AUG 2 0 1979

AF333343

Nonreciprocal Microwave Junctions and Circulators

Nonreciprocal Microwave Junctions and Circulators

J. HELSZAJN

Department of Electrical
and Electronic Engineering
Heriot-Watt University
Edinburgh, United Kingdom

A WILEY-INTERSCIENCE PUBLICATION

John Wiley & Sons

New York · London · Sydney · Toronto

Copyright © 1975 by John Wiley & Sons, Inc.

All rights reserved. Published simultaneously in Canada.

No part of this book may be reproduced by any means, nor transmitted, nor translated into a machine language without the written permission of the publisher.

Library of Congress Cataloging in Publication Data

Helszajn, Joseph.
   Nonreciprocal microwave junctions and circulators.

   Includes bibliographical references and index.
   1. Circulators, Wave-guide. 2. Junctions, Wave-guide. I. Title.
TK7871.65.H44   621.381′331   75–5588
ISBN 0–471–36935–7

Printed in the United States of America

10 9 8 7 6 5 4 3 2 1

550

Preface

The theory of junction circulators is given in terms either of a linear combination of the properties of one-port reactive networks or of a one-port equivalent circuit. The former description is particularly suitable for analysis, whereas the latter approach is more convenient for design purposes. The object of this book is to summarize the available theory of standard circulators, using both these approaches in a unified notation based on *ABCD* matrices, ideal transformers, and junction resonator eigennetworks. I hope the text will provide a useful introduction to the topic of junction circulators to both research and other workers in the field.

Chapter 1 defines the tensor permeability of a magnetized ferrite material on which the junction circulator relies for its operation. This chapter discusses the magnetic background required for the study of the text. Chapters 2 and 3 deal with scattering and immitance matrices of symmetrical junctions and their eigenvalues on which much of the material in this text rests. The basic nonreciprocal two-port gyrator network is defined in Chapter 4 and is used to construct equivalent circuits of ideal circulators. Chapter 5 states the boundary conditions of ideal circulators in terms of the scattering and immittance matrices and their eigenvalues and also in terms of an equivalent one-port approximate network. The mode structure of ferrite disk resonators required for the construction of circulators is treated in Chapter 6. Scattering matrices and other properties of lossy junctions are next established in Chapter 7, and Chapter 8 gives a general adjustment procedure for circulators with more than three ports in terms of the location of the scattering matrix eigenvalues on the unit circle. Chapters 9, 10, and 11 present the theory of practical lumped element, stripline, and waveguide circulators, using the boundary conditions established earlier in the text. The circuit theory of the junction is treated in Chapter 12, which also derives a universal one-port gyrator equation for three-port circulators. Measurement techniques are dealt with next in Chapter 13. The text continues with Chapters 14, 15, and 16, which give the frequency behavior of quarter wave matched wideband reciprocal and nonreciprocal junctions with Chebyshev characteristics. Chapter 17 describes the use of circulators for switching. Chapter 18 treats the theory of stripline and waveguide four-port single junction devices in terms of junction modes. The final chapter discusses the magnetic variables of circulators such as choice of magnetization and direct magnetic field.

I would like to thank Sheila Murray, Helene Young, and Moira Tullis at Heriot-Watt University for their patience and good temper in typing the text.

J. Helszajn

Edinburgh, United Kingdom
November 1974

Contents

Nonreciprocal Microwave Junctions and Circulators

INTRODUCTION: APPLICATION OF JUNCTION CIRCULATORS

The ideal circulator discussed in this book is defined as an *m*-port lossless passive network, in which a signal introduced at one port is transferred wholly to an adjacent one while decoupling it from the other ports. Circulators with three or more ports have nonreciprocal properties that are of value in many microwave systems. It is the purpose of this chapter to discuss some of these. The schematic of the 3-port circulator is illustrated in Figure I.1. One important property of such devices is that the performance of any junction circulator is a function of the impedance match of its ports. For example, a 3-port circulator with direction of circulation from port 1 to port 2 to port 3 has isolation from port 2 to 1, which is essentially a function of the voltage standing wave ratio (VSWR) at port 3. Figure I.2 is a theoretical graph of circulator isolation as a function of *VSWR*. One way to obtain more than three ports with such junctions is to connect them in tandem. This is shown in Figure I.3 in the case of 4-, 5-, and *m*-port junctions. Some mechanical configurations in the case of the 4-port junction are depicted in Figure I.4.

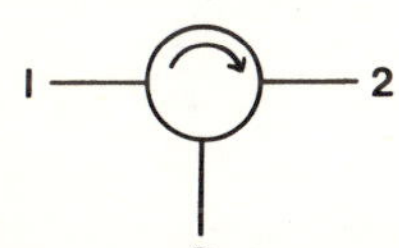

Figure I.1.   *Schematic of 3-port circulator.*

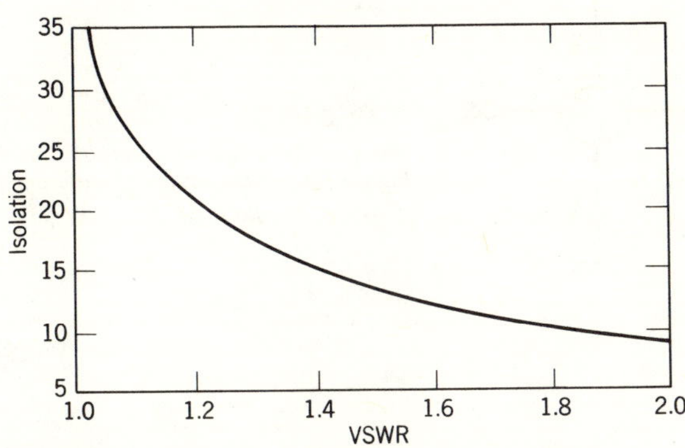

Figure I.2.   *Relation between VSWR and isolation of a 3-port circulator.*

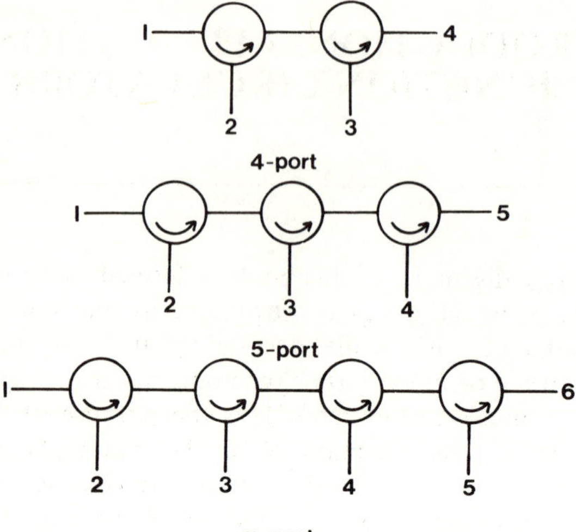

Figure I.3. *Schematics of 4-, 5-, and m-port circulators.*

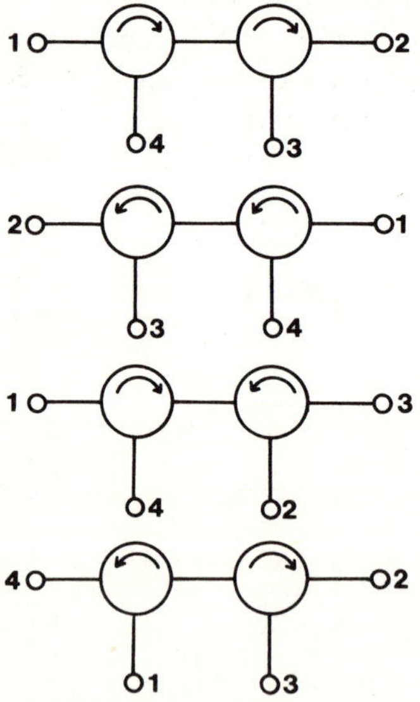

Figure I.4. *Schematics of different port arrangements of 4-port circulators.*

The junction device discussed here is, of course, only one possible type of circulator. They may also be constructed using either the principle of Faraday rotation in a longitudinally magnetized ferrite medium or nonreciprocal phase shifters in rectangular waveguides in conjunction with 3 *dB* hybrids and magic tees.

I.1   *ISOLATOR CONFIGURATION OF 3-PORT CIRCULATOR*

The most common application of the 3-port junction circulator is in the isolator configuration obtained by terminating port 3 by a matched load. This arrangement is shown schematically in Figure I.5. The isolator is the most commonly used of all microwave ferrite devices. It is a passive nonreciprocal 2-port transmission line permitting signals to pass in one direction but absorbing them in the other one with little attenuation. In the terminated circulator configuration the signal in the reverse direction is absorbed in the load placed at the third port of the device. One advantage of this isolator compared to other forms is that the average power at which this one can operate is determined by the rating of the load connected at the third port rather than by that of the junction itself. In such an isolator the input VSWR is determined by the isolation of the device and the VSWR of the load. Figure I.6 depicts the VSWR reduction within the ferrite isolator. This unilateral isolation nomogram can be used for determining the required attenuation in any isolation problem. With any two of the three scale values known, the third is readily determined. When reducing a terminating VSWR by a given value of isolation, the VSWR looking into the isolator must be multiplied by the VSWR of the isolator itself. The product is the maximum VSWR that can appear at the isolator input. One way to increase the isolation of the single junction is to use a number of them in tandem. A schematic representation of an *m*-port junction is illustrated in Figure I.7. A typical application of the ferrite isolator, which is shown in Figure I.8, is as a buffer stage between a microwave tube, such as a magnetron or klystron, and the rest of the microwave system. Such a buffer stage prevents frequency pulling in these tubes due to long line effects since the tube essentially works into a constant impedance.

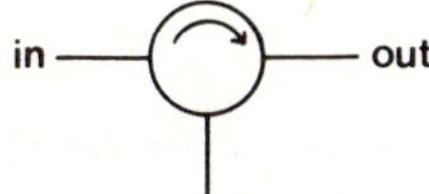

Figure I.5.   *Schematic of isolator using terminated 3-port circulator.*

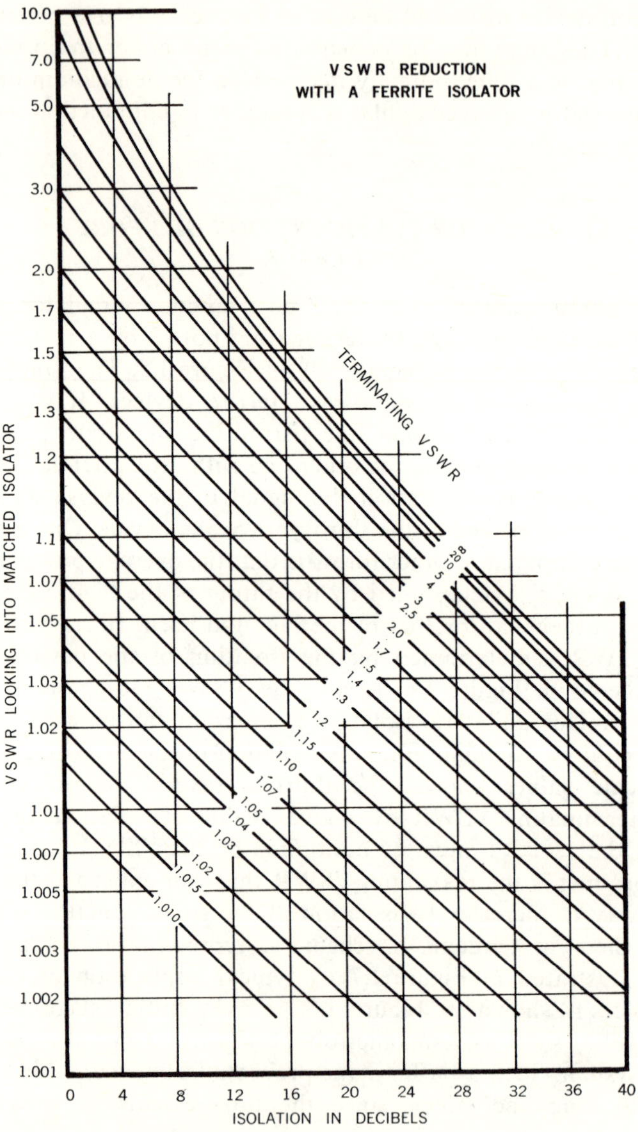

Figure I.6. *Unilateral isolation nomogram.*

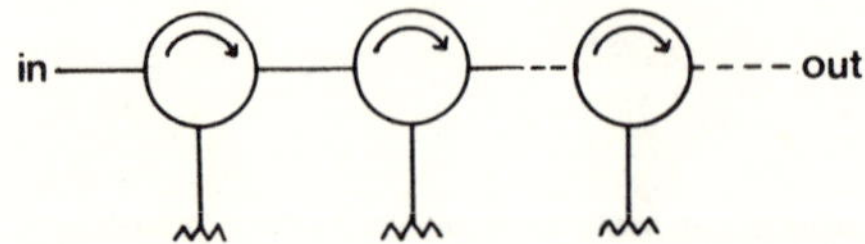

Figure I.7. *Tandem connection of terminated circulators to form high-ratio isolator.*

4

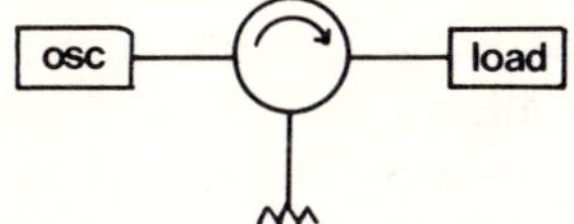

Figure I.8.   *Decoupling of oscillator and load by isolator.*

I.2.   *SWITCHABLE CIRCULATORS*

Since the direction of circulation of a circulator is determined by that of
the direct magnetic field used to magnetize it such a junction may be used
to switch an input signal at one port to either one of the other two.
Switching is here achieved by replacing the permanent magnet by an
electromagnet or by latching the microwave ferrite directly by inserting a
current carrying wire loop inside the ferrite shape. The schematic of the
switched junction is shown in Figure I.9. It is particularly useful in the
construction of Butler type matrices in phased array systems. A single pole
four-throw switch is illustrated schematically in Figure I.10.

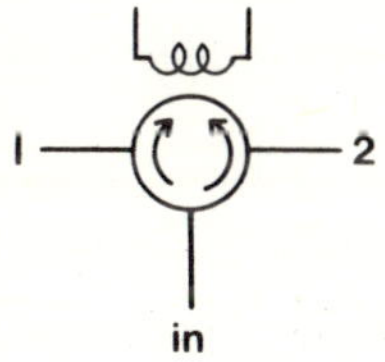

Figure I.9.   *Schematic of circulator switch.*

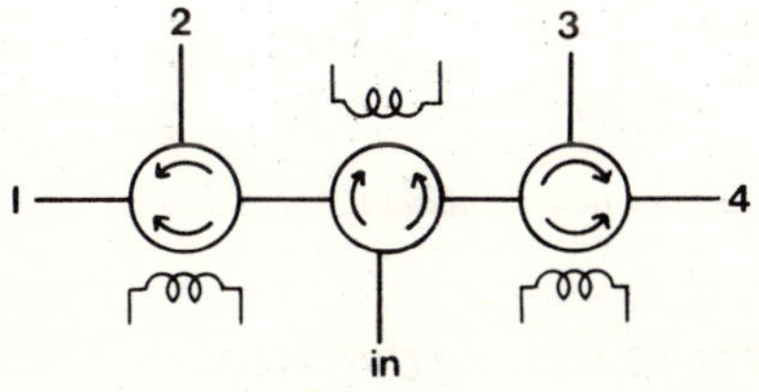

Figure I.10. *SP4T Buttler switch using circulators.*

I.3.   *SINGLE PORT AMPLIFIERS USING JUNCTION*
CIRCULATORS

A simplified schematic of a reflection tunnel diode amplifier (TDA) which
utilizes a circulator to separate the input signal from the amplified one is
shown in Figure I.11. Since the gain of these amplifiers may be comparable
to the isolation of the junction, such amplifiers are often used in conjunction with 5-port circulators, as depicted in Figure I.12 where the input and
output junctions are connected as isolators in order to minimize gain

variations due to source and load impedance variations. The magnetic field for the input circulator is here sometimes supplied by an electromagnet. The reversal of the magnetic field in this circuit allows the TDA to be protected from radio frequency leakage during the transmitting period by reversing the direction of circulation during this interval. Figure I.13 outlines an *m*-port amplifier chain using junction circulators.

Figure I.11.   *Single port amplifier using 3-port circulator.*

Figure I.12.   *Single port amplifier using 5-port circulator.*

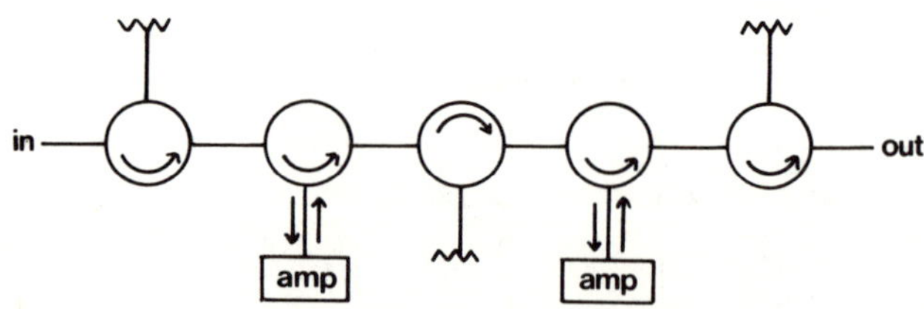

Figure I.13.   *m-Port amplifier chain using junction circulators.*

I.4   *MICROWAVE PHASE SHIFTING USING JUNCTION CIRCULATORS*

Two common arrangements in which ferrite circulators are used to obtain microwave phase shifting are outlined in Figure I.14 and I.15. The first uses a fixed circulator in conjunction with a *p–i–n* diode switch to vary the short circuit plane terminating port 2. A transmission analog phase shifter is therefore obtained between ports 1 and 3 with this mode of operation. The second arrangement is also a transmission configuration but here a switchable circulator is used to control the path between ports 1 and 3 of the circulator. The switching speed of the *p–i–n* device is normally the faster one.

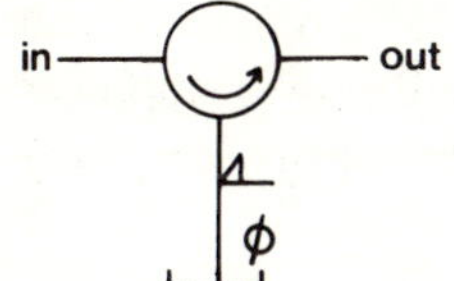

Figure I.14. *Microwave phase shifter using p.i.n. dioded switch and fixed circulator.*

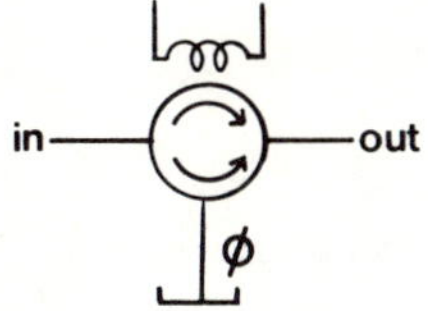

Figure I.15. *Microwave phase shifter using switched circulator.*

I.5.   *DUPLEXING USING JUNCTION CIRCULATORS*

The ferrite circulator may also be used in duplexing systems for simultaneous transmission and reception of microwave energy with a single antenna. Ferrite circulators here replace conventional types of duplexing and are suited for both high- and low-power systems. The arrangement used here is shown in Figure I.16. Circulators may also be used in communication systems to eliminate mutual interference between closely separated transmitters. Figure I.17 gives an example of a single antenna

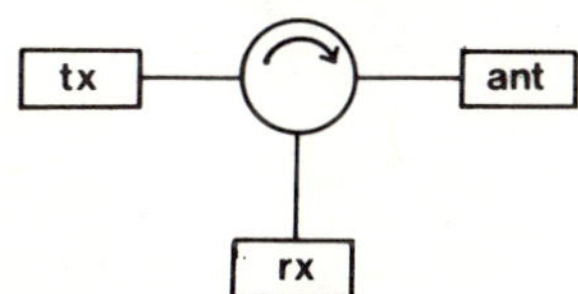

Figure I.16.   *Low-power duplexer using 3-port circulator.*

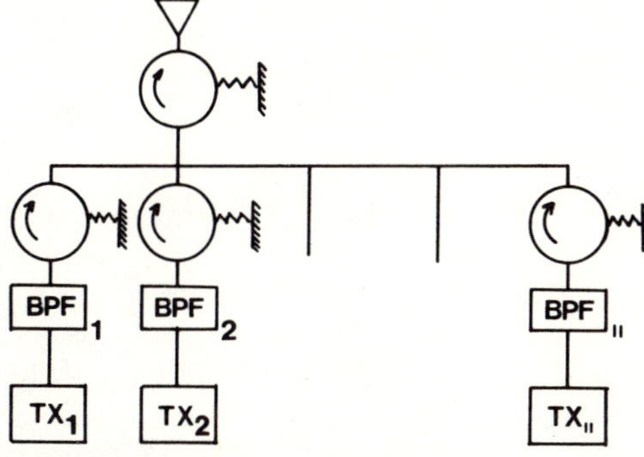

Figure I.17. *Duplexing between closely spaced transmitters.*

being shared by a number of transmitters with the help of circulators and bandpass filters (BPF). Figure I.18 illustrates a 4-port high-power duplexer which uses a reflection limiter to protect the receiver during the transmission interval.

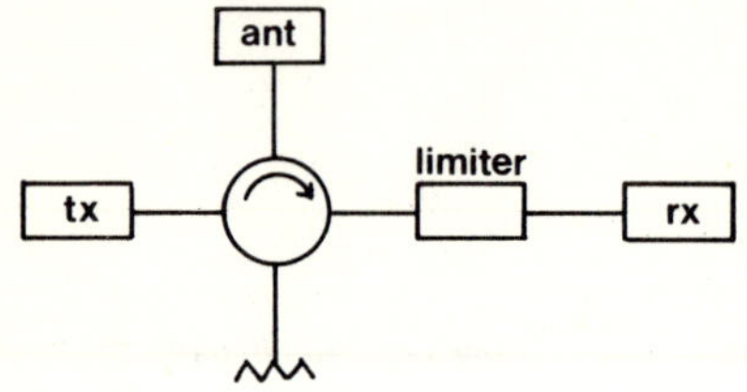

Figure I.18. *High-power duplexer using 4-port circulators.*

CHAPTER ONE

Tensor Permeability of Magnetized Ferrite

The characteristic property of a magnetized ferrite is that while it has a scalar dielectric constant, it has a tensor permeability at microwave frequencies. Inside a ferrite medium Maxwell's equations must therefore be solved in conjunction with this tensor permeability. The presence of imaginary off-diagonal components having opposite signs in this tensor gives rise to a number of important nonreciprocal effects not usually encountered in a medium for which the permeability is a scalar. The tensor permeability is derived from the linearized equation of motion of the magnetization vector that will be studied in detail in this first chapter under various driving fields. In its most simple form, the equation of motion gives the motion of the total number of magnetic dipoles per unit volume due to electron spins in the presence of a constant magnetic field. The motion of the magnetization vector is found to precess about the constant magnetic field at a natural precession frequency determined by the magnetic field. The direction of the precession depends on the direction of the steady magnetic field. Because damping is present the amplitude of the precession decreases until the magnetization comes into line with the magnetic field. The precession can, however, be maintained by superimposing a small r.f. magnetic field in the plane transverse to the steady magnetic field. The tensor permeability can be obtained from this simple arrangement. If the frequency of the r.f. magnetic field coincides with the natural precession frequency the amplitude of the precession becomes particularly large and the energy absorbed from the r.f. magnetic field peaks through a maximum. In this chapter, we shall study the equation of motion with this simple arrangement, and also in the presence of additional driving fields, such as demagnetizing fields, which occur in finite ellipsoidal shapes.

1.1. *EQUATION OF MOTION OF MAGNETIZATION VECTOR*

Much of the macroscopic theory of microwave ferrite devices is based on the equation of motion of the magnetization vector. The equation of motion can be derived by considering an elementary magnetic dipole having a dipole moment μ placed in a direct current (d.c.) magnetic field $\overline{H}_0$. Under equilibrium conditions, the dipole moment vector $\overline{\mu}$ lies in the direction $\overline{H}_0$, which is usually assumed to be in the z-direction. Now let us assume that the magnetic dipole is tilted by a small external force so that it makes an angle θ with H_0 as shown in Figure 1.1. Since the only field acting on $\overline{\mu}$ is $\overline{H}_0$ the torque excerted on $\overline{\mu}$ is

$$\overline{T} = \overline{\mu} \times \overline{H}_0 \tag{1.1}$$

Associated with the magnetic dipole $\overline{\mu}$ there is an angular momentum $\overline{J}$

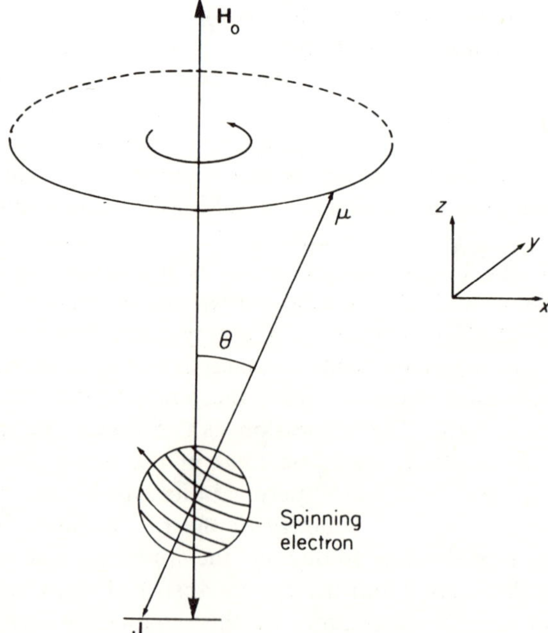

Figure 1.1. *Magnetic moment precession about a static magnetic field.*

given by

$$\bar{\mu} = -\gamma \bar{J} \qquad (1.2)$$

where γ is the gyromagnetic ratio given by $2 \cdot 21 \times 10^5 (\text{rad/sec})/(\text{At/m})$. Since the torque is the time rate of change of the angular momentum, it can be written from Eq. 1.2 as

$$\bar{T} = -\frac{1}{\gamma}\frac{d\bar{\mu}}{dt} \qquad (1.3)$$

Combining Eq. 1.1 and 1.3 gives the equation of motion for a single dipole

$$\frac{d\bar{\mu}}{dt} = -\gamma\left(\bar{\mu} \times \bar{H}_0\right) \qquad (1.4)$$

The total magnetic moment per unit volume in Webers per square meter is

$$M_0 = N\mu \qquad (1.5)$$

where N is the number of unbalanced spins per unit volume. Equation 1.4 now becomes

$$\frac{d\bar{M}_0}{dt} = -\gamma\left(\bar{M}_0 \times \bar{H}_0\right) \qquad (1.6)$$

Much of the classical theory of microwave ferrites is based on the equation of motion of the magnetization vector given by Eq. 1.6.

1.2. SUSCEPTIBILITY TENSOR IN INFINITE MEDIUM

In the most simple microwave case the total effective magnetic field in Eq. 1.6 consists of the d.c. magnetic field $\bar{H}_0$ and the r.f. magnetic field $\bar{h}$

$$\bar{H} = \bar{H}_0 + \bar{h} \qquad (1.7)$$

The total magnetization consists of the d.c. magnetization $\bar{M}_0$ and the r.f. magnetization $\bar{m}$

$$\bar{M} = \bar{M}_0 + \bar{m} \qquad (1.8)$$

In component form the above equations are

$$\bar{H}_0 = \begin{bmatrix} 0 \\ 0 \\ H_0 \end{bmatrix} \qquad (1.9)$$

$$\bar{h} = \begin{bmatrix} h_x \\ h_y \\ h_z \end{bmatrix} \tag{1.10}$$

$$\overline{M}_0 = \begin{bmatrix} 0 \\ 0 \\ M_0 \end{bmatrix} \tag{1.11}$$

$$\overline{m} = \begin{bmatrix} m_x \\ m_y \\ m_z \end{bmatrix} \tag{1.12}$$

Equation 1.6 can now be expanded to give

$$\frac{dm_x}{dt} = -m_y \gamma (H_0 + h_z) + h_y \gamma (M_0 + m_z) \tag{1.13}$$

$$\frac{dm_y}{dt} = m_x \gamma (H_0 + h_z) - h_x \gamma (M_0 + m_z) \tag{1.14}$$

$$\frac{dm_z}{dt} = -m_x \gamma h_y + m_y \gamma h_x \tag{1.15}$$

In the small signal approximation higher order terms of m and h are neglected. The small signal approximation is therefore

$$\frac{dm_x}{dt} = -m_y \gamma H_0 + h_y \gamma M_0 \tag{1.16}$$

$$\frac{dm_y}{dt} = m_x \gamma H_0 - h_x \gamma M_0 \tag{1.17}$$

$$\frac{dm_z}{dt} = 0 \tag{1.18}$$

Rewriting the last equations we have

$$\ddot{m}_x + \omega_0^2 m_x = \mu_0 \omega_m \omega_0 h_x + \mu_0 \omega_m \dot{h}_y \tag{1.19}$$

$$\ddot{m}_y + \omega_0^2 m_y = -\mu_0 \omega_m \dot{h}_x + \mu_0 \omega_m \omega_0 h_y \tag{1.20}$$

$$m_z \approx 0 \tag{1.21}$$

where

$$\omega_m = \frac{\gamma M_0}{\mu_0} \tag{1.22}$$

$$\omega_0 = \gamma H_0 \tag{1.23}$$

If the time dependence of the r.f. quantities is of the form $\exp(j\omega t)$, a susceptibility tensor $[\chi]$ can be defined that relates the r.f. magnetization to the r.f. magnetic field

$$\overline{m} = \mu_0 [\chi] \overline{h} \tag{1.24}$$

where

$$[\chi] = \begin{bmatrix} \chi_{xx} & \chi_{xy} & 0 \\ \chi_{yx} & \chi_{yy} & 0 \\ 0 & 0 & 0 \end{bmatrix} \tag{1.25}$$

and

$$\chi_{xx} = \chi_{yy} = \frac{\omega_m \omega_0}{-\omega^2 + \omega_0^2} \tag{1.26}$$

$$-\chi_{yx} = \chi_{xy} = \frac{j\omega_m \omega}{-\omega^2 + \omega_0^2} = -jK \tag{1.27}$$

The components of the susceptibility tensor have a singularity at $\omega = \omega_0$. This is defined as the resonance condition.

1.3. *DAMPING*

To stabilize the motion of the magnetization vector at the resonance a damping term must be introduced into the equation of motion. One form of phenomenological damping often used is the Bloch–Bloembergen one given by

$$\left(\frac{d\overline{M}}{dt}\right)_{xy} = -\gamma(\overline{M}\times\overline{H})_{xy} - \frac{\overline{M}_{xy}}{T_2} \tag{1.28}$$

$$\left(\frac{d\overline{M}}{dt}\right)_{z} = -\gamma(\overline{M}\times\overline{H})_{z} - \frac{M_z - M_0}{T_1} \tag{1.29}$$

where T_1 and T_2 are longitudinal and transverse relaxation times.

Another phenomenological damping term is due to Landau–Lifshitz. Using this form the equation of motion becomes

$$\frac{d\overline{M}}{dt} = -\gamma(\overline{M}\times\overline{H}) + \frac{\gamma\alpha}{|M|}\left[\overline{M}\times(\overline{M}\times\overline{H})\right] \tag{1.30}$$

where α is a dimensionless parameter that determines the damping. Taking the cross product of $\overline{M}$ with the last equation gives

$$\frac{\gamma\alpha}{|M|}\left[\overline{M}\times(\overline{M}\times\overline{H})\right] = -\frac{\alpha}{|M|}\left(\overline{M}\times\frac{d\overline{M}}{dt}\right) + \frac{\gamma\alpha^2}{|M|^2}\overline{M}\times\left[\overline{M}\times(\overline{M}\times\overline{H})\right]$$

$$\tag{1.31}$$

Introducing this last equation back into Eq. 1.30 we have

$$\frac{d\overline{M}}{dt} = -\gamma(\overline{M}\times\overline{H}) - \frac{\alpha}{|M|}\left(\overline{M}\times\frac{d\overline{M}}{dt}\right) + \frac{\gamma\alpha^2}{|M|^2}\overline{M}\times\left[\overline{M}\times(\overline{M}\times\overline{H})\right]$$

$$\tag{1.32}$$

For most ferrites the last term is negligible, therefore Eq. 1.32 can be written as

$$\frac{d\overline{M}}{dt} = -\gamma(\overline{M}\times\overline{H}) - \frac{\alpha}{|M|}\left(\overline{M}\times\frac{d\overline{M}}{dt}\right) \tag{1.33}$$

This last equation, which is the Gilbert form of the original Landau–Lifshitz phenomenological damping term, is the one that will be adopted in the topics to be discussed. This is shown in Figure 1.2. The components of

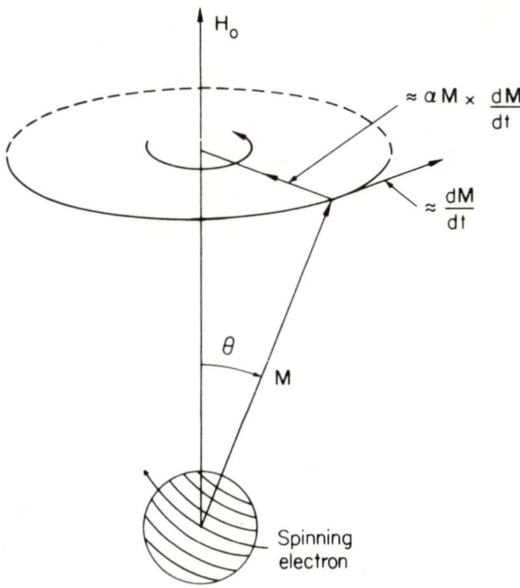

Figure 1.2. Relation between $\overline{M}$, $d\overline{M}/dt$, and $\overline{M} \times (d\overline{M}/dt)$. The loss term tends to decrease θ and damp out the precession.

the tensor susceptibility obtained from the small signal solution to Eq. 1.33 are

$$\chi_{xx} = \chi_{yy} = \frac{\omega_m(\omega_0 + j\omega\alpha)}{-\omega^2 + (\omega_0 + j\omega\alpha)^2} \tag{1.34}$$

$$-\chi_{yx} = \chi_{xy} = \frac{j\omega_m\omega}{-\omega^2 + (\omega_0 + j\omega\alpha)^2} = -jK \tag{1.35}$$

The only difference between Eq. 1.26 and 1.27 and Eqs. 1.34 and 1.35 is that an imaginary frequency term $j\omega\alpha$ has been added to the resonant frequency ω_0. Hence, the damping term can always be introduced by replacing ω_0 by $\omega_0 + j\omega\alpha$ in the loss free components. The real and imaginary parts of the components of the susceptibility tensor are given by

$$\chi'_{xx} = \frac{\omega_m\omega_0(\omega_0^2 - \omega^2) + \omega_m\omega_0\omega^2\alpha^2}{[\omega_0^2 - \omega^2(1 + \alpha^2)]^2 + 4\omega_0^2\omega^2\alpha^2} \tag{1.36}$$

$$\chi''_{xx} = \frac{-\omega_m\omega\alpha\left[\omega_0^2+\omega^2(1+\alpha^2)\right]}{\left[\omega_0^2-\omega^2(1+\alpha^2)\right]^2+4\omega_0^2\omega^2\alpha^2} \qquad (1.37)$$

$$K' = \frac{-\omega_m\omega\left[\omega_0^2-\omega^2(1+\alpha^2)\right]}{\left[\omega_0^2-\omega^2(1+\alpha^2)\right]^2+4\omega_0^2\omega^2\alpha^2} \qquad (1.38)$$

$$K'' = \frac{-2\omega_m\omega_0\omega^2\alpha}{\left[\omega_0^2-\omega^2(1+\alpha^2)\right]^2+4\omega_0^2\omega^2\alpha^2} \qquad (1.39)$$

The real and imaginary parts of the susceptibility tensor are plotted in Figures 1.3 through 1.6 as a function of ω_m/ω, ω_0/ω, and α.

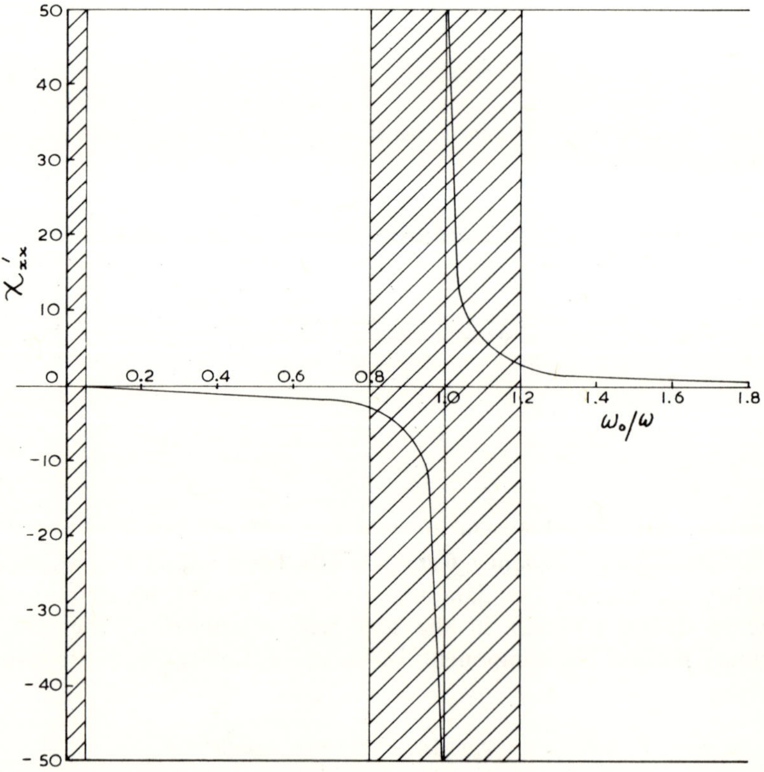

Figure 1.3. *Real part of* χ_{xx} *component, for* $\omega_m/\omega=1$ *and* $\alpha=0.01$.

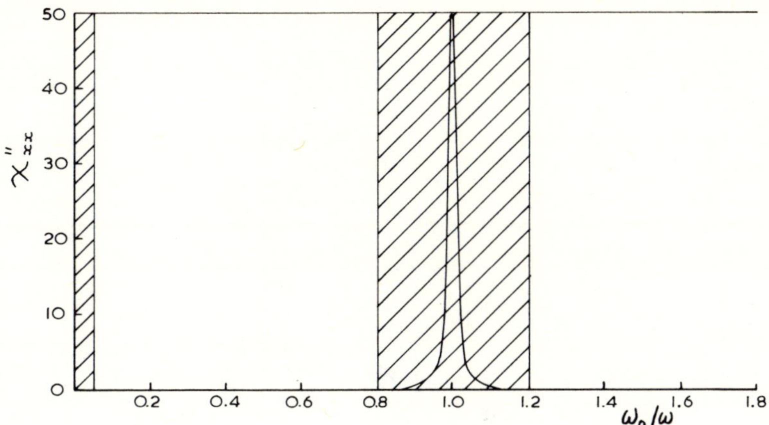

Figure 1.4. *Imaginary part of χ_{xx} component for $\omega_m/\omega = 1$ and $\alpha = 0.01$.*

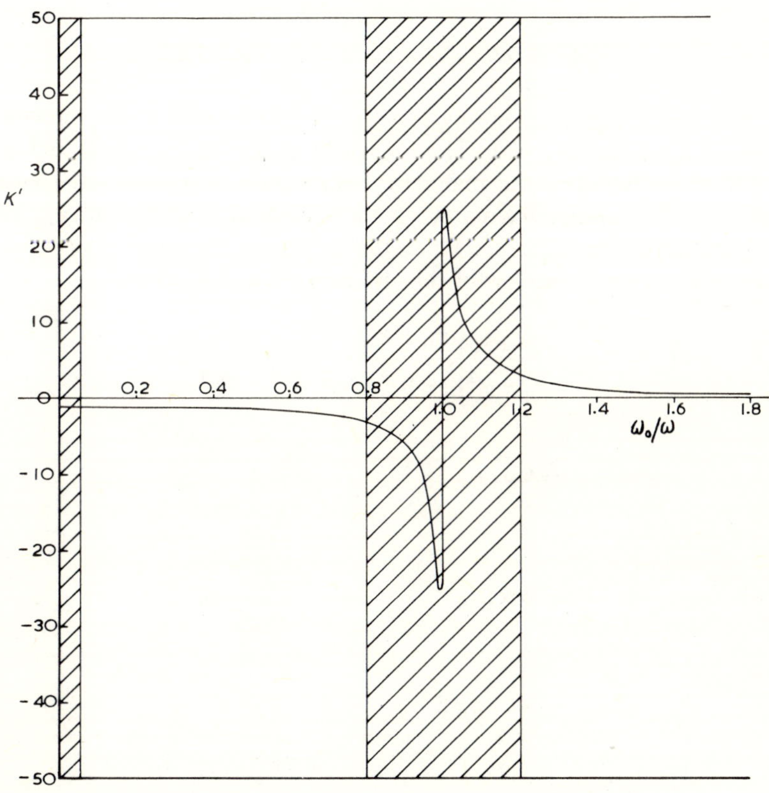

Figure 1.5. *Real part of K component for $\omega_m/\omega = 1$ and $\alpha = 0.01$.*

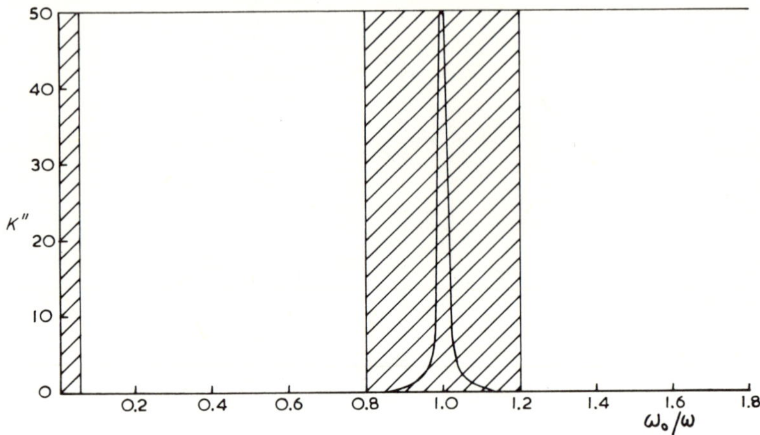

Figure 1.6.   *Imaginary part of K component for $\omega_m/\omega = 1$ and $\alpha = 0.01$.*

1.4.   *THE LINEWIDTH OF FERRITES*

The phenomenological damping factor α can be related to the linewidth of ferrites in the following way. The linewidth of ferrites is usually defined as the difference between the magnetic field values at a constant frequency where χ''_{xx}, the imaginary parts of the diagonal component χ_{xx} of the susceptibility tensor, attains a value that is half of its value at resonance as shown in Figure 1.7. In view of this definition Eq. 1.37 gives

$$\frac{\omega_m \omega \alpha (\omega_{01}^2 + \omega^2)}{(\omega_{01}^2 - \omega^2)^2 + 4\omega_{01}^2 \omega^2 \alpha^2} = \frac{1}{2}\left(\frac{\omega_m}{2\omega\alpha}\right) \tag{1.40}$$

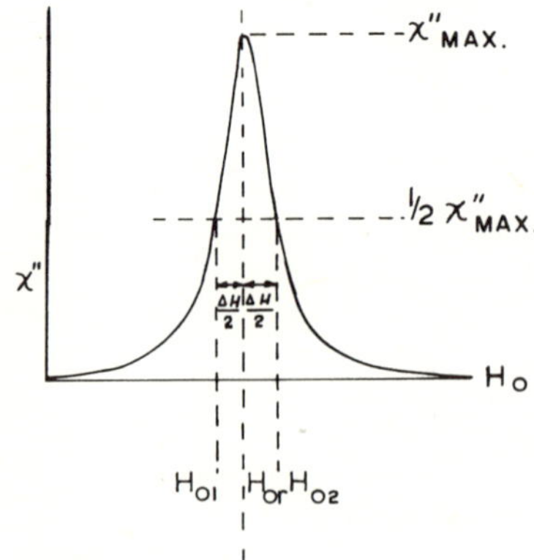

Figure 1.7.   *Linewidth of ferrites.*

Rearranging Eq. 1.40 gives a quadratic in ω_{01}^2

$$\omega_{01}^4 - 2\omega^2\omega_{01}^2 + \omega^4(1-4\alpha^2)=0 \tag{1.41}$$

The solution to Eq. 1.41 is

$$\omega_{01,2}=\omega(1\pm 2\alpha)^{1/2}\approx\omega\pm\omega\alpha \tag{1.42}$$

Hence,

$$\omega\alpha = \frac{\omega_{02}-\omega_{01}}{2} \tag{1.43}$$

The full linewidth of the material is usually defined by

$$\Delta H = \frac{\omega_{02}-\omega_{01}}{\gamma} \tag{1.44}$$

Combining Eqs. 1.43 and 1.44 gives

$$\alpha\omega = \frac{\gamma\Delta H}{2} \tag{1.45}$$

For engineering purposes the imaginary frequency $j\alpha\omega$ is usually written in terms of ΔH from the last equation.

1.5. SCALAR SUSCEPTIBILITY

The relation between $\overline{m}$ and $\overline{h}$ is a scalar quantity, provided $\overline{h}$ is adjusted to correspond to one of the normal modes of the system. These modes may be found by first finding the eigenvalues (scalar susceptibilities) of the tensor susceptibility. The eigenvalue equation to be solved is

$$\chi\overline{H}=[\chi]\overline{H} \tag{1.46}$$

where χ is an eigenvalue, $\overline{H}$ is an eigenvector, and $[\chi]$ is the susceptibility tensor. The magnetic fields are proportional to the eigenvectors.

Equation 1.46 has a nonvanishing value for $\overline{H}$ provided

$$\begin{vmatrix} (\chi_{xx}-\chi) & \chi_{xy} & 0 \\ -\chi_{xy} & (\chi_{xx}-\chi) & 0 \\ 0 & 0 & -\chi \end{vmatrix} = 0 \tag{1.47}$$

The three eigenvalues are

$$\chi_1 = \chi_- = \chi_{xx} + j\chi_{xy} \tag{1.48}$$

$$\chi_2 = \chi_+ = \chi_{xx} - j\chi_{xy} \tag{1.49}$$

$$\chi_3 = 0 \tag{1.50}$$

The eigenvectors are now found by substituting Eqs. 1.48 through 1.50 into Eq. 1.46. Each eigenvalue corresponds to one of the eigenvectors:

$$\overline{H}_1 = \overline{h}_- = \frac{1}{\sqrt{2}} \begin{bmatrix} h_0 \\ jh_0 \\ 0 \end{bmatrix} \tag{1.51}$$

$$\overline{H}_2 = \overline{h}_+ = \frac{1}{\sqrt{2}} \begin{bmatrix} h_0 \\ -jh_0 \\ 0 \end{bmatrix} \tag{1.52}$$

$$\overline{H}_3 = \overline{h}_z = \begin{bmatrix} 0 \\ 0 \\ h_0 \end{bmatrix} \tag{1.53}$$

The magnetic fields corresponding to the last three equations are orthonormal

$$\overline{H}_i \overline{H}_j^* = \delta_{ij} h_0^2 \tag{1.54}$$

δ_{ij} is the Kronecker delta

$$\delta_{ij} = 1 \text{ if } i = j$$

$$= 0 \text{ if } i \neq j$$

The two normal modes $\overline{h}_+$ and $\overline{h}_-$ correspond to circularly polarized r.f. fields in the transverse plane. The normal mode $\overline{h}_z$ corresponds to the excitation where the r.f. magnetic field is parallel to the biasing field H_0, in which case the ferrite does not exhibit any gyromagnetic properties.

In terms of the original variables, one also has

$$\chi_{\pm} = \frac{\omega_m}{\omega_0 \mp \omega} \tag{1.55}$$

Damping may be introduced into the last equation by adding an imaginary frequency to the resonant one in the usual way

$$\chi_{\pm} = \frac{\omega_m}{(\omega_0 + j\alpha\omega) \mp \omega} \tag{1.56}$$

The real and imaginary parts of the two scalar susceptibilities corresponding to the two possible circularly polarized waves are shown in Figure 1.8 in terms of ω_m/ω, ω_0/ω and α. The positive susceptibility has a

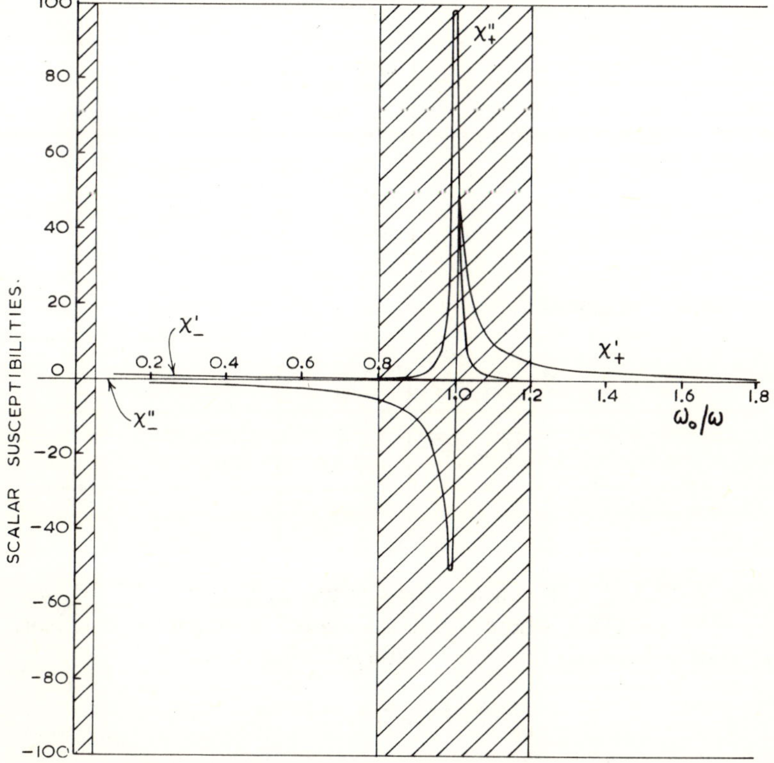

Figure 1.8.   Real and imaginary parts of scalar susceptibilities with $\omega_m/\omega = 1$ and $\alpha = 0.01$.

singularity at $\omega = \omega_0$ and corresponds to the circularly polarized wave that rotate in the same sense as the precessional motion of the magnetization vector. The negative susceptibility describes the response of the system to a circularly polarized field that rotates in the opposite direction. In this instance there is no singularity at $\omega = \omega_0$. This result is responsible for a number of the microwave ferrite devices.

1.6. *SUSCEPTIBILITY TENSOR IN ELLIPSOIDAL MEDIUM*

To evaluate the magnetic fields inside a finite ferrite medium it is necessary, in principle, to solve Maxwell's equations in conjunction with the Polder tensor permeability subject to the boundary conditions. However, this approach is not always simple or convenient. For an ellipsoidal sample in a uniform field the problem can be simplified by introducing demagnetizing fields as is done in the d.c. case. For the r.f. magnetic field the demagnetizing field is given by

$$\bar{h}_{\text{dem}} = - \begin{bmatrix} N_x \dfrac{m_x}{\mu_0} \\ N_y \dfrac{m_y}{\mu_0} \\ N_z \dfrac{m_z}{\mu_0} \end{bmatrix} \tag{1.57}$$

and for the d.c. magnetic field by

$$\overline{H}_{\text{dem}} = - \begin{bmatrix} 0 \\ 0 \\ \dfrac{N_z M_0}{\mu_0} \end{bmatrix} \tag{1.58}$$

Equations 1.57 and 1.58 can also be written as $\bar{h}_{\text{dem}} = -[N] \cdot (\bar{m}/\mu_0)$ and $H_{\text{dem}} = -[N] \cdot (\overline{M}_0/\mu_0)$, where $[N]$ is a diagonal demagnetizing tensor.

The relation between the shape demagnetizing factors is

$$N_x + N_y + N_z = 1 \tag{1.59}$$

There are several shapes that are of particular interest in the discussion of microwave ferrite devices. For a sphere

$$N_x = N_y = N_z = \tfrac{1}{3} \tag{1.60}$$

while for a long cylinder of radius R and length L

$$N_z = 1 - \left[1 + \left(\frac{2R}{L} \right)^2 \right]^{-1/2} \tag{1.61}$$

$$N_x = N_y = \tfrac{1}{2}(1 - N_z) \tag{1.62}$$

and for a thin disk

$$N_z = 1 - \left(\frac{L}{2R} \right) \left[1 + \left(\frac{L}{(2R)} \right)^2 \right]^{-1/2} \tag{1.63}$$

$$N_x = N_y = \tfrac{1}{2}(1 - N_z) \tag{1.64}$$

Figure 1.9 depicts the demagnetizing factors of some simple shapes.

The small signal approximation of the equation of motion in terms of the external fields is now given with

$$\overline{H} = \overline{H}_0 + \overline{H}_{\text{dem}} + \bar{h}^e + \bar{h}_{\text{dem}} \tag{1.65}$$

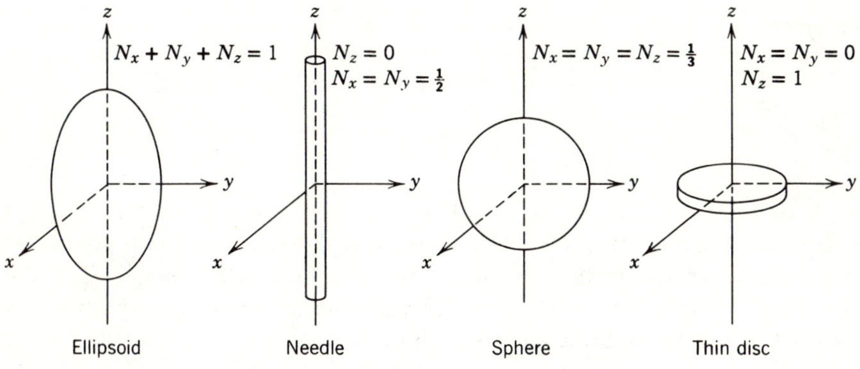

Figure 1.9. *Demagnetizing factors.*

by

$$\ddot{m}_x + \omega_r^2 m_x = \mu_0 \omega_m \omega_y h_x^e + \mu_0 \omega_m \dot{h}_y^e \tag{1.66}$$

$$\ddot{m}_y + \omega_r^2 m_y = -\mu_0 \omega_m \dot{h}_x^e + \mu_0 \omega_m \omega_x h_y^e \tag{1.67}$$

$$m_z \approx 0 \tag{1.68}$$

where

$$\omega_x = (\omega_0 - N_z \omega_m + N_x \omega_m) \tag{1.69}$$

$$\omega_y = (\omega_0 - N_z \omega_m + N_y \omega_m) \tag{1.70}$$

and

$$\omega_r = \sqrt{\omega_x \omega_y} \tag{1.71}$$

The last equation is the well-known Kittel resonance relation for an ellipsoid. The singularity of the susceptibility components now occurs at $\omega = \omega_r$, instead of $\omega = \omega_0$, and involves the demagnetizing factors.

The components of the external susceptibility tensor are given by

$$\chi_{xx}^e = e_0 \left(\frac{\omega_m \omega_r}{-\omega^2 + \omega_r^2} \right) \tag{1.72}$$

$$\chi_{yy}^e = \frac{1}{e_0} \left(\frac{\omega_m \omega_r}{-\omega^2 + \omega_r^2} \right) \tag{1.73}$$

$$-\chi_{yx}^e = \chi_{xy}^e = \left(\frac{j\omega_m \omega}{-\omega^2 + \omega_r^2} \right) = -jK \tag{1.74}$$

where

$$e_0 = \sqrt{\frac{\omega_y}{\omega_x}} \tag{1.75}$$

e_0 is defined as the ellipticity of the normal modes of the uniform precession. For the infinite medium $e_0 = 1$ and $\omega_r = \omega_0$. The ellipticity is also unity whenever the transverse demagnetizing factors are equal. The advantage of writing the components of the external susceptibility tensor in terms of the uniform mode ellipticity as is done above is that the quantities inside the brackets are the same as for the components of the internal susceptibility except that the resonant frequency is now given by ω_r instead of ω_0. In the presence of damping the loss term can be represented by introducing an imaginary frequency term $j\omega\alpha$ to ω_x and ω_y. This is equivalent to introducing an imaginary frequency $j\omega\alpha$ to ω_r provided

$$\frac{\omega_x+\omega_y}{2}\approx\sqrt{\omega_x\omega_y} \tag{1.76}$$

The components of the external susceptibility tensor with damping is therefore approximately given by

$$\chi_{xx}^e=e_0\left[\frac{\omega_m(\omega_r+j\omega\alpha)}{-\omega^2+(\omega_r+j\omega\alpha)^2}\right] \tag{1.78}$$

$$\chi_{yy}^e=\frac{1}{e_0}\left[\frac{\omega_m(\omega_r+j\omega\alpha)}{-\omega^2+(\omega_r+j\omega\alpha)^2}\right] \tag{1.79}$$

$$-\chi_{yx}^e=\chi_{xy}^e=\left[\frac{j\omega_m\omega}{-\omega^2+(\omega_r+j\omega\alpha)^2}\right]=-jK \tag{1.80}$$

The real and imaginary parts inside the brackets of Eqs. 1.78 and 1.79 are the same as those given by Eqs. 1.36 and 1.37, with ω_0 replaced by ω_r. The real and imaginary parts of Eq. 1.80 are the same as those given by Eqs. 1.38 and 1.39 with ω_0 replaced by ω_r.

The external susceptibility can also be diagonalized. In this instance, the normal modes of the magnetization vector are elliptically polarized. This requires that the r.f. excitation be elliptically polarized also. The nature of the normal modes is obtained in a similar way to that used in Section 1.5.

1.7.   *TENSOR PERMEABILITY*

It is always necessary to account for the d.c. demagnetizing field in a finite medium. In using the tensor permeability to be derived now it is not necessary to account for the r.f. demagnetizing fields. The boundary conditions automatically satisfy the demagnetizing factors of the magnetic fields. This means that only the internal components of the permeability tensor appear in the final answer. However, the d.c. field is shifted from H_0 to $H_0-N_zM_0/\mu_0$.

A tensor permeability can also be defined by relating the r.f. flux density $\bar{b}$ to the r.f. magnetic field $\bar{h}$:

$$\bar{b} = \mu_0 \bar{h} + \bar{m} \tag{1.81}$$

or

$$\bar{b} = \mu_0 [\mu_r] \bar{h} \tag{1.82}$$

where

$$[\mu_r] = [1] + [\chi] \tag{1.83}$$

In component form, the tensor permeability is given from Eq. 1.25 by

$$[\mu_r] = \begin{bmatrix} \mu & -jK & 0 \\ jK & \mu & 0 \\ 0 & 0 & 1 \end{bmatrix} \tag{1.84}$$

where

$$\mu = 1 + \chi_{xx} \tag{1.85}$$

$$-jK = \chi_{xy} \tag{1.86}$$

The variation of K with the magnetic parameters has already been given in Figure 1.5. The variation of μ with the magnetic parameters is obtained from Figure 1.3 by noting that $\chi_{xx} = \mu - 1$. Equation 1.84 is the well-known Polder tensor permeability. It is usually employed in conjunction with Maxwell's equations to solve microwave ferrite boundary problems. Damping may be introduced into Eq. 1.84 in the usual way by adding an imaginary frequency term $j\alpha\omega$ to ω_0.

1.8. SCALAR PERMEABILITY

The permeability is also a scalar for the r.f. magnetic fields defined by Eqs. 1.51 through 1.53.

$$\mu_1 = \mu_+ = \mu - K \tag{1.87}$$

$$\mu_2 = \mu_- = \mu + K \tag{1.88}$$

$$\mu_3 = \mu_z = 1 \tag{1.89}$$

This result is obtained quite simply by determining the eigenvalues of the permeability tensor given by Eq. 1.84. The damping can again be introduced into Eqs. 1.87 and 1.88 in the usual way. In keeping with the behavior of the scalar susceptibilities the positive scalar permeability has a singularity at $\omega = \omega_0$, whereas the negative scalar susceptibility does not exhibit any resonance. Figures 1.10 and 1.11 depict the real and imaginary parts of $\mu \pm K$.

1.9.   *EFFECTIVE PERMEABILITY FOR PROPAGATION IN INFINITE MEDIUM*

When the direction of propagation makes an angle ϕ with that of the direction of magnetization, and when the tensor permeability is that given by Eq. 1.84, the solution to Maxwell's equations gives two values for the propagation coefficient for which the effective permeabilities are

$$\mu_\pm(\phi) = \left[\frac{(\mu^2 - K^2 - \mu)\sin^2\phi + 2\mu \mp \left\{ (\mu^2 - K^2 - \mu)^2 \sin^4\phi + 4K^2 \cos^2\phi \right\}^{1/2}}{2\left\{ 1 + (\mu - 1)\sin^2\phi \right\}} \right]$$

$$(1.90)$$

The two values of μ correspond to elliptically polarized waves traveling in the same direction but with different velocity, and of positive and negative sense, respectively. The positive component is that which rotates in the direction of the current that creates the steady magnetic field. If ϕ is made zero in Eq. 1.90 propagation is along the direction of the direct magnetic field. For the two permeabilities Equation 1.90 then gives

$$\mu_\pm = (\mu \mp K) \tag{1.91}$$

when they now correspond to circularly polarized waves. This result is the same as in the previous section.

If ϕ is made 90° in Eq. 1.90, the two values for the permeabilities become

$$\mu_- = \mu_e = \frac{\mu^2 - K^2}{\mu} \tag{1.92}$$

and

$$\mu_+ = \mu = 1 \tag{1.93}$$

The two permeabilities now refer to waves that are polarized with their magnetic vectors normal with and parallel to the applied field direction.

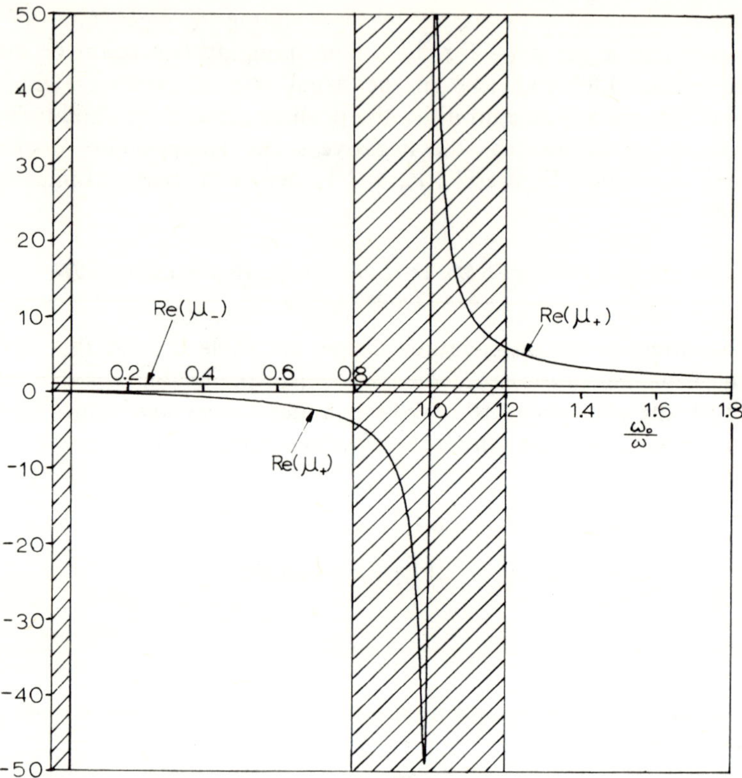

Figure 1.10.   *Real parts of* $(\mu \pm K)$ *for* $\omega_m/\omega = 1$ *and* $\alpha = 0.01$.

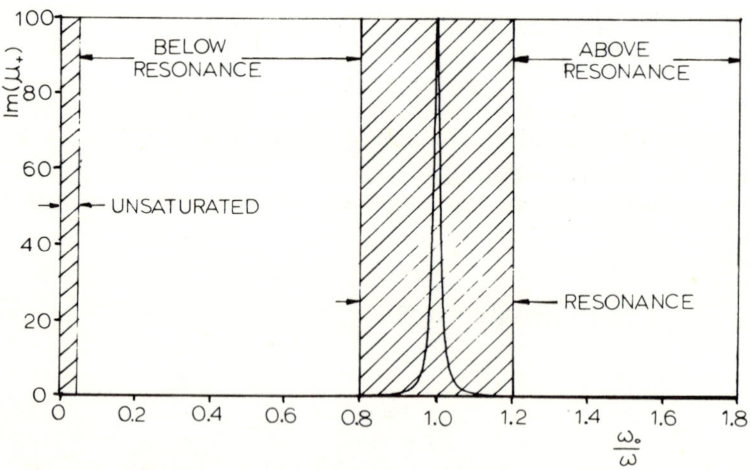

Figure 1.11.   *Imaginary parts of* $(\mu \pm K)$ *for* $\omega_m/\omega = 1$ *and* $\alpha = 0.01$.

The value of μ_+ is independent of applied field since the exciting torque on the electrons is zero. The medium appears isotropic to this wave with a permeability equal to unity. The value of μ_- depends on the magnetic field and to such a wave the medium is birefringent. Figures 1.12 and 1.13 show the real and imaginary parts of μ_e.

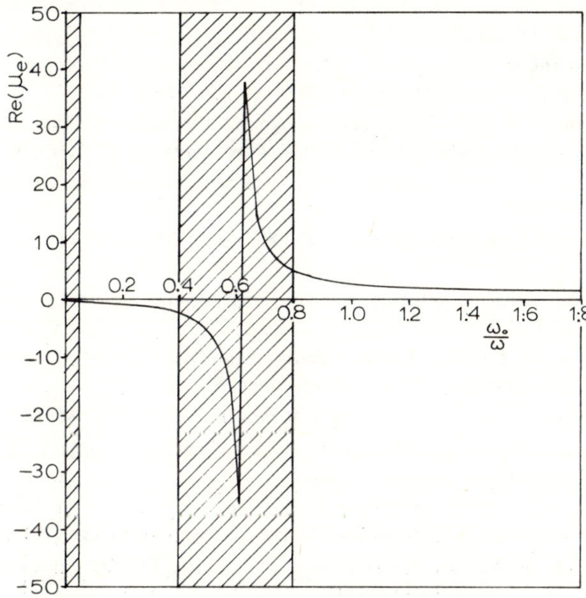

Figure 1.12.   *Real part of μ_e for $\omega_m/\omega = 1$ and $\alpha = 0.01$.*

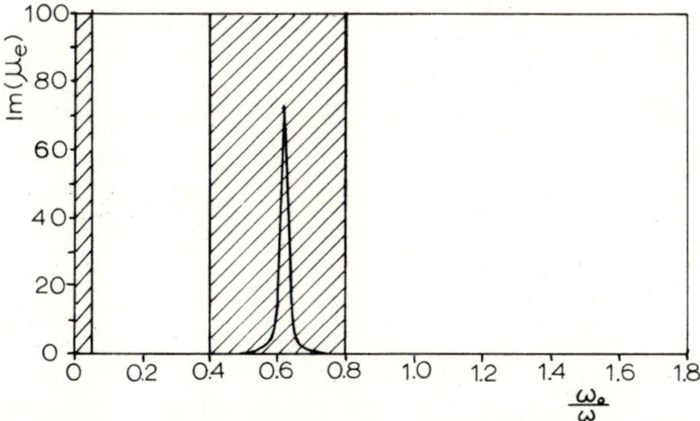

Figure 1.13.   *Imaginary part of μ_e for $\omega_m/\omega = 1$ and $\alpha = 0.01$.*

REFERENCES

1. D. Polder, "On the Theory of Ferromagnetic Resonance," *Phil. Mag.*, **40**, 100–115 (1949).

2. G. Rado, "Theory of Microwave Permeability Tensor and Faraday Effect in Non-Saturated Ferromagnetic Materials," *Phys. Rev.*, **89**, 529 (1953).

3. C. Kittel, "On the Theory of Ferromagnetic Resonance Absorption," *Phys. Rev.*, **73**, 155–161 (1948).

4. C. R. Buffler, "Resonance Properties of Single Crystal Hexagonal Ferrites," *J. Appl. Phys. Suppl.* **33** (1962).

5. H. S. Belson and C. J. Kreissman, "Microwave Resonance in Hexagonal Ferromagnetic Single Crystals," *J. Appl. Phys. Suppl.*, **30** (4), 175S (1959).

6. E. Schlömann and R. V. Jones, "Ferromagnetic Resonance in Polycrystalline Ferrites with Hexagonal Crystal Structure," *J. Appl. Phys. Suppl.* **30**, 177S (1959).

7. B. Lax, "Frequency and Loss Characteristics of Microwave Ferrite Devices," *Proc. IRE*, **44**, 1368–1386 (1956).

8. J. A. Osborn, "Demagnetizing Factors of the General Ellipsoid," *Phys. Rev.*, **67**, 351–357 (1945).

9. J. O. Artman, "Microwave Resonance Relations in Anisotropic Single-Crystal Ferrites," *Phys. Rev.*, **105**, 62 (1957).

10. L. Landau and E. Lifshitz, "On the Theory of the Dispersion of Magnetic Permeability in Ferromagnetic Bodies," *Phys. Z. Sowjetunion*, **8**, 153 (1935).

11. N. Bloembergen, "Magnetic Resonance in Ferrites," *Proc. IRE*, **44**, 1259–1250 (1956).

12. F. C. Rossol, "Subsidiary Resonance in the Coincidence Region in Yttrium Iron Garnet," *J. Appl. Phys.*, **31**, 2273 (1960).

13. J. McStay and J. Helszajn, "External Susceptibility Tensor of Magnetized ferrite ellipsoid in terms of Uniform-Mode Ellipticity," *Proc. IEEE*, **116** (1969) pp 2088–2092.

14. H. Suhl and L. R. Walker, "Guided Wave Propagation Through Gyromagnetic Media," *Bell Syst. Tech. J.*, **33**, 579–659 (1954).

15. D. M. Bolle and L. Lewin, "On the Definitions of Parameters in Ferrite Electromagnetic Wave Interactions," *IEEE Trans. on MTT*, **MTT-21**, 118 (1973).

CHAPTER TWO

Scattering Matrix of m-Port Junction

The entries of the scattering matrix of an *m*-port junction are a set of quantities that relate incident and reflected waves at the ports or terminal planes of the junction. It describes the performance of a network under any specified terminating conditions. The coefficients along the main diagonal of this matrix are reflection coefficients and those along the off-diagonal are transmission coefficients. The scattering matrix is modified if one or more of the terminal planes are moved. A scattering matrix exists for every linear, passive, time invariant network.

Figure 2.1 depicts a generalized junction enclosed by a surface S, which

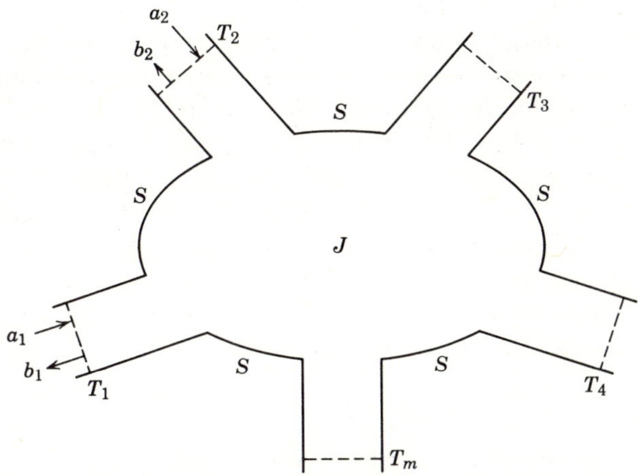

Figure 2.1. *Schematic of m-port junction showing incident and reflected waves.*

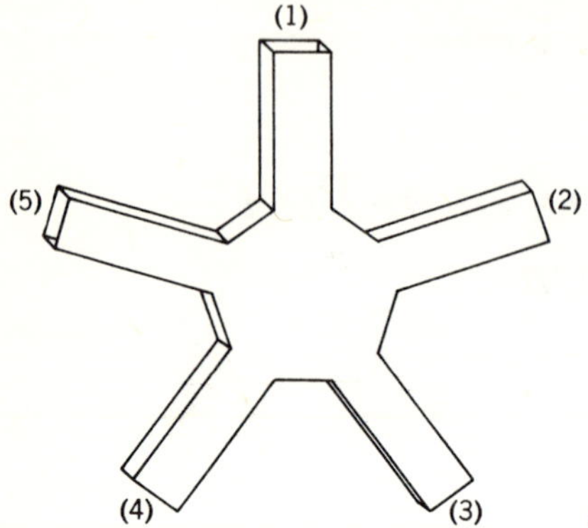

Figure 2.2. Schematic of 5-port waveguide junction.

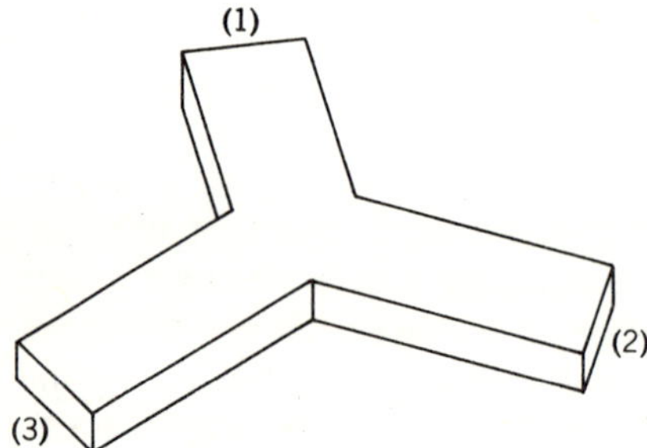

Figure 2.3.   Schematic of 3-port waveguide junction.

cuts the various transmission lines perpendicular to the axes and provides a definition for the ports or terminal planes of the junction. However, the text is primarily concerned with *m*-port junctions with the symmetry illustrated in Figures 2.2 and 2.3. It is possible to deduce important general properties of junctions containing a number of ports by invoking such properties as symmetry, reciprocity, and energy conservation. The *m*-eigenvalues of the scattering matrix that are reflection coefficients are derived. These are obtained by forming the characteristic equation. A set of *m*-eigenvectors, which are the possible *m*-ways in which the junction can be excited, is also given. A linear combination of the *m*-eigenvectors is equivalent to exciting a single port only. The eigenvectors are determined by symmetry properties of the junction only. If the input waves at the *m*-ports are proportional to one of the eigenvectors, the reflection coefficient at any port is the corresponding eigenvalue.

2.1.   *THE SCATTERING MATRIX*

The scattering matrix of the *m*-port junction is defined by

$$\bar{b} = \bar{\bar{S}}\bar{a}$$

$$(2.1)$$

where $\bar{S}$ relates the complex amplitude and phase of the incident wave a_q at the qth port to the emergent complex wave b_q at the same port defined below. It is assumed that a_q and b_q are normalized in such a way that $\frac{1}{2}a_q^* a_q$ is the average input power and $\frac{1}{2}b_q^* b_q$ is the average emergent power. For a lossless junction the principle of the conservation of energy requires that the scattering matrix be unitary. The matrix $\bar{S}$ is a square matrix, which has the form

$$\bar{S} = \begin{bmatrix} S_{11} & S_{12} & \cdots & S_{1m} \\ S_{21} & S_{22} & \cdots & S_{2m} \\ \vdots & \vdots & & \vdots \\ S_{m1} & S_{m2} & \cdots & S_{mm} \end{bmatrix} \tag{2.2}$$

The elements along the main diagonal are reflection coefficients and those along the off-diagonal are transmission coefficients. The vectors $\bar{a}$ and $\bar{b}$ are column vectors

$$\bar{a} = \begin{bmatrix} a_1 \\ a_2 \\ \vdots \\ a_m \end{bmatrix} \tag{2.3}$$

$$\bar{b} = \begin{bmatrix} b_1 \\ b_2 \\ \vdots \\ b_m \end{bmatrix} \tag{2.4}$$

a_q and b_q can be related to voltage V_q and current i_q at the port q in the following way:

$$a_q = \frac{1}{2} \left(\frac{V_q}{\sqrt{R_0}} + i_q \sqrt{R_0} \right) \tag{2.5}$$

$$b_q = \frac{1}{2} \left(\frac{V_q}{\sqrt{R_0}} - i_q \sqrt{R_0} \right) \tag{2.6}$$

where R_0 is the characteristic impedance of the termination at each port. In terms of the incident and reflected waves V_q and i_q are given by

$$V_q = V_q^+ + V_q^- \tag{2.7}$$

$$i_q = i_q^+ + i_q^- \tag{2.8}$$

Substituting Eqs. 2.7 and 2.8 into Eqs. 2.5 and 2.6 gives

$$a_q = \frac{V_q^+}{\sqrt{R_0}} \tag{2.9}$$

$$b_q = \frac{V_q^-}{\sqrt{R_0}} \tag{2.10}$$

2.2. CIRCULATOR DEFINITION BY MEANS OF CYCLIC SUBSTITUTION

In the theory of finite groups, an operation, called cyclic substitution, is defined. This operation is usually illustrated as being performed on a sequence of letters or numbers. For example, the operation of cyclic substitution ($abcd$) means that a be replaced by b, b by c, c by d, and d by a. If this operation is performed on the sequence $bdac$ the result is

$$(abcd) \rightarrow bdac = cabd \tag{2.11}$$

In the scattering relationship between incident and reflected waves in Eq. 2.1, the scattering matrix may be considered as indicating an operation performed on the incident waves, the result of which yields the reflected waves.

If the operator $\bar{S}$ corresponds to a cyclic substitution, then it can be seen that the device having such a scattering matrix corresponds to the intuitive notion of the circulator property. Consequently, a circulator may be defined as: "A device, whose scattering matrix operates on the incident waves so as to produce the same result as the operation of a cyclic substitution on the incident voltages, is called a circulator."

For a given m-port circulator different systems of port numbering will lead to ($m-1$) distinct scattering matrices. In the case of a structurally symmetrical circulator the number of possibilities may be reduced, however, by specifying a standard numbering system. For an m-port junction having the symmetry in Figures 2.2 and 2.3, it is always possible to number the ports to represent the cyclic substitution $(1, 2, 3, \ldots, m)$.

2.3.   *SYMMETRY AND GROUP PROPERTIES OF CIRCULATOR*

In the preceding section, a concept of group theory was used in formulating an exact definition of a circulator. In this section, those properties of circulators that follow directly from this definition will be examined. Primarily, it will be shown how the various symmetries that a circulator may possess are derived from its definition.

It has been shown by Dicke that the scattering matrix $\overline{S}$ of any symmetrical junction must satisfy a set of commutation relations,

$$\overline{F}\overline{S} = \overline{S}\overline{F} \tag{2.12}$$

which determine the restrictions imposed on $\overline{S}$ by the junction symmetry, where $\overline{F}$ is a "symmetry operator." The symmetry operators are matrices that indicate how the terminal fields transform under the operations of the symmetry group of the junction, that is, rotations and reflections that carry the junction into itself. The problem of finding all the symmetries of $\overline{S}$ reduces to finding all the symmetry operators that commute with $\overline{S}$. Fortunately, this can be done.

Since the $\overline{S}$ matrices of circulators are, by definition, equivalent to cyclic substitutions, an established theorem for cyclic substitutions will apply. This theorem states: "The only substitutions on m letters which are commutative with a cyclic substitution are the powers of this cyclic substitution." Since the $\overline{F}$ operators of Dicke are essentially substitutions, these $\overline{F}$ operators may be found by taking the powers of $\overline{S}$ as indicated by the theorem. Accordingly, taking a 3-port circulator as an example,

$$\overline{F}_1 = \overline{S}_0 = \begin{bmatrix} 0 & 1 & 0 \\ 0 & 0 & 1 \\ 1 & 0 & 0 \end{bmatrix} \tag{2.13}$$

$$\overline{F}_2 = \overline{S}_0^2 = \begin{bmatrix} 0 & 0 & 1 \\ 1 & 0 & 0 \\ 0 & 1 & 0 \end{bmatrix} \tag{2.14}$$

$$\bar{F}_3 = \bar{S}_0^3 = \begin{bmatrix} 1 & 0 & 0 \\ 0 & 1 & 0 \\ 0 & 0 & 1 \end{bmatrix} \qquad (2.15)$$

where $\bar{S}_0$ is the scattering matrix of an ideal circulator represented by the cyclic substitution $(1, 2, 3)$. Since $(\bar{S}_0)^3$ is equal to $\bar{I}$, the identity matrix, the powers of $\bar{S}_0$ are said to form a cyclic group with $\bar{S}_0$ as the "generator" of the group.

The $\bar{F}$ operators listed above would comprise a mathematically complete list of the symmetries of the 3-port circulator, except for the fact that negative $\bar{F}$ operators, which are permissible, also satisfy Eq. 2.12. Therefore, the complete and mathematically exhaustive list of symmetries of the 3-port circulator is contained in the following group whose order is 6:

$$F_1, F_2, I, -F_2, -I, -F_1 \qquad (2.16)$$

If a structure possesses a symmetry not contained in this group, then it cannot be a circulator.

For the 3-port junction the most general form for the scattering matrix is

$$\bar{S} = \begin{bmatrix} S_{11} & S_{12} & S_{13} \\ S_{21} & S_{22} & S_{23} \\ S_{31} & S_{32} & S_{33} \end{bmatrix} \qquad (2.17)$$

Applying the commutation relation

$$\bar{F}_1 \bar{S} = \bar{S}\bar{F}_1 \qquad (2.18)$$

gives

$$\begin{bmatrix} S_{21} & S_{22} & S_{23} \\ S_{31} & S_{32} & S_{33} \\ S_{11} & S_{12} & S_{13} \end{bmatrix} = \begin{bmatrix} S_{13} & S_{11} & S_{12} \\ S_{23} & S_{21} & S_{22} \\ S_{33} & S_{31} & S_{32} \end{bmatrix} \qquad (2.19)$$

which is possible only if

$$S_{11} = S_{22} = S_{33} \tag{2.20}$$

$$S_{23} = S_{31} = S_{12} \tag{2.21}$$

$$S_{21} = S_{32} = S_{13} \tag{2.22}$$

Hence, the general form for the scattering matrix of a 3-port circulator is

$$\bar{S} = \begin{bmatrix} S_{11} & S_{12} & S_{13} \\ S_{13} & S_{11} & S_{12} \\ S_{12} & S_{13} & S_{11} \end{bmatrix} \tag{2.23}$$

Applying the commutation relation $\bar{F}_2\bar{S} = \bar{S}\bar{F}_2$ yields a similar result.
Taking now a 4-port circulator as an example gives

$$\bar{F}_1 = \bar{S}_0 = \begin{bmatrix} 0 & 1 & 0 & 0 \\ 0 & 0 & 1 & 0 \\ 0 & 0 & 0 & 1 \\ 1 & 0 & 0 & 0 \end{bmatrix} \tag{2.24}$$

$$\bar{F}_2 = \bar{S}_0^2 = \begin{bmatrix} 0 & 0 & 1 & 0 \\ 0 & 0 & 0 & 1 \\ 1 & 0 & 0 & 0 \\ 0 & 1 & 0 & 0 \end{bmatrix} \tag{2.25}$$

$$\bar{F}_3 = \bar{S}_0^3 = \begin{bmatrix} 0 & 0 & 0 & 1 \\ 1 & 0 & 0 & 0 \\ 0 & 1 & 0 & 0 \\ 0 & 0 & 1 & 0 \end{bmatrix} \tag{2.26}$$

$$\bar{F}_4 = \bar{S}_0^4 = \begin{bmatrix} 1 & 0 & 0 & 0 \\ 0 & 1 & 0 & 0 \\ 0 & 0 & 1 & 0 \\ 0 & 0 & 0 & 1 \end{bmatrix} \tag{2.27}$$

For the 4-port junction the most general form for the scattering matrix is

$$\bar{S} = \begin{bmatrix} S_{11} & S_{12} & S_{13} & S_{14} \\ S_{21} & S_{22} & S_{23} & S_{24} \\ S_{31} & S_{32} & S_{33} & S_{34} \\ S_{41} & S_{42} & S_{43} & S_{44} \end{bmatrix} \tag{2.28}$$

Applying the commutation relation with $\bar{F}_1$ given by Eq. 2.24 gives

$$\begin{bmatrix} S_{14} & S_{11} & S_{12} & S_{13} \\ S_{24} & S_{21} & S_{22} & S_{23} \\ S_{34} & S_{31} & S_{32} & S_{33} \\ S_{44} & S_{41} & S_{42} & S_{43} \end{bmatrix} = \begin{bmatrix} S_{21} & S_{22} & S_{23} & S_{24} \\ S_{31} & S_{32} & S_{33} & S_{34} \\ S_{41} & S_{42} & S_{43} & S_{44} \\ S_{11} & S_{12} & S_{13} & S_{14} \end{bmatrix} \tag{2.29}$$

which is possible only if

$$S_{11} = S_{22} = S_{33} = S_{44} \tag{2.30}$$

$$S_{12} = S_{23} = S_{34} = S_{41} \tag{2.31}$$

$$S_{13} = S_{31} = S_{24} = S_{42} \tag{2.32}$$

$$S_{14} = S_{43} = S_{32} = S_{21} \tag{2.33}$$

The general form for the scattering matrix is therefore

$$\bar{S} = \begin{bmatrix} S_{11} & S_{12} & S_{13} & S_{14} \\ S_{14} & S_{11} & S_{12} & S_{13} \\ S_{13} & S_{14} & S_{11} & S_{12} \\ S_{12} & S_{13} & S_{14} & S_{11} \end{bmatrix} \tag{2.34}$$

2.4. *UNITARY PROPERTIES OF SCATTERING MATRIX*

For a lossless junction the principle of the conservation of energy requires that the scattering matrix be unitary

$$\bar{S}(\bar{S}*)^{T} = \bar{I} \tag{2.35}$$

The above equation will now be used to demonstrate that it is not possible to match a lossless reciprocal 3-port junction. If the junction is reciprocal it has the property that the scattering matrix is symmetric

$$S_{ij} = S_{ji} \tag{2.36}$$

If the junction is matched the diagonal elements must be zero.

The scattering matrix for a matched reciprocal 3-port junction is, therefore,

$$\bar{S} = \begin{bmatrix} 0 & S_{12} & S_{12} \\ S_{12} & 0 & S_{12} \\ S_{12} & S_{12} & 0 \end{bmatrix} \tag{2.37}$$

Combining the last equation with the unitary condition gives

$$2S_{12}S_{12}^{*} = 1 \tag{2.38}$$

and

$$S_{12}S_{12}^{*} = 0 \tag{2.39}$$

It is therefore impossible to match a reciprocal lossless 3-port junction, since it is not possible to satisfy Eqs. 2.38 and 2.39 simultaneously. It will be shown later on that the best possible match for a reciprocal 3-port junction occurs when $S_{11} = 1/3$.

It will now be demonstrated that a matched symmetric 3-port junction is necessarily a circulator. The scattering matrix of this junction is given from Eq. 2.23 by

$$\bar{S} = \begin{bmatrix} 0 & S_{12} & S_{13} \\ S_{13} & 0 & S_{12} \\ S_{12} & S_{13} & 0 \end{bmatrix} \tag{2.40}$$

Applying the unitary condition to the scattering matrix gives

$$|S_{12}|^2 + |S_{13}|^2 = 1 \tag{2.41}$$

and

$$S_{12}S_{13}^* = 0 \tag{2.42}$$

The last two equations are satisfied provided

$$|S_{12}| = 1 \tag{2.43}$$

and

$$|S_{13}| = 0 \tag{2.44}$$

or

$$|S_{12}| = 0 \tag{2.45}$$

$$|S_{13}| = 1 \tag{2.46}$$

It is obvious from the above relations that if $S_{11} = 0$ (matched condition), then either $|S_{12}| = 0$ and $|S_{13}| = 1$, or $|S_{13}| = 0$ and $|S_{12}| = 1$. These are the conditions for a perfect circulator.

It is also possible to obtain the relation between the scattering coefficients of a lossless but nearly matched circulator. The scattering matrix for this network is

$$S = \begin{bmatrix} S_{11} & S_{12} & S_{13} \\ S_{13} & S_{11} & S_{12} \\ S_{12} & S_{13} & S_{11} \end{bmatrix} \tag{2.47}$$

Applying the unitary condition gives

$$|S_{11}|^2 + |S_{12}|^2 + |S_{13}|^2 = 1 \tag{2.48}$$

and

$$S_{12}S_{13}^* + S_{12}S_{11}^* + S_{13}S_{11}^* = 0 \tag{2.49}$$

For a lossless, but not perfectly matched symmetrical 3-port junction with S_{12} close to unity and S_{11} and S_{13} small, the unitary relation shows that

$$|S_{11}| \approx |S_{13}| \tag{2.50}$$

and

$$|S_{12}| \approx 1 - 2|S_{11}|^2 \tag{2.51}$$

Thus, minimum insertion loss corresponds to both maximum isolation and minimum VSWR looking into any of the three ports.

By virtue of Eq. 2.48 $|S_{12}|$ can be expressed in terms of $|S_{11}|$ and $|S_{13}|$. The three terms in Eq. 2.49 can be interpreted as three vectors that span a triangle. Then inequality relations such as

$$|S_{12}||S_{13}| \leqslant |S_{11}|(|S_{12}|+|S_{13}|) \tag{2.52}$$

are valid. From these considerations Butterweck deduced that in the $|S_{11}|$, $|S_{12}|$ diagram the possible three ports are restricted to a region that is bounded by three ellipses

$$|S_{11}|^2 + |S_{11}||S_{13}| + |S_{13}|^2 - |S_{11}| - |S_{13}| = 0 \tag{2.53}$$

$$|S_{11}|^2 - |S_{11}||S_{13}| + |S_{13}|^2 + |S_{11}| - |S_{13}| = 0 \tag{2.54}$$

$$|S_{11}|^2 - |S_{11}||S_{13}| + |S_{13}|^2 - |S_{11}| + |S_{13}| = 0 \tag{2.55}$$

In Figure 2.4 this region is given by the shaded area. The origin of the diagram $|S_{11}|=0$, $|S_{13}|=0$, represents an ideal clockwise circulator.

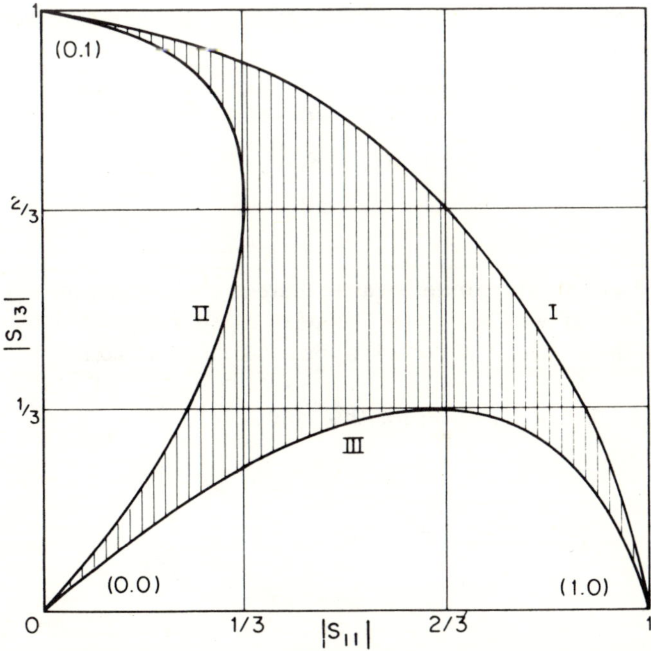

Figure 2.4.   The close region of possible lossless and cyclic–symmetric 3 ports (Ref. 2).

2.5. *THE SCATTERING MATRIX EIGENVALUES*

The relation between the scattering matrix and its eigenvalues can be obtained from the eigenvalue equation of the square matrix $\bar{S}$

$$\bar{S}\bar{U}_n = s_n \bar{U}_n \tag{2.56}$$

where $\bar{U}_n$ is an eigenvector and s_n is an eigenvalue. Through comparison with Eq. 2.1, it can be seen that $\bar{U}_n$ represents a possible excitation in the junction with the fields at the terminal planes proportional to the elements of the eigenvector, and s_n represents a reflection coefficient measured at any terminal plane. Equation 2.56 has a nonvanishing value for $\bar{U}_n$ provided

$$\det|\bar{S} - s_n \bar{I}| = 0 \tag{2.57}$$

where $\bar{I}$ is a unit vector.

Equation 2.57 is known as the characteristic equation. The determinant given by the last equation is a polynomial of degree m. The m roots of this equation are the m eigenvalues, of $\bar{S}$, some of which may be equal (degenerate). For a lossless junction, they lie in the complex plane with unit amplitude. These eigenvalues can be obtained once the coefficients of the scattering matrix are stated.

The $\bar{S}$ matrix for a 3-port junction with no transmission between the three ports is

$$\bar{S} = \begin{bmatrix} S_{11} & 0 & 0 \\ 0 & S_{11} & 0 \\ 0 & 0 & S_{11} \end{bmatrix} \tag{2.58}$$

where S_{11} has unit amplitude because there is total reflection at each port. The terminals at which $S_{11} = -1$ appear as short-circuited transmission lines, and those where $S_{11} = +1$ appear as open circuited transmission lines.

The characteristic equation for this matrix is

$$(S_{11} - s_n)^3 = 0 \tag{2.59}$$

The result for $S_{11} = -1$ is

$$s_0 = s_{+1} = s_{-1} = -1 \tag{2.60}$$

The above eigenvalues lie on a unit circle as shown in Figure 2.5.

For a reciprocal 3-port junction, the $\bar{S}$ matrix is

$$\bar{S} = \begin{bmatrix} S_{11} & S_{12} & S_{12} \\ S_{12} & S_{11} & S_{12} \\ S_{12} & S_{12} & S_{11} \end{bmatrix} \tag{2.61}$$

The characteristic equation for this matrix is

$$(S_{11} - s_n)^3 - 3(S_{11} - s_n)S_{12}^2 + 2S_{12}^3 = 0 \tag{2.62}$$

The three eigenvalues are

$$s_0 = S_{11} + 2S_{12} \tag{2.63}$$

$$s_{+1} = s_{-1} = S_{11} - S_{12} \tag{2.64}$$

This result indicates that two of the eigenvalues in a reciprocal 3-port junction are degenerate.

The entries of the $\bar{S}$ matrix can also be written in terms of the eigenvalues with the help of the last two equations:

$$S_{11} = \frac{s_0 + 2s_{+1}}{3} \tag{2.65}$$

$$S_{12} = \frac{s_0 - s_{+1}}{3} \tag{2.66}$$

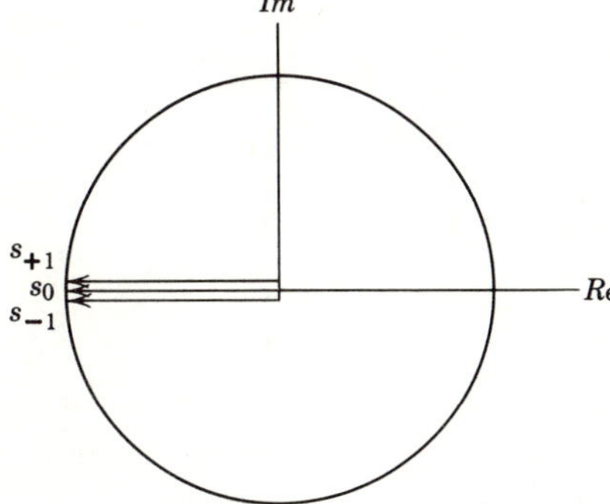

Figure 2.5.   *Eigenvalue diagram of a 3-port junction having a unit matrix as its scattering matrix.*

It is observed from the first of these two equations that S_{11} is a minimum equal to $|1/3|$ when

$$s_{+1} = -s_0 \tag{2.67}$$

The second of these equations indicates that the minimum value for S_{11} coincides with the maximum value of S_{12}. The eigenvalue diagram for maximum power transfer through a reciprocal 3-port network is depicted in Figure 2.6.

The $\bar{S}$ matrix for an ideal circulator is

$$\bar{S} = \begin{bmatrix} 0 & S_{12} & 0 \\ 0 & 0 & S_{12} \\ S_{12} & 0 & 0 \end{bmatrix} \tag{2.68}$$

where S_{12} has unit amplitude.

The characteristic equation for this matrix is

$$-s_n^3 + S_{12}^3 = 0 \tag{2.69}$$

For a circulator with $S_{12} = -1$ the result is

$$s_n = -e^{-j2\pi n/3} \tag{2.70}$$

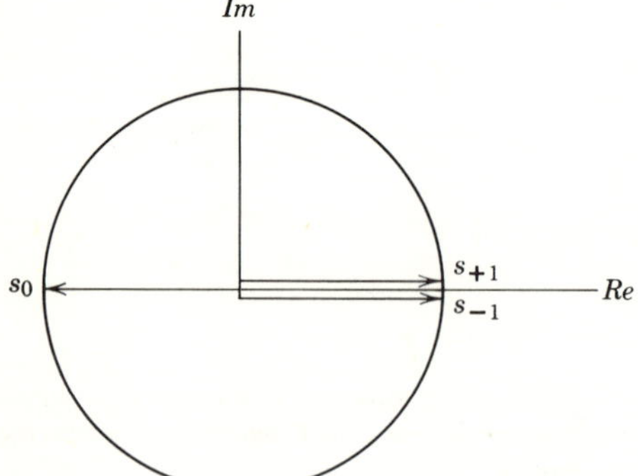

Figure 2.6. Eigenvalue diagram of reciprocal 3-port junction for maximum power transfer.

which gives

$$s_0 = -1$$

$$s_{+1} = s_0 e^{-j2\pi/3}$$

$$s_{-1} = s_0 e^{j2\pi/3}$$

$$(2.71)$$

The eigenvalues for an ideal circulator lie equally spaced on a unit circle. This is illustrated in Figure 2.7.

The characteristic equation for an ideal m-port circulator is

$$-s_n^m + S_{12}^m = 0 \tag{2.72}$$

The result is

$$s_n = s_0 e^{-j2\pi n/m} \tag{2.73}$$

where

$$n = 0, \pm 1, \pm 2, \ldots, \frac{m}{2}$$

For m odd there is one real eigenvalue and $(m-1)$ complex ones. For m even there are two real eigenvalues and $(m-2)$ complex ones. Hence, for an ideal circulator the eigenvalues lie equally spaced on a unit circle. In terms of the eigenvalue arrangement given by Eq. 2.73 the synthesis procedure consists of adjusting the phases of $(m-1)$ of them on the unit circle and hence requires $(m-1)$ independent physical variables. The phase angle of S_{12} is determined by that of s_0. For $S_{12}=1$ one has $s_0=1$, and for $S_{12}=-1$ one has $s_0=-1$.

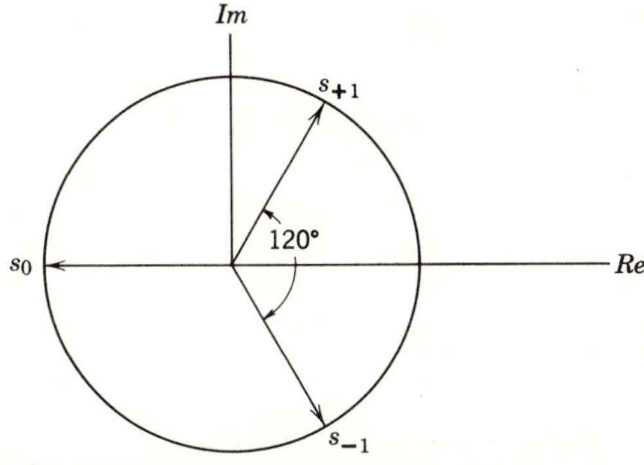

Figure 2.7. Eigenvalue diagram of ideal 3-port circulator.

The eigenvalues for an ideal 3-port circulator have already been given by Eq. 2.71. For the 4- and 5-port junctions they are

$$s_0 = -1$$

$$s_{+1} = s_0 e^{-j2\pi/4}$$

$$s_{-1} = s_0 e^{j2\pi/4}$$

$$s_{+2} = s_0 e^{-j4\pi/4}$$

$$(2.74)$$

and

$$s_0 = -1$$

$$s_{+1} = s_0 e^{-j2\pi/5}$$

$$s_{-1} = s_0 e^{j2\pi/5}$$

$$s_{+2} = s_0 e^{-j4\pi/5}$$

$$s_{-2} = s_0 e^{j4\pi/5}$$

$$(2.75)$$

Figures 2.8 and 2.9 depict the eigenvalue diagrams for the 4- and 5-port circulators. The above eigenvalues can always be rotated through an arbitrary phase angle by a suitable adjustment of the reference planes.

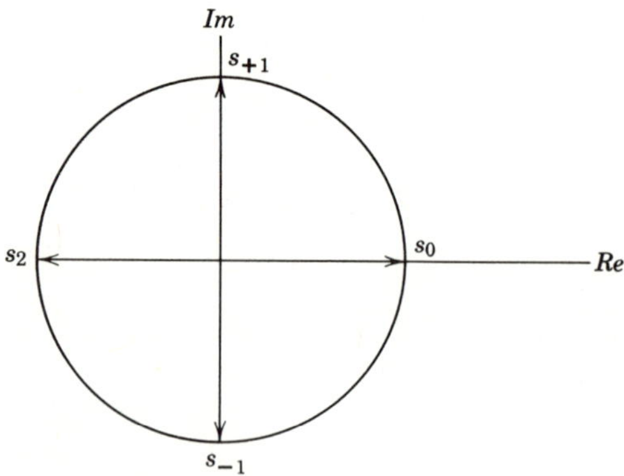

Figure 2.8.   *Eigenvalue diagram of ideal 4-port circulator.*

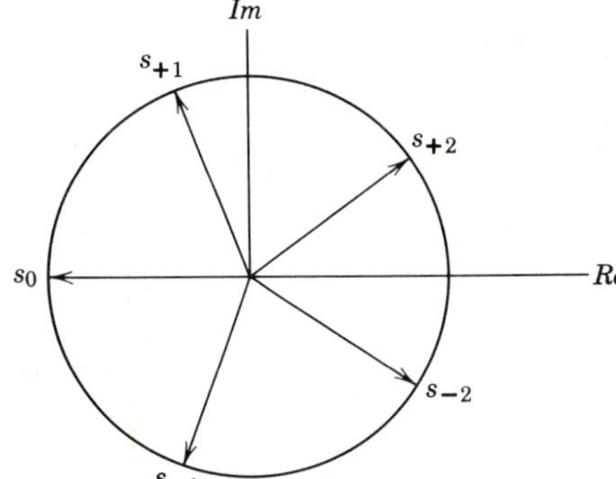

Figure 2.9.   *Eigenvalue diagram of ideal 5-port circulator.*

2.6.   *SCATTERING MATRIX EIGENVECTORS*

The eigenvectors of the junction are determined by symmetry properties of the junction only. In the case of the star junction circulator in Figure 2.2 a set of eigenvectors are given by the set of column matrices $\overline{U}_n$

$$\left(\overline{U}_n\right)_q = \frac{e^{-j2\pi n(q-1)/m}}{\sqrt{m}}$$

$$n = 0, \pm 1, \pm 2, \ldots, \frac{m}{2} \qquad\qquad (2.76)$$

$$q = 1, 2, 3, \ldots, m$$

That the above eigenvectors satisfy the eigenvalue equation can be demonstrated by direct substitution into Equation 2.56. Each eigenvalue and eigenvector represents a unique eigensolution to the junction boundary value problem which can be solved. The above eigenvectors are not orthogonal because the symmetry matrices in Section 2.3 are not symmetrical. Since the eigenvectors are completely determined by the junction symmetry, a symmetric perturbation of the junction alters the phases of the eigenvalues but leaves the eigenvectors unchanged.

For a 3-port symmetric junction Eq. 2.76 gives for the eigenvectors

$$\overline{U}_0 = \frac{1}{\sqrt{3}} \begin{bmatrix} 1 \\ 1 \\ 1 \end{bmatrix} \qquad\qquad (2.77)$$

$$\overline{U}_{+1} = \frac{1}{\sqrt{3}} \begin{bmatrix} 1 \\ e^{-j2\pi/3} \\ e^{-j4\pi/3} \end{bmatrix} \tag{2.78}$$

$$\overline{U}_{-1} = \frac{1}{\sqrt{3}} \begin{bmatrix} 1 \\ e^{j2\pi/3} \\ e^{j4\pi/3} \end{bmatrix} \tag{2.79}$$

The above three eigenvectors represent the three possible ways of exciting the junction which will give identical reflection coefficients at each port. A linear combination of these individual excitations is equivalent to a single input at any port.

The eigensolution corresponding to the set of eigenvectors have been described in the literature. For those corresponding to the eigenvector $\overline{U}_0$ the electromagnetic field at the center of the junction has only components parallel to the axis of the junction. For the eigensolutions corresponding to the eigenvectors $\overline{U}_{+1}$ and $\overline{U}_{-1}$, the axial fields vanish at the center of the junction, but the transverse components of the electric and magnetic field give rise to circularly-polarized waves rotating in one sense for one eigenvector and in the opposite sense for the other. Corresponding to these three excitations, one obtains the reflection coefficients s_0, s_{+1}, and s_{-1} at any port.

In the case of the 4-port symmetric junction the eigenvectors are

$$\overline{U}_0 = \frac{1}{\sqrt{4}} \begin{bmatrix} 1 \\ 1 \\ 1 \\ 1 \end{bmatrix} \tag{2.80}$$

$$\overline{U}_{+1} = \frac{1}{\sqrt{4}} \begin{bmatrix} 1 \\ e^{-j2\pi/4} \\ e^{-j4\pi/4} \\ e^{-j6\pi/4} \end{bmatrix} \tag{2.81}$$

$$\overline{U}_{-1} = \frac{1}{\sqrt{4}} \begin{bmatrix} 1 \\ e^{j2\pi/4} \\ e^{j4\pi/4} \\ e^{j6\pi/4} \end{bmatrix} \qquad (2.82)$$

$$\overline{U}_{+2} = \frac{1}{\sqrt{4}} \begin{bmatrix} 1 \\ e^{-j4\pi/4} \\ e^{-j8\pi/4} \\ e^{-j12\pi/4} \end{bmatrix} \qquad (2.83)$$

For the 5-port they are

$$\overline{U}_{0} = \frac{1}{\sqrt{5}} \begin{bmatrix} 1 \\ 1 \\ 1 \\ 1 \\ 1 \end{bmatrix} \qquad (2.84)$$

$$\overline{U}_{+1} = \frac{1}{\sqrt{5}} \begin{bmatrix} 1 \\ e^{-j2\pi/5} \\ e^{-j4\pi/5} \\ e^{-j6\pi/5} \\ e^{-j8\pi/5} \end{bmatrix} \qquad (2.85)$$

$$\overline{U}_{-1} = \frac{1}{\sqrt{5}} \begin{bmatrix} 1 \\ e^{+j2\pi/5} \\ e^{+j4\pi/5} \\ e^{+j6\pi/5} \\ e^{+j8\pi/5} \end{bmatrix} \qquad (2.86)$$

$$\overline{U}_{+2} = \frac{1}{\sqrt{5}} \begin{bmatrix} 1 \\ e^{-j4\pi/5} \\ e^{-j8\pi/5} \\ e^{-j12\pi/5} \\ e^{-j16\pi/5} \end{bmatrix} \qquad (2.87)$$

$$\overline{U}_{-2} = \frac{1}{\sqrt{5}} \begin{bmatrix} 1 \\ e^{+j4\pi/5} \\ e^{+j8\pi/5} \\ e^{+j12\pi/5} \\ e^{+j16\pi/5} \end{bmatrix} \qquad (2.88)$$

2.7. DIAGONALIZATION OF SCATTERING MATRIX

If the eigenvalues are known it is possible to form the coefficients of the matrix $\overline{S}$. The relation between the two is obtained by diagonalizing $\overline{S}$. This can be done by a matrix $\overline{U}$ having for its columns the eigenvectors of $\overline{S}$

$$\overline{S} = \overline{U}\overline{\lambda}\overline{U}^{-1} \qquad (2.89)$$

where $\overline{\lambda}$ is a diagonal matrix with the eigenvalues of $\overline{S}$ along its main diagonal and $\overline{U}^{-1}$ is the inverse of $\overline{U}$. If the eigenvectors of $\overline{S}$ are obtained from the eigenvalue equation of an ideal lossless circulator the following applies:

$$\overline{U}^{-1} = (\overline{U}^*)^T \qquad (2.90)$$

where $(\overline{U}^*)^T$ is the transpose of the complex conjugate of $\overline{U}$. The relation between the eigenvalues and the coefficients of the scattering matrix is obtained by multiplying out Eq. 2.89.

The diagonalization procedure will now be developed for a 3-port junction. This gives the relation between the eigenvalues and the scattering coefficients of the scattering matrix.

The matrix $\overline{U}$ having the eigenvectors of $\overline{S}$ as columns is from Eqs. 2.77 through 2.79

$$\overline{U}=\frac{1}{\sqrt{3}}\begin{bmatrix} 1 & 1 & 1 \\ 1 & e^{-j2\pi/3} & e^{j2\pi/3} \\ 1 & e^{-j4\pi/3} & e^{j4\pi/3} \end{bmatrix} \qquad (2.91)$$

The matrix $(\overline{U}*)^{T}$ is

$$(\overline{U}*)^{T}=\frac{1}{\sqrt{3}}\begin{bmatrix} 1 & 1 & 1 \\ 1 & e^{j2\pi/3} & e^{j4\pi/3} \\ 1 & e^{-j2\pi/3} & e^{-j4\pi/3} \end{bmatrix} \qquad (2.92)$$

In the case of the 3-port junction, there are a degenerate pair of eigenvalues and one nondegenerate one. The diagonal matrix $\overline{\lambda}$ with the eigenvalues of $\overline{S}$ is, therefore,

$$\overline{\lambda}=\begin{bmatrix} s_0 & 0 & 0 \\ 0 & s_{+1} & 0 \\ 0 & 0 & s_{-1} \end{bmatrix} \qquad (2.93)$$

Substituting the above quantities into Eq. 2.89 gives the following relation between the scattering coefficients and the eigenvalues

$$3S_{11}=s_0+s_{+1}+s_{-1} \qquad (2.94)$$

$$3S_{12}=s_0+s_{+1}e^{j2\pi/3}+s_{-1}e^{-j2\pi/3} \qquad (2.95)$$

$$3S_{13}=s_0+s_{+1}e^{-j2\pi/3}+s_{-1}e^{j2\pi/3} \qquad (2.96)$$

The eigenvalues given by Eq. 2.71 give the conditions for an ideal circulator. If s_{+1} and s_{-1} are interchanged, the sense of circulation is reversed. The eigenvalues become interchanged when the biasing direct magnetic field is reversed. It can be seen from Eq. 2.94 that the spur of the scattering matrix is equal to the sum of the eigenvalues. This is a general result.

If only the reflection coefficient S_{11} is required, it is not necessary to diagonalize the matrix $\bar{S}$, since it can be obtained quite simply by equating the spur of the scattering matrix to the sum of the eigenvalues.

In the case of the 4-port junction the matrix U having the eigenvectors of $\bar{S}$ as columns is

$$\bar{U} = \frac{1}{\sqrt{4}} \begin{bmatrix} 1 & 1 & 1 & 1 \\ 1 & e^{-j2\pi/4} & e^{j2\pi/4} & e^{-j4\pi/4} \\ 1 & e^{-j4\pi/4} & e^{j4\pi/4} & e^{-j8\pi/4} \\ 1 & e^{-j6\pi/4} & e^{j6\pi/4} & e^{-j12\pi/4} \end{bmatrix} \tag{2.97}$$

The diagonal matrix $\bar{\lambda}$ has one degenerate pair of eigenvalues and two nondegenerate ones

$$\bar{\lambda} = \begin{bmatrix} s_0 & 0 & 0 & 0 \\ 0 & s_{+1} & 0 & 0 \\ 0 & 0 & s_{-1} & 0 \\ 0 & 0 & 0 & s_2 \end{bmatrix} \tag{2.98}$$

The relation between the scattering coefficients and the eigenvalues is, therefore,

$$\begin{aligned} 4S_{11} &= s_0 + s_{+1} + s_{-1} + s_2 \\ 4S_{12} &= s_0 + js_{+1} - js_{-1} - s_2 \\ 4S_{13} &= s_0 - s_{+1} - s_{-1} + s_2 \\ 4S_{14} &= s_0 - js_{+1} + js_{-1} - s_2 \end{aligned} \tag{2.99}$$

For the 5-port junction one has

$$\bar{U} = \frac{1}{\sqrt{5}} \begin{bmatrix} 1 & 1 & 1 & 1 & 1 \\ 1 & e^{-j2\pi/5} & e^{+j2\pi/5} & e^{-j4\pi/5} & e^{+j4\pi/5} \\ 1 & e^{-j4\pi/5} & e^{+j4\pi/5} & e^{-j8\pi/5} & e^{+j8\pi/5} \\ 1 & e^{-j6\pi/5} & e^{+j6\pi/5} & e^{-j12\pi/5} & e^{+j12\pi/5} \\ 1 & e^{-j8\pi/5} & e^{+j8\pi/5} & e^{-j16\pi/5} & e^{+j16\pi/5} \end{bmatrix} \tag{2.100}$$

In this case there are two degenerate pairs of eigenvalues and one nondegenerate one. The diagonal matrix $\bar{\lambda}$ with the eigenvalues of $\bar{S}$ is, therefore,

$$\bar{\lambda} = \begin{bmatrix} s_0 & 0 & 0 & 0 & 0 \\ 0 & s_{+1} & 0 & 0 & 0 \\ 0 & 0 & s_{-1} & 0 & 0 \\ 0 & 0 & 0 & s_{+2} & 0 \\ 0 & 0 & 0 & 0 & s_{-2} \end{bmatrix} \tag{2.101}$$

Substituting the above quantities into Eq. 2.89 and making use of Eq. 2.90 gives the following relation between the scattering coefficients and the eigenvalues:

$$5S_{11} = s_0 + s_{+1} + s_{-1} + s_{+2} + s_{-2} \tag{2.102}$$

$$5S_{12} = s_0 + s_{+1}e^{+j2\pi/5} + s_{-1}e^{-j2\pi/5} + s_{+2}e^{+j4\pi/5} + s_{-2}e^{-j4\pi/5} \tag{2.103}$$

$$5S_{13} = s_0 + s_{+1}e^{+j4\pi/5} + s_{-1}e^{-j4\pi/5} + s_{+2}e^{+j8\pi/5} + s_{-2}e^{-j8\pi/5} \tag{2.104}$$

$$5S_{14} = s_0 + s_{+1}e^{+j6\pi/5} + s_{-1}e^{-j6\pi/5} + s_{+2}e^{+j12\pi/5} + s_{-2}e^{-j12\pi/5} \tag{2.105}$$

$$5S_{15} = s_0 + s_{+1}e^{+j8\pi/5} + s_{-1}e^{-j8\pi/5} + s_{+2}e^{+j16\pi/5} + s_{-2}e^{-j16\pi/5} \tag{2.106}$$

If the scattering coefficients are known instead of the eigenvalues one has by postmultiplying and premultiplying Eq. (2.89) by $\bar{U}$ and $\bar{U}^{-1}$,

$$\bar{\lambda} = \bar{U}^{-1}\bar{S}\bar{U} \tag{2.107}$$

This last equation gives the eigenvalues in terms of the scattering coefficients. For the 3-port junction the result is

$$s_0 = S_{11} + S_{12} + S_{13} \tag{2.108}$$

$$s_{+1} = S_{11} + S_{12}e^{j2\pi/3} + S_{13}e^{-j2\pi/3} \tag{2.109}$$

$$s_{-1} = S_{11} + S_{12}e^{-j2\pi/3} + S_{13}e^{j2\pi/3} \tag{2.110}$$

Adding the last three equations again gives the general result that the spur of the scattering matrix is equal to the sum of the eigenvalues. For a reciprocal junction $S_{12} = S_{13}$ and the above equations give

$$s_{+1} = s_{-1} = s_1 \tag{2.111}$$

This means that two of the eigenvalues in a reciprocal junction are equal.

2.8. *EIGENSOLUTIONS*

In this section it is desired to demonstrate that if the input wave at each terminal of a junction correspond to a particular eigenvector the reflection coefficient at any port is the corresponding eigenvalue. For the 3-port junction, the incident and reflected waves are related by the scattering matrix in the following way:

$$
\begin{bmatrix} b_1 \\ b_2 \\ b_3 \end{bmatrix} = \begin{bmatrix} S_{11} & S_{12} & S_{13} \\ S_{13} & S_{11} & S_{12} \\ S_{12} & S_{13} & S_{11} \end{bmatrix} \begin{bmatrix} a_1 \\ a_2 \\ a_3 \end{bmatrix}
\tag{2.112}
$$

For the eigensolution corresponding to the eigenvector $\overline{U}_0$ Eq. 2.112 gives

$$
\frac{b_1}{a_1} = \frac{b_2}{a_2} = \frac{b_3}{a_3} = s_0
\tag{2.113}
$$

This last equation is obtained by writing the scattering coefficients in terms of the eigenvalues with the help of Eqs. 2.91 through 2.93. The eigensolutions corresponding to the eigenvectors $\overline{U}_{+1}$ and $\overline{U}_{-1}$ are

$$
\frac{b_1}{a_1} = \frac{b_2}{a_2} = \frac{b_3}{a_3} = s_{+1}
\tag{2.114}
$$

and

$$
\frac{b_1}{a_1} = \frac{b_2}{a_2} = \frac{b_3}{a_3} = s_{-1}
\tag{2.115}
$$

These eigensolutions are illustrated in Figure 2.10.

2.9. *SCATTERING MATRIX OF 3-PORT ASYMMETRIC CIRCULATOR IN TERMS OF 2-PORT NETWORKS*

One model for a 3-port junction circulator consists of an ideal 3-port circulator with 2-port networks connected in each port. This equivalent network will be derived in Chapter 5. In this model the frequency behavior of the overall circulator is contained in the 2-port networks. The overall

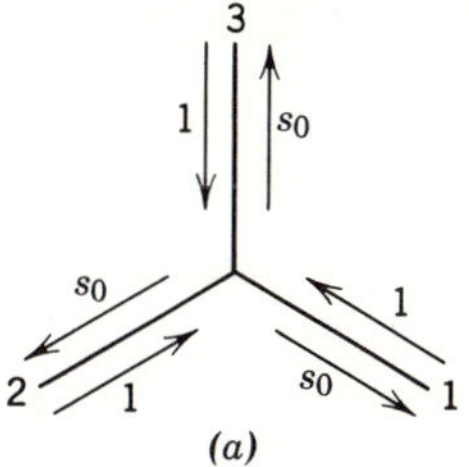

(a)

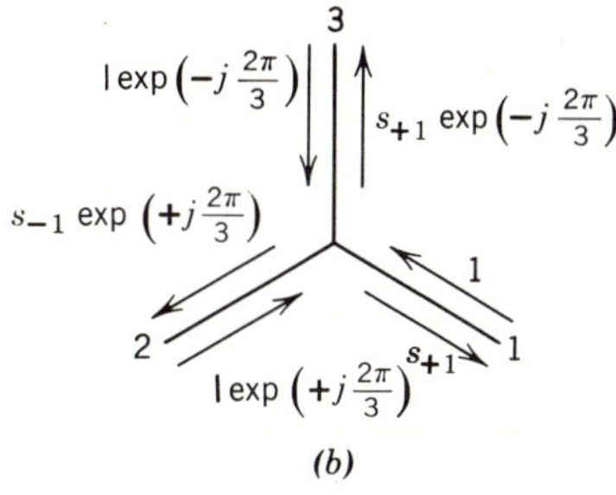

(b)

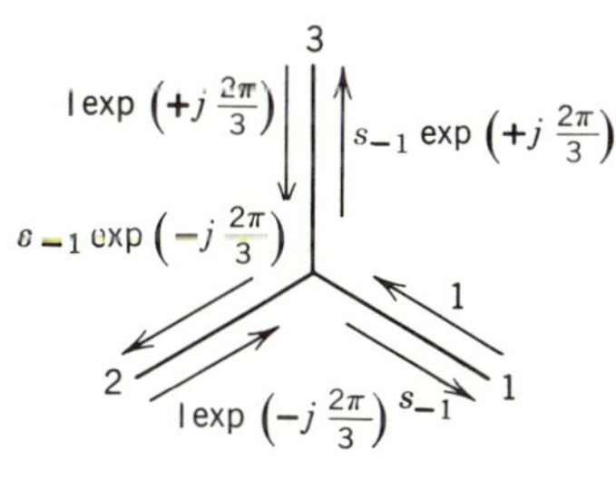

(c)

Figure 2.10. (a) *Eigensolution for* $\overline{U}_0$ *excitation.* (b) *Eigensolution for* $\overline{U}_{+1}$ *excitation.* (c) *Eigensolution for* $\overline{U}_{-1}$ *excitation.*

scattering matrix can be constructed from the known scattering matrices of the individual networks. This can be done using Mason's nontouching loop rule. In Figure 2.11 the flow graph of the 3-port circulator model is shown.

Here the 2-port scattering matrices are given by

$$\overline{S}^{(i)} = \begin{bmatrix} S_{11}^{(i)} & S_{12}^{(i)} \\ S_{21}^{(i)} & S_{22}^{(i)} \end{bmatrix} \tag{2.116}$$

where $i = 1, 2,$ or 3 denotes the port where the network is connected.

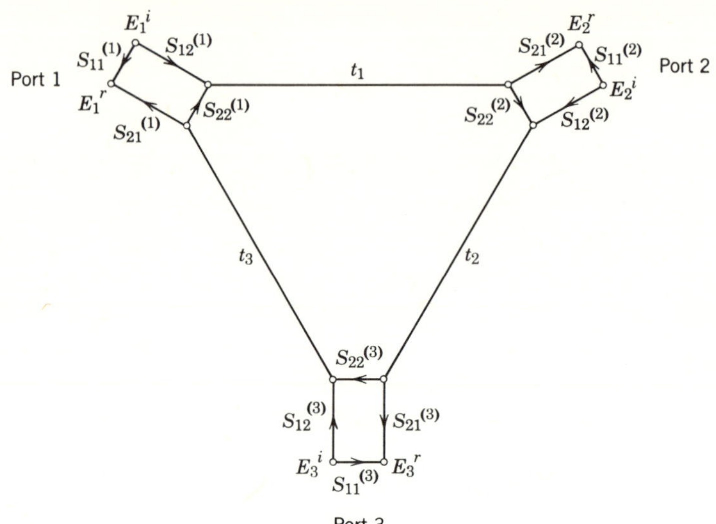

Figure 2.11. *Schematic of 3-port circulator in terms of ideal 3-port circulator with 2-port networks connected to each port (Ref. 1).*

The following expressions for the scattering coefficients of the overall network can now be obtained:

$$S_{11} = S_{11}^{(1)} + \frac{S_{21}^{(1)} S_{12}^{(1)} S_{22}^{(2)} S_{22}^{(3)} t_1 t_2 t_3}{N} \tag{2.117}$$

$$S_{21} = \frac{S_{21}^{(1)} S_{12}^{(2)} t_1}{N} \tag{2.118}$$

$$S_{31} = \frac{S_{21}^{(1)} S_{12}^{(3)} S_{22}^{(2)} t_1 t_2}{N} \tag{2.119}$$

$$S_{12} = \frac{S_{21}^{(2)} S_{12}^{(1)} S_{22}^{(3)} t_2 t_3}{N} \tag{2.120}$$

$$S_{22} = S_{11}^{(2)} + \frac{S_{21}^{(2)} S_{12}^{(2)} S_{22}^{(1)} S_{22}^{(3)} t_1 t_2 t_3}{N} \tag{2.121}$$

$$S_{32} = \frac{S_{21}^{(2)} S_{12}^{(3)} t_2}{N} \tag{2.122}$$

$$S_{13} = \frac{S_{21}^{(3)} S_{12}^{(1)} t_3}{N} \tag{2.123}$$

$$S_{23} = \frac{S_{21}^{(3)} S_{12}^{(2)} S_{22}^{(1)} t_1 t_3}{N} \tag{2.124}$$

$$S_{33} = S_{11}^{(3)} + \frac{S_{21}^{(3)} S_{12}^{(3)} S_{22}^{(1)} S_{22}^{(2)} t_1 t_2 t_3}{N} \tag{2.125}$$

where

$$N = 1 - S_{22}^{(1)} S_{22}^{(2)} S_{22}^{(3)} t_1 t_2 t_3 \tag{2.126}$$

and $t_i = e^{j\theta_i}$ and θ_i is the phase shift through the junction. The overall scattering matrix of the device is, therefore, determined by the entries of the 2-port matrices. A number of separate cases may be considered. For symmetric circulators with ideal transformers connected at each port, the 2-port scattering matrix is

$$\bar{S} = \begin{bmatrix} 0 & 1 \\ 1 & 0 \end{bmatrix} \tag{2.127}$$

and the overall matrix reduces to that of an ideal circulator with circulation in the direction 1 to 2 to 3.

For a symmetric circulator with pure shunt admittances the scattering matrix of the 2-port is

$$\bar{S} = \frac{1}{y+2} \begin{bmatrix} -y & 2 \\ 2 & -y \end{bmatrix} \tag{2.128}$$

This type of model can be used to predict the overall response of a circulator, if the nature of the admittance function in the vicinity of the center frequency is known.

Another case of interest is where the two ports represent a transformer with a turns ratio $a:1$ for which the $\bar{S}$ matrix is

$$\bar{S} = \begin{bmatrix} \dfrac{a^2-1}{a^2+1} & \dfrac{2a}{a^2+1} \\ \dfrac{2a}{a^2+1} & \dfrac{1-a^2}{a^2+1} \end{bmatrix} \tag{2.129}$$

Figures 2.12a through d depict the $\bar{S}$ matrices for the 2-port networks discussed.

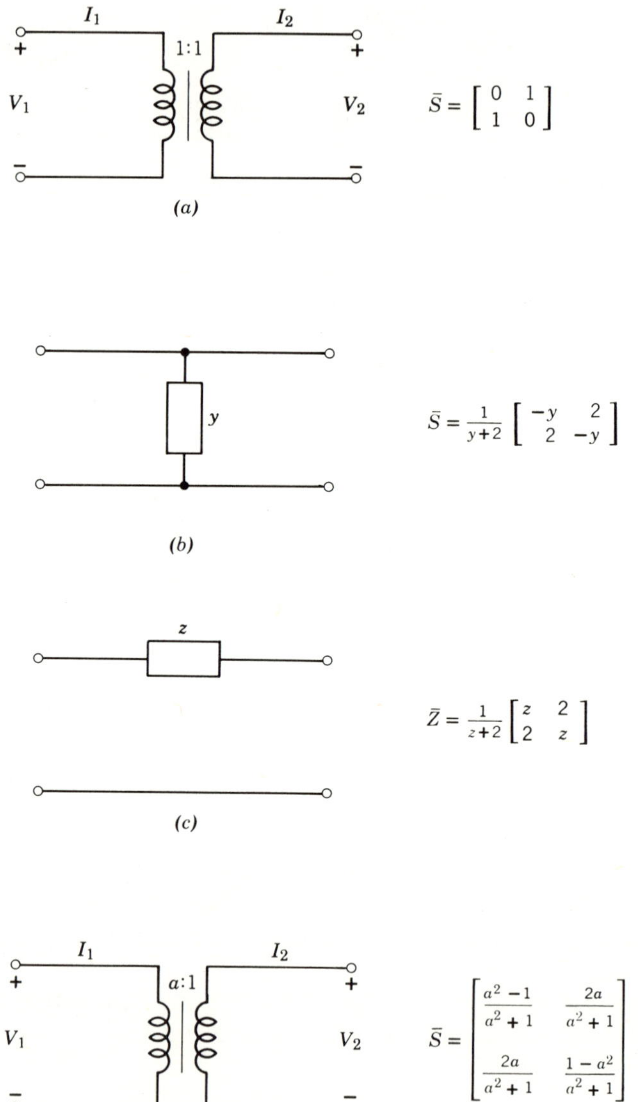

$$\bar{S} = \begin{bmatrix} 0 & 1 \\ 1 & 0 \end{bmatrix}$$

(a)

$$\bar{S} = \frac{1}{y+2} \begin{bmatrix} -y & 2 \\ 2 & -y \end{bmatrix}$$

(b)

$$\bar{Z} = \frac{1}{z+2} \begin{bmatrix} z & 2 \\ 2 & z \end{bmatrix}$$

(c)

$$\bar{S} = \begin{bmatrix} \dfrac{a^2-1}{a^2+1} & \dfrac{2a}{a^2+1} \\[2mm] \dfrac{2a}{a^2+1} & \dfrac{1-a^2}{a^2+1} \end{bmatrix}$$

(d)

Figure 2.12. (a) Schematic of ideal 2-port one-to-one transformer, and its scattering matrix. (b) Schematic of 2-port shunt admittance and its scattering matrix. (c) Schematic of 2-port series impedance and its scattering matrix. (d) Schematic of 2-port a-to-one transformer and its scattering matrix.

REFERENCES

1. S. Hagelin, "A Flow Graph Analysis of 3- and 4-port Junction Circulators," *IEEE Trans. Microwave Theory Tech.* **MTT-14** (5), 243–249 (1966).

2. D. M. Kerns, "Analysis of Symmetrical Waveguide Junctions," *J. Res. Nat. Bur. Stand. Sect.* 000 **46**, 000, (1951).

3. H. J. Butterweck, "The Y Circulator," *Arch. Elek. Ubertragung*, 17 (4), 163–76 (1963).

4. B. A. Auld, "The Synthesis of Symmetrical Waveguide Circulators," *IRE Trans. Microwave Theory Tech.*, **MTT-7**, 137–46, (1962).

5. C. G. Montgomery, R. H. Dicke, and E. M. Purcell, *Principles of Microwave Circuits*, McGraw-Hill, New York, 1948.

6. M. A. Treuhaft, "Network Properties of Circulators Based on the Scattering Concept," *Proc. IRE*, **44**, 1394–1402 (1956).

7. B. L. Humphreys and J. B. Davies, "The Synthesis of N-Port Circulators," *IRE Trans. Microwave Theory Tech.*, **MTT-10**, 551–554 (1962).

CHAPTER THREE

Imittance Matrix

In addition to the scattering matrix description of the m-port junction another useful way in which a junction may be described is in terms of its immittance matrix. By the immittance of a junction it is meant either the impedance or admittance matrices.

The immittance matrix can be diagonalized in a similar way to the scattering matrix by noting that the eigenvectors of $\bar{S}$ are also those of $\bar{Z}$ and $\bar{Y}$. This is so because the matrices $\bar{Z}$ and $\bar{Y}$ commute with $\bar{S}$ and the symmetry operators. Another property of these matrices is that the eigenvalues of all three are related. The derivation of the immittance matrix allows equivalent networks of junctions to be determined.

The procedure used in this book for relating the $\bar{S}$, $\bar{Z}$, and $\bar{Y}$ matrices will consist of first finding the eigenvalues of, for instance, the $\bar{S}$ matrix. The eigenvalues of the latter matrix will then be related to those of $\bar{Z}$ and $\bar{Y}$, and these will be used to construct the immittances of $\bar{Z}$ and $\bar{Y}$ directly.

It is sometimes possible that a scattering matrix for a junction has no impedance or admittance matrix (the elements becoming infinitely large). This situation is determined by the entries of the $\bar{S}$ matrix. If it is not possible to determine directly an immittance matrix corresponding to a given scattering matrix, a new scattering matrix can be defined by shifting the terminal planes. An immittance matrix may then be possible at the new reference plane.

3.1. *IMPEDANCE MATRIX*

The complete relation between voltages and currents of an m-port junction is defined through an impedance matrix given by

$$\bar{V} = \bar{Z}\bar{I} \tag{3.1}$$

where the impedance matrix $\overline{Z}$ is defined by the square matrix

$$\overline{Z} = \begin{bmatrix} Z_{11} & Z_{12} & \cdots & Z_{1m} \\ Z_{21} & Z_{22} & \cdots & Z_{2m} \\ \cdot\cdot\cdot\cdot\cdot\cdot\cdot\cdot\cdot\cdot\cdot\cdot\cdot\cdot \\ Z_{m1} & Z_{m2} & \cdots & Z_{mm} \end{bmatrix} \qquad (3.2)$$

and the voltage and current matrices are column matrices

$$\overline{V} = \begin{bmatrix} V_1 \\ V_2 \\ V_m \end{bmatrix} \qquad (3.3)$$

$$\bar{i} = \begin{bmatrix} i_1 \\ i_2 \\ i_m \end{bmatrix} \qquad (3.4)$$

which are defined in Chapter 2. There is a one-to-one correspondence between the scattering and impedance matrices that will now be derived.

The derivation starts with the normalized scattering waves given by Eqs. 2.5 and 2.6

$$\bar{a} = \tfrac{1}{2}(\overline{V} + \bar{i}) \qquad (3.5)$$

$$\bar{b} = \tfrac{1}{2}(\overline{V} - \bar{i}) \qquad (3.6)$$

Rewriting the above equations gives

$$\overline{V} = \bar{a} + \bar{b} \qquad (3.7)$$

$$\bar{i} = \bar{a} - \bar{b} \qquad (3.8)$$

Substituting the relation between $\bar{a}$ and $\bar{b}$ given by Eq. 2.1 into the last two equations gives

$$\overline{V} = (\overline{I} + \overline{S})\overline{a} \tag{3.9}$$

$$\overline{i} = (\overline{I} - \overline{S})\overline{a} \tag{3.10}$$

If now Eq. 3.1 is used to eliminate $\overline{V}$ from Eq. 3.9 the result is

$$\overline{Z}\overline{i} = (\overline{I} + \overline{S})\overline{a} \tag{3.11}$$

$$\overline{i} = (\overline{I} - \overline{S})\overline{a} \tag{3.12}$$

Hence

$$(\overline{I} + \overline{S}) = \overline{Z}(\overline{I} - \overline{S}) \tag{3.13}$$

Postmultiplying both sides by $(\overline{I} - \overline{S})^{-1}$ gives

$$\overline{Z} = (\overline{I} + \overline{S})(\overline{I} - \overline{S})^{-1} \tag{3.14}$$

A $\overline{Z}$ matrix exists provided the inverse matrix $(\overline{I} - \overline{S})^{-1}$ exists. This condition is satisfied if the matrix $(\overline{I} - \overline{S})$ is nonsingular. This requires that the determinant of the matrix should not be zero.

$$|\overline{I} - \overline{S}| \neq 0 \tag{3.15}$$

This last condition is determined by the entries of the scattering matrix.

If the $\overline{Z}$ matrix is known the $\overline{S}$ matrix can also be formed. To obtain this relation it is first necessary to eliminate $\overline{V}$ in Eqs. 3.5 and 3.6 by using Eq. 3.1

$$\overline{a} = \tfrac{1}{2}(\overline{Z} + \overline{I})\overline{i} \tag{3.16}$$

$$\overline{b} = \tfrac{1}{2}(\overline{Z} - \overline{I})\overline{i} \tag{3.17}$$

Using Eq. 2.1 to eliminate $\overline{b}$ in terms of $\overline{a}$ in Eq. 3.16 gives

$$\overline{a} = \tfrac{1}{2}(\overline{Z} + \overline{I})\overline{i} \tag{3.18}$$

$$\overline{S}\overline{a} = \tfrac{1}{2}(\overline{Z} - \overline{I})\overline{i} \tag{3.19}$$

Hence

$$\bar{S}(\bar{Z}+\bar{I})=(\bar{Z}-\bar{I}) \tag{3.20}$$

Postmultiplying both sides of the last equation by $(\bar{Z}+\bar{I})^{-1}$ gives

$$\bar{S}=(\bar{Z}-\bar{I})(\bar{Z}+\bar{I})^{-1} \tag{3.21}$$

In order not to have to invert matrices, the procedure used in this book for relating the matrices $\bar{S}$, $\bar{Y}$, and $\bar{Z}$ will consist of first finding the eigenvalues of, for instance, $\bar{S}$. These eigenvalues will then be related to those of $\bar{Y}$ and $\bar{Z}$, and they will be used to directly construct these matrices.

3.2.   THE ADMITTANCE MATRIX

It is also possible to define an admittance matrix $\bar{Y}$ for a junction provided the inverse of the impedance matrix $\bar{Z}$ exists.

$$\bar{Y}=\bar{Z}^{-1} \tag{3.22}$$

The admittance matrix relates currents and voltages at the m-ports of the junction by

$$\bar{i}=\bar{Y}\bar{V} \tag{3.23}$$

where the admittance matrix $\bar{Y}$ is a square matrix given by

$$\bar{Y}=\begin{bmatrix} Y_{11} & Y_{12} & \cdots & Y_{1m} \\ Y_{21} & Y_{22} & \cdots & Y_{2m} \\ Y_{m1} & Y_{m2} & \cdots & Y_{mm} \end{bmatrix} \tag{3.24}$$

and the voltage and current matrices $\bar{V}$ and $\bar{i}$ are column matrices given by Eqs. 3.3 and 3.4.

The admittance matrix can also be related to the scattering matrix. To derive this relation it is necessary to substitute for $\bar{i}$ in Eqs. 3.9 and 3.10 instead of for $\bar{V}$

$$\bar{V}=(\bar{I}+\bar{S})\bar{a} \tag{3.25}$$

$$\bar{Y}\bar{V}=(\bar{I}-\bar{S})\bar{a} \tag{3.26}$$

Hence

$$\overline{Y}(\overline{I}+\overline{S})=(\overline{I}-\overline{S}) \tag{3.27}$$

or

$$\overline{Y}=(\overline{I}-\overline{S})(\overline{I}+\overline{S})^{-1} \tag{3.28}$$

The $\overline{Y}$ matrix exists provided $(\overline{I}+\overline{S})$ is nonsingular. The determinant formed by the matrix must therefore not be zero. Starting with Eqs. 3.16 and 3.17, the following relation is readily established:

$$\overline{S}=(\overline{I}-\overline{Y})(\overline{I}+\overline{Y})^{-1} \tag{3.29}$$

3.3.   EIGENVALUES OF IMMITTANCE MATRICES

The eigenvalues of the $\overline{Z}$ or $\overline{Y}$ matrices can be found in the usual way by forming the eigenvalue equation. If the immittance matrices are constructed in terms of the $\overline{S}$ matrix, it is possible to avoid the need to invert matrices by first finding the eigenvalues of $\overline{S}$. These eigenvalues are then used to construct the immittance and thereafter the immittance matrices.

The eigenvalues of the $\overline{S}$, $\overline{Z}$, and $\overline{Y}$ matrices, which have common eigenvectors, may be related by making use of the following theorem: If

$$\overline{S}\,\overline{U}_n = s_n \overline{U}_n \tag{3.30}$$

then

$$f(\overline{S})\overline{U}_n = f(s_n)\overline{U}_n \tag{3.31}$$

This theorem will first be applied to obtain the relation between the eigenvalues of the $\overline{S}$ and $\overline{Z}$ matrices.

From Eq. 3.21 $\overline{S}$ is related to $\overline{Z}$ by

$$\overline{S}=(\overline{Z}-I)(\overline{Z}+I)^{-1} \tag{3.32}$$

Using Eq. 3.31 the relation between s_n and z_n is

$$s_n = \frac{z_n-1}{z_n+1} \tag{3.33}$$

Writing z_n in terms of s_n gives

$$z_n = \frac{1+s_n}{1-s_n} \tag{3.34}$$

where z_n is a normalized eigenvalue that satisfies the eigenvalue equation given by

$$\overline{Z}\,\overline{U}_n = z_n \overline{U}_n \tag{3.35}$$

To obtain the relation between the eigenvalues of the scattering and admittance matrices, one starts with Eq. 3.29

$$\overline{S} = (\overline{I} - \overline{Y})(I + \overline{Y})^{-1} \tag{3.36}$$

The relation between the eigenvalues is now given from Eq. 3.31 by

$$s_n = \frac{1 - y_n}{1 + y_n} \tag{3.37}$$

Writing y_n in terms of s_n gives

$$y_n = \frac{1 - s_n}{1 + s_n} \tag{3.38}$$

where y_n is a normalized eigenvalue that satisfies the eigenvalues equation given by

$$\overline{Y}\,\overline{U}_n = y_n \overline{U}_n \tag{3.39}$$

From Eqs. 3.34 and 3.38, it is also observed that

$$y_n = z_n \tag{3.40}$$

3.4. *DIAGONALIZATION OF IMPEDANCE MATRIX*

This matrix can be diagonalized in a similar way to that of the scattering one by noting that the eigenvectors of $\overline{S}$ are also those of $\overline{Z}$. The two have common eigenvectors because they commute. Hence,

$$\overline{Z} = \overline{U}\bar{z}\,\overline{U}^{-1} \tag{3.41}$$

where $\bar{z}$ is a diagonal matrix with the eigenvalues of the impedance matrix $\overline{Z}$.

Since the impedance eigenvalues are related to the scattering ones, an impedance matrix can be constructed after each symmetrical adjustment of the junction. This means that it is possible to form $(m-1)$ equivalent networks that correspond to the $(m-1)$ adjustments of the matrix eigenvalues to be described in Chapter 8.

It is assumed in this section that a $\bar{Z}$ matrix exists. This statement requires that the matrix $(\bar{I} - \bar{S})$ in Eq. 3.14 remains nonsingular. Following the procedure used in Chapter 2 to derive the relation between the scattering coefficients and its eigenvalues one obtains in the case of the $\bar{Z}$ matrix

$$3Z_{11} = z_0 + z_{+1} + z_{-1} \tag{3.42}$$

$$3Z_{12} = z_0 + z_{+1}e^{j2\pi/3} + z_{-1}e^{-j2\pi/3} \tag{3.43}$$

$$3Z_{13} = z_0 + z_{+1}e^{-j2\pi/3} + z_{-1}e^{j2\pi/3} \tag{3.44}$$

The relation between the coefficients of the $\bar{Z}$ matrix and its eigenvalues for the 4-port junction is

$$4Z_{11} = z_0 + z_{+1} + z_{-1} + z_2 \tag{3.45}$$

$$4Z_{12} = z_0 + jz_{+1} - jz_{-1} - z_2 \tag{3.46}$$

$$4Z_{13} = z_0 - z_{+1} - z_{-1} + z_2 \tag{3.47}$$

$$4Z_{14} = z_0 - jz_1 + jz_{-1} - z_2 \tag{3.48}$$

For the 5-port junction the result is

$$5Z_{11} = z_0 + z_{+1} + z_{-1} + z_{+2} + z_{-2} \tag{3.49}$$

$$5Z_{12} = z_0 + z_{+1}e^{+j2\pi/5} + z_{-1}e^{-j2\pi/5} + z_{+2}e^{+j4\pi/5} + z_{-2}e^{-j4\pi/5} \tag{3.50}$$

$$5Z_{13} = z_0 + z_{+1}e^{+j4\pi/5} + z_{-1}e^{-j4\pi/5} + z_{+2}e^{+j8\pi/5} + z_{-2}e^{-j8\pi/5} \tag{3.51}$$

$$5Z_{14} = z_0 + z_{+1}e^{+j6\pi/5} + z_{-1}e^{-j6\pi/5} + z_{+2}e^{+j12\pi/5} + z_{-2}e^{-j12\pi/5} \tag{3.52}$$

$$5Z_{15} = z_0 + z_{+1}e^{+j8\pi/5} + z_{-1}e^{-j8\pi/5} + z_{+2}e^{+j16\pi/5} + z_{-2}e^{-j16\pi/5} \tag{3.53}$$

Using the symmetry operators developed in Chapter 2, it is found that the $\bar{Z}$ matrix has the symmetry of $\bar{S}$. For a reciprocal junction, the $\bar{Z}$ matrix is symmetrical,

$$Z_{ij} = Z_{ji} \tag{3.54}$$

This can be demonstrated by noting that for a reciprocal junction $z_n = z_{-n}$, since $s_n = s_{-n}$.

3.5. *DIAGONALIZATION OF ADMITTANCE MATRIX*

This matrix can be diagonalized in a similar way to that of the scattering matrix by noting that the eigenvectors of $\bar{S}$ are also those of $\bar{Y}$,

$$\bar{Y} = \bar{U}\bar{y}\bar{U}^{-1} \tag{3.55}$$

where $\bar{y}$ is a diagonal matrix with the admittance matrix eigenvalues. An admittance matrix can, therefore, be constructed after each symmetrical adjustment of the junction, provided $(\bar{I}+\bar{S})$ in Eq. 3.28 remains nonsingular. This again leads to $(m-1)$ equivalent networks, which correspond to the $(m-1)$ adjustments of the scattering matrix eigenvalues.

The relation between the coefficients of the $\bar{Y}$ matrix and its eigenvalues is given by inspection from the previous section. For the 3-port junction the result is

$$3Y_{11} = y_0 + y_{+1} + y_{-1} \tag{3.56}$$

$$3Y_{12} = y_0 + y_{+1}e^{j2\pi/3} + y_{-1}e^{-j2\pi/3} \tag{3.57}$$

$$3Y_{13} = y_0 + y_{+1}e^{-j2\pi/3} + y_{-1}e^{j2\pi/3} \tag{3.58}$$

For the 4-port junction the result is

$$4Y_{11} = y_0 + y_{+1} + y_{-1} + y_2 \tag{3.59}$$

$$4Y_{12} = y_0 + jy_{+1} - jy_{-1} - y_2 \tag{3.60}$$

$$4Y_{13} = y_0 - y_{+1} - y_{-1} + y_2 \tag{3.61}$$

$$4Y_{14} = y_0 - jy_{+1} + jy_{-1} - y_2 \tag{3.62}$$

For the 5-port junction, the result is

$$5Y_{11} = y_0 + y_{+1} + y_{-1} + y_{+2} + y_{-2} \tag{3.63}$$

$$5Y_{12} = y_0 + y_{+1}e^{+j2\pi/5} + y_{-1}e^{-j2\pi/5} + y_{+2}e^{+j4\pi/5} + y_{-2}e^{-j4\pi/5} \tag{3.64}$$

$$5Y_{13} = y_0 + y_{+1}e^{+j4\pi/5} + y_{-1}e^{-j4\pi/5} + y_{+2}e^{+j8\pi/5} + y_{-2}e^{-j8\pi/5} \tag{3.65}$$

$$5Y_{14} = y_0 + y_{+1}e^{+j6\pi/5} + y_{-1}e^{-j6\pi/5} + y_{+2}e^{+j12\pi/5} + y_{-2}e^{-j12\pi/5} \tag{3.66}$$

$$5Y_{15} = y_0 + y_{+1}e^{+j8\pi/5} + y_{-1}e^{-j8\pi/5} + y_{+2}e^{+j16\pi/5} + y_{-2}e^{-j16\pi/5} \tag{3.67}$$

The $\overline{Y}$ matrix has the same symmetry as the $\overline{S}$ and $\overline{Z}$ matrices. For a reciprocal junction, the $\overline{Y}$ matrix is symmetrical,

$$Y_{ij} = Y_{ji} \tag{3.68}$$

3.6. IMMITTANCE EIGENSOLUTIONS

In this section, it is desired to demonstrate that if the input waves at each terminal of a junction correspond to a particular eigenvector, the input impedance at any terminal coincides with the corresponding impedance eigenvalue. For the 3-port junction, the voltages and current waves are related by the $\overline{Z}$ matrix in the following way:

$$
\begin{bmatrix} V_1 \\ V_2 \\ V_3 \end{bmatrix} =
\begin{bmatrix} Z_{11} & Z_{12} & Z_{13} \\ Z_{13} & Z_{11} & Z_{12} \\ Z_{12} & Z_{13} & Z_{11} \end{bmatrix}
\begin{bmatrix} i_1 \\ i_2 \\ i_3 \end{bmatrix} \tag{3.69}
$$

For the eigensolution corresponding to the eigenvector $\overline{U}_0$, Eq. 3.69 gives

$$\frac{V_1}{i_1} = \frac{V_2}{i_2} = \frac{V_3}{i_3} = z_0 \tag{3.70}$$

This last equation is obtained by writing the coefficients of the $\overline{Z}$ matrix in terms of its eigenvalues with the help of Eqs. 3.42 through 3.44.

For the eigensolutions corresponding to the eigenvectors $\overline{U}_{+1}$ and $\overline{U}_{-1}$ the results are

$$\frac{V_1}{i_1} = \frac{V_2}{i_2} = \frac{V_3}{i_3} = z_{+1} \tag{3.71}$$

and

$$\frac{V_1}{i_1} = \frac{V_2}{i_2} = \frac{V_3}{i_3} = z_{-1} \tag{3.72}$$

Figures 3.1a through c give the equivalent networks corresponding to the eigenvectors $\overline{U}_0$, $\overline{U}_{+1}$, and $\overline{U}_{-1}$. A similar result applies for the admittance eigenvalues.

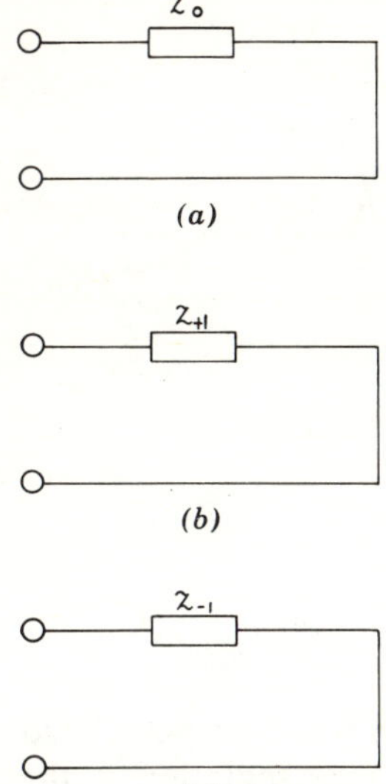

Figure 3.1. (a) Impedance eigenvalue z_0 for $\overline{U}_0$ excitation. (b) Impedance eigenvalue z_{+1} for $\overline{U}_{+1}$ excitation. (c) Impedance eigenvalue z_{-1} for $\overline{U}_{-1}$ excitation.

REFERENCES

1. C. G. Montgomery, R. H. Dicke, and E. M. Purcell, *Principles of Microwave Circuits*, McGraw-Hill, New York, 1948.

2. J. A. Altman, *Microwave Circuits*, van Nostrand, New York, 1964.

3. H. J. Carlin and A. B. Giordano, *Network Theory, an Introduction to Reciprocal and Nonreciprocal Circuits*, Prentice Hall, Englewood Cliffs, New Jersey, 1964.

CHAPTER FOUR

The Gyrator Network

The basic nonreciprocal element is the gyrator, which was first postulated by Tellegen. It is a 2-port network whose network representation is the topic of this chapter. Such a network exhibits 180° nonreciprocal phase shift between its input and output terminals. There are a number of different physical ways in which a gyrator network can be realized. The one that will be described in this chapter consists of two orthogonal coils coupled through a magnetized yttrium iron garnet (YIG) sphere. The importance of the gyrator lies in the fact that at any frequency, an arbitrary linear passive nonreciprocal network can always be represented by an equivalent circuit containing only conventional reciprocal elements and gyrators.

It is also shown that the minimum number of gyrator networks required to construct an m-port circulator corresponds to the number of degenerate eigenvalues whose degeneracy must be removed to construct the device. Chapter 5 deals with the construction of equivalent circuits of junction circulators in terms of such 2-port gyrator networks.

4.1. *SCATTERING MATRIX OF 2-PORT GYRATOR NETWORK*

The 2-port gyrator network has a scattering matrix given by

$$\bar{S} = \begin{bmatrix} 0 & -1 \\ 1 & 0 \end{bmatrix} \tag{4.1}$$

This matrix describes the basic nonreciprocal 2-port network defined by

Tellegen. Such a network exhibits 180° nonreciprocal phase shift between its input and output terminals. The schematic for this network is shown in Figure 4.1.

The characteristic equation for the above matrix is

$$s_n^2 + 1 = 0 \tag{4.2}$$

The two roots for this equation are

$$s_{+1} = j \tag{4.3}$$

$$s_{-1} = -j \tag{4.4}$$

The eigenvalues for the gyrator network are shown in Figure 4.2.

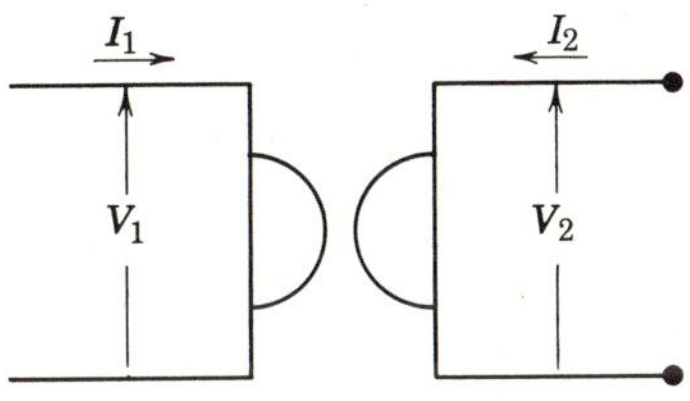

Figure 4.1.   Schematic of gyrator network.

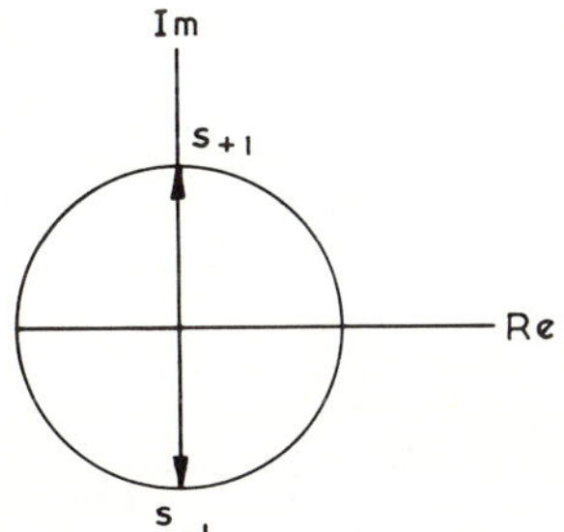

Figure 4.2.   Eigenvalues of gyrator network.

4.2.   IMMITTANCE MATRICES OF GYRATOR NETWORK

It will now be demonstrated that both the impedance and admittance matrices exist for such a network. The normalized impedance characteristic values are

$$z_{+1} = \frac{1 + s_{+1}}{1 - s_{+1}} = \frac{1 + j}{1 - j} = j \tag{4.5}$$

$$z_{-1} = \frac{1 + s_{-1}}{1 - s_{-1}} = \frac{1 - j}{1 + j} = -j \tag{4.6}$$

The normalized admittance characteristic values are the reciprocal of the impedance ones.

$$y_{+1} = \frac{1}{z_{+1}} = -j \tag{4.7}$$

$$y_{-1} = \frac{1}{z_{-1}} = j \tag{4.8}$$

The characteristic vectors are obtained from the characteristic equation defined in Chapter 2.

$$\overline{S}\,\overline{U}_n = s_n \overline{U}_n \tag{4.9}$$

For s_{+1} one has

$$\begin{bmatrix} 0 & -1 \\ 1 & 0 \end{bmatrix} \begin{bmatrix} a_1 \\ a_2 \end{bmatrix} = s_{+1} \begin{bmatrix} a_1 \\ a_2 \end{bmatrix} \tag{4.10}$$

Hence

$$\frac{a_1}{a_2} = j \tag{4.11}$$

A normalized characteristic vector is, therefore,

$$\overline{U}_{+1} = \frac{1}{\sqrt{2}} \begin{bmatrix} 1 \\ -j \end{bmatrix} \tag{4.12}$$

In a similar way one has

$$\overline{U}_{-1} = \frac{1}{\sqrt{2}} \begin{bmatrix} 1 \\ j \end{bmatrix} \tag{4.13}$$

The impedance matrix is obtained by diagonalizing the matrix $\overline{Z}$

$$\overline{Z} = \overline{U}\overline{\lambda}(\overline{U}^*)^T \tag{4.14}$$

where

$$\bar{U} = \frac{1}{\sqrt{2}} \begin{bmatrix} 1 & 1 \\ -j & j \end{bmatrix} \tag{4.15}$$

$$(\bar{U}^*)^T = \frac{1}{\sqrt{2}} \begin{bmatrix} 1 & j \\ 1 & -j \end{bmatrix} \tag{4.16}$$

$$\bar{\lambda} = \begin{bmatrix} z_{+1} & 0 \\ 0 & z_{-1} \end{bmatrix} \tag{4.17}$$

The result is

$$\bar{Z} = \frac{1}{2} \begin{bmatrix} (z_{+1} + z_{-1}), & j(z_{+1} - z_{-1}) \\ -j(z_{+1} - z_{-1}), & (z_{+1} + z_{-1}) \end{bmatrix} \tag{4.18}$$

In terms of the original variables, the above becomes

$$\bar{Z} = \begin{bmatrix} 0 & -1 \\ +1 & 0 \end{bmatrix} \tag{4.19}$$

The $\bar{Z}$-matrix schematic of the 2-port gyrator is shown in Figure 4.3. The result for the admittance matrix is obtained in exactly the same way as for the impedance one

$$\bar{Y} = \frac{1}{2} \begin{bmatrix} (y_{+1} + y_{-1}), & j(y_{+1} - y_{-1}) \\ -j(y_{+1} - y_{-1}), & (y_{+1} + y_{-1}) \end{bmatrix} \tag{4.20}$$

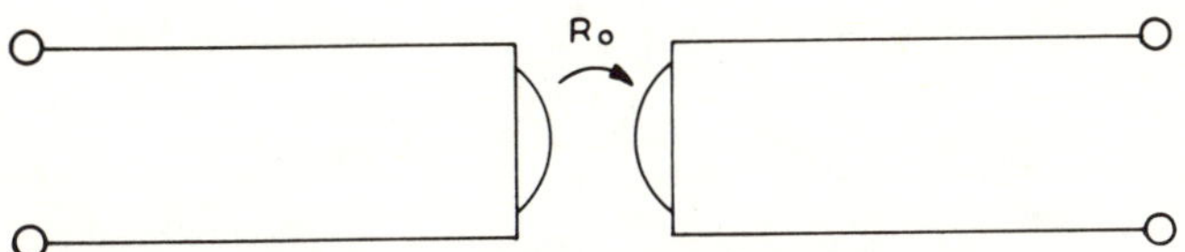

Figure 4.3. $\bar{Z}$ matrix equivalent circuit of ideal 2-port gyrator.

In terms of the original variables the result is

$$\overline{Y} = \begin{bmatrix} 0 & +1 \\ -1 & 0 \end{bmatrix} \qquad (4.21)$$

The $\overline{Y}$-matrix schematic of the 2-port gyrator is illustrated in Figure 4.4.

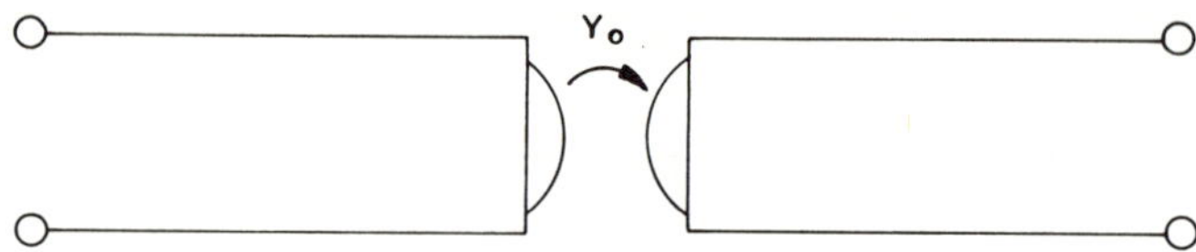

Figure 4.4.   *$\overline{Y}$ matrix equivalent circuit of ideal 2-port gyrator.*

4.3. *2-PORT GYRATOR USING ORTHOGONAL LOOPS COUPLED BY A YIG SPHERE*

There are a number of different physical ways in which gyrator networks can be constructed.

One circuit arrangement, which has the properties of a 2-port gyrator, is that of the bandpass filter in Figure 4.5. Two coils have their axes at right angles to each other, and a small ferrite sample is placed at the intersection of the coil axes. When the sample is not magnetized, no power is transferred between the coils because the loop axes are perpendicular to each other and there is no interaction with the ferrite. When a d.c. field is applied along the z-axis, the two coils are coupled through the transverse components of the dipolar field of the ferrite resonator. This coupling is largest at ferrimagnetic resonance. These orthogonal circuits can be crossed wires, loops, striplines, waveguides, cavity modes.

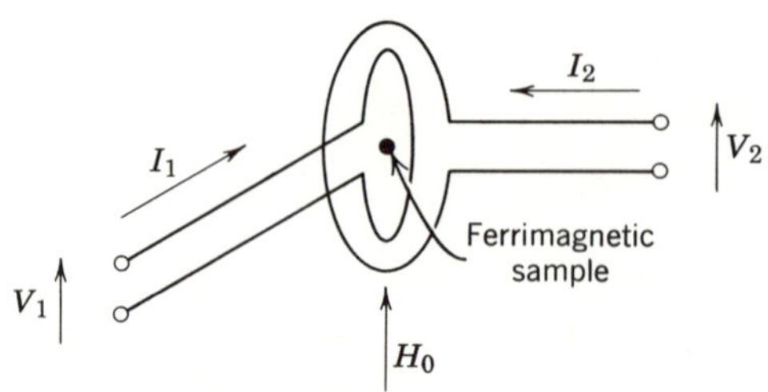

Figure 4.5.   *2-port gyrator network using YIG sphere.*

The equivalent circuit of the ferrimagnetic resonator at the intersection of the two loops is shown in Figure 4.6. It consists of a gyrator network of characteristic impedance R_0 connected between series impedances Z_{11} at the input and output ports. This equivalent circuit can be used in any of the available synthesis procedures for bandpass filters either in terms of a lowpass prototype or directly using the image parameter method.

In the following analysis, the ferrite resonator is assumed to be so small that the r.f. field is substantially uniform throughout the volume of the sample, and, therefore, higher-order modes are not excited.

The equivalent circuit can be obtained by assuming currents to exist in the two circuits (shown in Figure 4.5) that produce magnetic fields at the ferrite sample. The resultant external magnetic field at the sample is then

$$h_x^e = c_x I_1 \tag{4.22}$$

$$h_y^e = c_y I_2 \tag{4.23}$$

where the coupling components c_x and c_y depend only on the geometry of the circuit and the position of the ferrite sample.

The transverse magnetization in the sample is given from Eq. 1.66 and 1.67 by

$$m_x = \mu_0 \chi_{xx}^e h_x^e + \mu_0 \chi_{xy}^e h_y^e \tag{4.24}$$

$$m_y = \mu_0 \chi_{yx}^e h_x^e + \mu_0 \chi_{yy}^e h_y^e \tag{4.25}$$

The voltages induced in the two orthogonal circuits are given by

$$V_1 = j\omega \int_{\text{circuit } x} b_x \cdot da \tag{4.26}$$

$$V_2 = j\omega \int_{\text{circuit } y} b_y \cdot da \tag{4.27}$$

The inductions at the two loops due to the equivalent magnetic dipolar

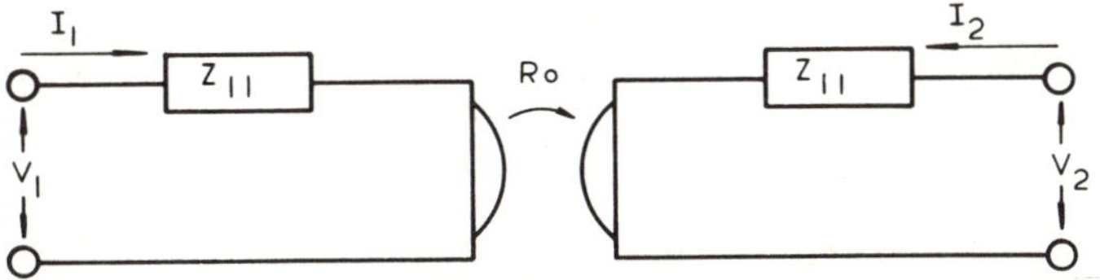

Figure 4.6.   *Equivalent circuit of 2-port gyrator network using YIG sphere.*

fields of the ferrite resonator are

$$b_x = \mu_0 h_x = \frac{m_x v}{2\pi r^3} \tag{4.28}$$

$$b_y = \mu_0 h_y = \frac{m_y v}{2\pi r^3} \tag{4.29}$$

where v is the volume of the ferrite resonator and r is the radial distance from the magnetic dipole, Substituting for m_x and m_y

$$b_x = \frac{\mu_0 v}{2\pi r^3} (\chi_{xx}^e c_x I_1 + \chi_{xy}^e c_y I_2) \tag{4.30}$$

$$b_y = \frac{\mu_0 v}{2\pi r^3} (\chi_{yx}^e c_x I_1 + \chi_{yy}^e c_y I_2) \tag{4.31}$$

Finally, introducing the above equations into Eq. 4.26 and 4.27 one obtains

$$V_1 = \frac{j\mu_0 \omega v}{2\pi} \int_{\text{circuit } x} \frac{(\chi_{xx}^e c_x I_1 + \chi_{xy}^e c_y I_2)}{r^3} \, da \tag{4.32}$$

$$V_2 = \frac{j\mu_0 \omega v}{2\pi} \int_{\text{circuit } y} \frac{(\chi_{yx}^e c_x I_1 + \chi_{yy}^e c_y I_2)}{r^3} \, da \tag{4.33}$$

As an example of the use of this result, consider the coupling to two orthogonal loops (radius r_0) by a ferrite resonator in the geometrical center as shown in Figure 4.5. In this case

$$h_x^e = \frac{I_1}{2r_0} \tag{4.34}$$

$$h_y^e = \frac{I_2}{2r_0} \tag{4.35}$$

and

$$c_x = c_y = \frac{1}{2r_0} \tag{4.36}$$

When integrated over the area of the loops, Eq. 4.32 and 4.33 become

$$V_1 = \frac{j\mu_0\omega\upsilon}{2r_0^2}(\chi_{xx}^e I_1 + \chi_{xy}^e I_2) \tag{4.37}$$

$$V_2 = \frac{j\mu_0\omega\upsilon}{2r_0^2}(\chi_{yx}^e I_1 + \chi_{yy}^e I_2) \tag{4.38}$$

where V_1 and V_2 are the voltages developed around the loops and I_1 and I_2 are the currents flowing in the loops.

In matrix notation the result is now given by

$$\begin{bmatrix} V_1 \\ V_2 \end{bmatrix} = \begin{bmatrix} Z_{11} & -R_0 \\ R_0 & Z_{11} \end{bmatrix} \begin{bmatrix} I_1 \\ I_2 \end{bmatrix} \tag{4.39}$$

where

$$Z_{11} = \frac{j\mu_0\omega\upsilon\chi_{xx}^e}{2r_0^2} \tag{4.40}$$

$$R_0 = \frac{-j\mu_0\omega\upsilon\chi_{xy}^e}{2r_0^2} \tag{4.41}$$

Using simple matrix addition the impedance matrix defined by Eq. 4.39 becomes

$$\bar{Z} = \begin{bmatrix} Z_{11} & 0 \\ 0 & 0 \end{bmatrix} + \begin{bmatrix} 0 & -R_0 \\ R_0 & 0 \end{bmatrix} + \begin{bmatrix} 0 & 0 \\ 0 & Z_{11} \end{bmatrix} \tag{4.42}$$

The equivalent circuit for this impedance matrix is indicated in Figure 4.6.

4.4.   SYNTHESIS OF CIRCULATOR NETWORKS USING MINIMUM NUMBER OF GYRATOR NETWORKS

It will be shown in Chapter 5 that the equivalent circuit for symmetrical 3-port circulators is readily realized by a network of three gyrators, but this is not the minimum number.

This minimum number is equal to 1/2 the rank of the $\bar{Z}$ matrix (the rank is the order of the highest order nonvanishing determinant, an even number for every skew–symmetric matrix) *or 1/2 the rank of the skew–symmetric part of the scattering matrix.* Such a minimum gyrator network is formed by performing linear transformations on $\bar{Z}$ until it has a form in which the only nonzero elements are those on the skew diagonal that stretches from the upper right-hand corner to the lower left-hand corner.

The minimum circuit requirements for m-port circulators have been expressed by Carlin in terms of the number of interconnected gyrator networks, as follows:

$$\tfrac{1}{2}(m-1) \text{ for } m \text{ odd} \tag{4.43}$$

$$\tfrac{1}{2}(m-2) \text{ for } m \text{ even} \tag{4.44}$$

Figure 4.7 depicts a possible circuit connection using a single gyrator for such a $\bar{Z}$ matrix. Figure 4.8 gives the dual circuit obtained from the $\bar{Y}$ matrix. Chapter 17 gives one example of such a circulator.

The minimum number of gyrators can also be obtained from a knowledge of the number of degenerate eigenvalues. Each pair of degenerate eigenvalues corresponding to one ideal gyrator.

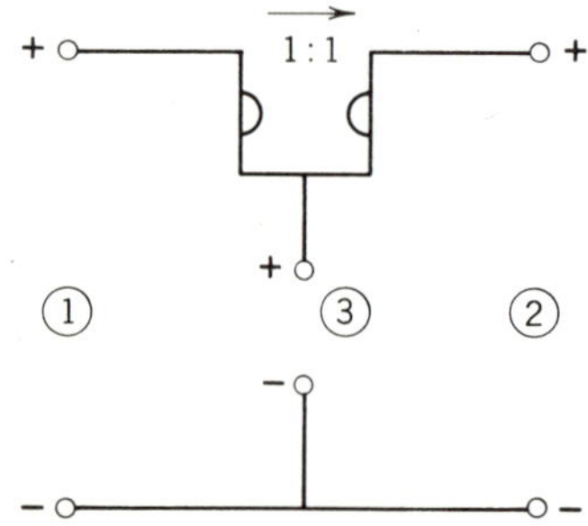

Figure 4.7. Circulator network constructed from $\bar{Z}$ matrix using single gyrator network.

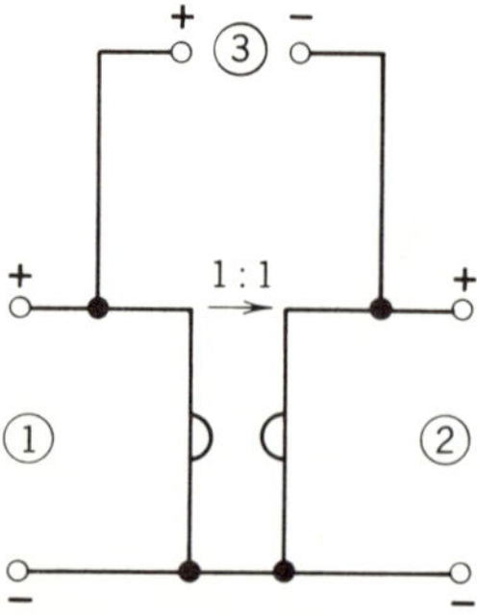

Figure 4.8. Circulator network constructed from $\bar{Y}$ matrix using single gyrator network.

4.5. *NONRECIPROCAL 2-PORT NETWORK*

It will now be shown that it is not possible to have a one way transmission line. This is done by applying the unitary condition to $\bar{S}$. The result is

$$|S_{11}|^2 + |S_{21}|^2 = 1 \qquad (4.45)$$

$$|S_{12}|^2 + |S_{22}|^2 = 1 \qquad (4.46)$$

$$S_{11}^* S_{12} + S_{21}^* S_{22} = 0 \qquad (4.47)$$

$$S_{11} S_{12}^* + S_{21} S_{22}^* = 0 \qquad (4.48)$$

If there is no transmission between ports 1 and 2, one has

$$S_{12} = 0 \qquad (4.49)$$

Substituting this last equation into Eq. 4.46 gives

$$|S_{22}|^2 = 1 \qquad (4.50)$$

Combining the last two equations with Eq. 4.48 gives

$$S_{21} = 0 \qquad (4.51)$$

Hence

$$S_{12} = S_{21} = 0 \qquad (4.52)$$

which is in contradiction of the nonreciprocity condition. It is, therefore, not possible to have a one-way transmission line.

REFERENCES

1. C. L. Hogan, "The Microwave Gyrator," *Bell Sys. Tech. J.*, **31**, 1 (1952).
2. B. D. H. Tellegen, "The Gyrator, a new Electric Network Element," *Phillips Res. Rep.*, **3**, 81–101 (1948).
3. R. W. Grasse, "Low-Loss Gyrometric Coupling through Single Crystal Garnets," *J. Appl. Phys. Suppl.* **30**, 155S (1959).
4. P. S. Carter, Jr., "Magnetically-Tunable Microwave Filters using Single-Crystal Yttrium-Garnet Resonators," *IRE Trans. Microwave Theory Tech.*, **MTT-9**, 252 (1962).
5. H. Carlin, Synthesis of nonreciprocal networks, *Proc. Brooklyn Symp. Modern Network Synthesis*, **5**, 11–42, 1955.

CHAPTER FIVE

The Boundary Conditions of the Ideal Circulator

The boundary conditions of an ideal circulator may be applied through the $\overline{S}$, $\overline{Z}$, and $\overline{Y}$ matrices or through their eigenvalues. The chapter starts by establishing the boundary conditions of the circulator in terms of the $\overline{S}$ matrix and its eigenvalues. The eigenvalues of the $\overline{Z}$ or $\overline{Y}$ matrices are then obtained from those of the $\overline{S}$ matrix by the method of Chapter 3, and the $\overline{Z}$ and $\overline{Y}$ matrices are then constructed in terms of their own eigenvalues. The $\overline{S}$ matrix and all sets of eigenvalues are always realizable, while it is possible that the $\overline{Z}$ and $\overline{Y}$ matrices do not exist. The 3-port circuit associated with the demagnetized circulator is one example of a circuit that has neither $\overline{Z}$ nor $\overline{Y}$ matrices. The latter two matrices are also helpful in constructing equivalent circuits of circulators in terms of ideal gyrator networks.

One approximate boundary condition which is often convenient is obtained by deriving a 1-port equivalent circuit that applies in the vicinity of the circulation frequency. The boundary conditions of the demagnetized reciprocal 3-port network, which is closely related to that of the ideal circulator, is also derived.

The discussion of the boundary conditions is simplified by relating the eigenvalues to shortcircuited and opencircuited distributed transmission lines having the characteristic impedance of the input lines, for which the electrical lengths satisfy the phase angles of the reflection eigenvalues. In this chapter the immittance eigenvalues and the other boundary conditions are unnormalized. In general, whether a quantity is normalized or not will be understood in the context used.

5.1. *EIGENNETWORKS*

The *m*-eigensolutions of the *m*-port junction are 1-port junctions for which the usual relations between *s*, *z*, and *y* apply. A suitable coordinate system for these eigennetworks may be obtained with reference to the section of uniform transmission line showing positive directions of *I* and *V* in Figure 5.1.

A standing wave solution to the wave equation is

$$V(z) = A\cos kz + B\sin kz \tag{5.1}$$

$$I(z) = \frac{-j}{Z_0}(A\sin kz - B\cos kz) \tag{5.2}$$

The constants *A* and *B* are obtained by applying the boundary conditions at $z = z_0$

$$V(z_0) = A\cos kz_0 + B\sin kz_0 \tag{5.3}$$

$$I(z_0) = \frac{-j}{Z_0}(A\sin kz_0 - B\cos kz_0) \tag{5.4}$$

The complete solution is

$$V(z) = V(z_0)\cos k(z_0 - z) + jZ_0 I(z_0)\sin k(z_0 - z) \tag{5.5}$$

$$I(z) = jY_0 V(z_0)\sin k(z_0 - z) + I(z_0)\cos k(z_0 - z) \tag{5.6}$$

These last two equations define the *ABCD* matrix of the network

$$\begin{bmatrix} V(z) \\ I(z) \end{bmatrix} = \begin{bmatrix} A & jB \\ jC & D \end{bmatrix} \begin{bmatrix} V(z_0) \\ I(z_0) \end{bmatrix} \tag{5.7}$$

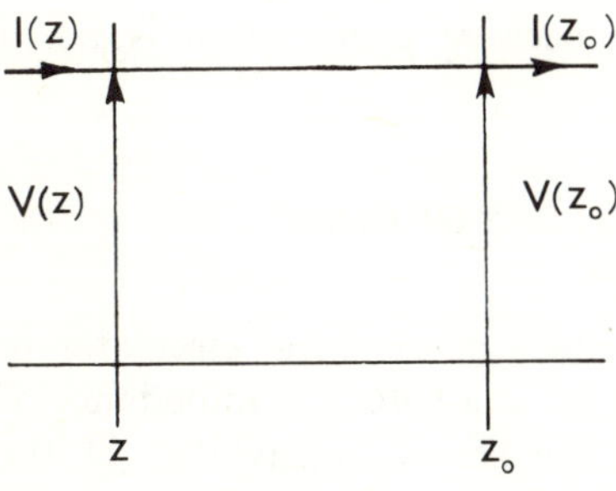

Figure 5.1. *Transmission line showing positive direction of propagation.*

which will be discussed further in this section.

The input admittance of the network in terms of this $ABCD$ matrix is

$$Y(z) = \frac{I(z)}{V(z)} \tag{5.8}$$

Its reflection coefficient is

$$s(z) = \frac{Y_0 - Y(z)}{Y_0 + Y(z)} = s(z_0)e^{j2k(z-z_0)} \tag{5.9}$$

where $s(z)$ represents the phase shift of $s(z_0)$ along the transmission line of characteristic admittance Y_0, where

$$s(z_0) = \frac{Y_0 - Y(z_0)}{Y_0 + Y(z_0)} \tag{5.10}$$

Two special cases that are of interest in this text are $Y(z_0) = \infty$ and $Y(z_0) = 0$ at $z_0 = 0$.
In the first case,

$$s(z_0) = -1 = e^{-j\pi} \tag{5.11}$$

for which

$$s(-z) = e^{-j2(kz + \pi/2)} \tag{5.12}$$

In the second situation,

$$s(z_0) = +1 \tag{5.13}$$

for which

$$s(-z) = e^{-j2(kz)} \tag{5.14}$$

The input terminals are taken at $z = -z$ because propagation is taken along the positive z direction.

5.2.   IMMITTANCE EIGENVALUES OF IDEAL CIRCULATOR

This section derives the immitance eigenvalues of the ideal circulator in terms of distributed transmission lines with the characteristic impedance of the input lines. This is done by writing the set of eigenvalues of the scattering matrix of the ideal circulator in Section 2.5 in the following form:

$$s_0 = e^{-j2\theta_0} \tag{5.15}$$

$$s_{+1} = e^{-j2(\theta_1 + \theta_{+1} + \pi/2)} \tag{5.16}$$

$$s_{-1} = e^{-j2(\theta_1 + \theta_{-1} + \pi/2)} \tag{5.17}$$

Here s_0 is the reflection coefficient of an open circuited transmission line of length θ_0, and $s_{\pm 1}$ are the reflection coefficients of short circuited transmission lines of length $\theta_1 + \theta_{\pm 1}$. The angles θ_1 and θ_0 represent the electrical lengths of the demagnetized transmission lines, and $\theta_{\pm 1}$ gives the change in the electrical length θ_1 of the demagnetized transmission lines when the networks are magnetized. These eigenvalues are depicted on the unit circle in Figure 5.2.

The admittance eigenvalues are related to the angles that the scattering matrix eigenvalues make by Eq. 3.38.

$$y_0 = jY_0 \tan\theta_0 \tag{5.18}$$

$$y_{+1} = jY_0 \tan\left(\theta_1 + \theta_{+1} + \frac{\pi}{2}\right) \tag{5.19}$$

$$y_{-1} = jY_0 \tan\left(\theta_1 + \theta_{-1} + \frac{\pi}{2}\right) \tag{5.20}$$

The impedance eigenvalues are related to those of the scattering ones by Eq. 3.34

$$z_0 = -jR_0 \cot\theta_0 \tag{5.21}$$

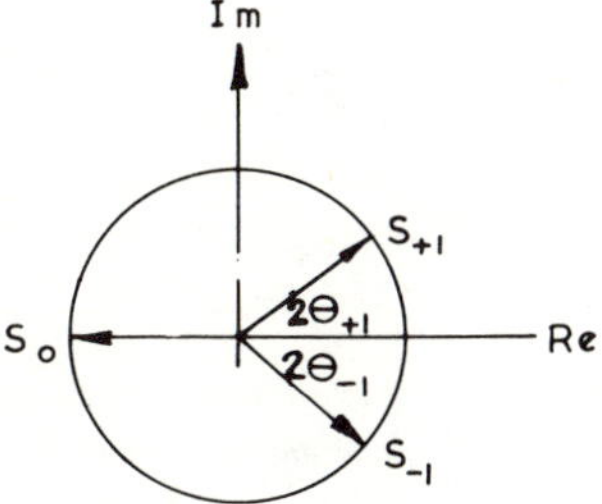

Figure 5.2. *Eigenvalues diagram of ideal 3-port circulator.*

$$z_{+1} = -jR_0 \cot\left(\theta_1 + \theta_{+1} + \frac{\pi}{2}\right)$$
(5.22)

$$z_{-1} = -jR_0 \cot\left(\theta_1 + \theta_{-1} + \frac{\pi}{2}\right)$$
(5.23)

The three eigennetworks are given in Figure 5.3.

For an ideal circulator

$$\theta_0 = \frac{\pi}{2}$$
(5.24)

$$\theta_1 = \frac{\pi}{2}$$
(5.25)

$$\theta_{\pm 1} = \mp \frac{\pi}{6}$$
(5.26)

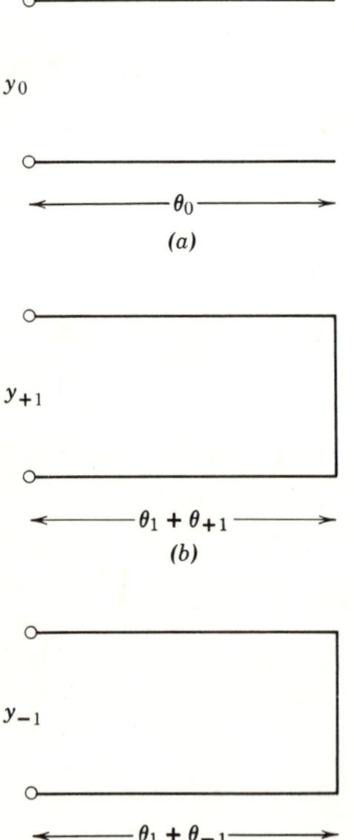

Figure 5.3.   (a) Eigennetwork for $\overline{U}_0$ excitation; (b) Eigennetwork for $\overline{U}_{+1}$ excitation; (c) Eigennetwork for $\overline{U}_{-1}$ excitation.

which satisfies the eigenvalues of the ideal circulator derived in Chapter 2.

Using the above boundary conditions, the admittance eigenvalues of the ideal circulator assume the following forms:

$$y_0 = \infty \tag{5.27}$$

$$y_{+1} = \frac{-jY_0}{\sqrt{3}} \tag{5.28}$$

$$y_{-1} = \frac{jY_0}{\sqrt{3}} \tag{5.29}$$

while the impedance eigenvalues of the ideal circulator are

$$z_0 = 0 \tag{5.30}$$

$$z_{+1} = j\sqrt{3}\, R_0 \tag{5.31}$$

$$z_{-1} = -j\sqrt{3}\, R_0 \tag{5.32}$$

Circulation boundary conditions may be stated in terms of either of the above three representations.

The eigennetwork connected with the s_0 reflection coefficient is known as an in-phase eigennetwork because the fields at the center of the junction are in phase for this excitation. The eigennetworks associated with the $s_{\pm 1}$ reflection coefficients are known as counter-rotating ones because the transverse components of the electric and magnetic fields are circularly polarized waves rotating in one sense for one eigenvector and in the other sense for the other eigenvector. Since the fields are circularly polarized for these latter modes the scalar permeabilities introduced in Chapter 1 apply.

5.3.   *IMPEDANCE MATRIX OF IDEAL 3-PORT CIRCULATOR*

The impedance matrix of an ideal 3-port circulator can be constructed with the help of Eqs. 3.42 through 3.44 in terms of the impedance eigenvalues. The impedance eigenvalues can be formed once those of the scattering matrix are known. For an ideal circulator with $S_{12} = -1$ the matrix $(\bar{I} - \bar{S})$ is nonsingular. The impedance matrix, therefore, exists. Notice that the sign of S_{12} is determined by that of s_0.

The coefficients of the impedance matrix are now obtained by substituting the last equations into Eqs. 3.42 through 3.44

$$Z_{11} = 0 \tag{5.33}$$

$$Z_{12} = -R_0 \tag{5.34}$$

$$Z_{13} = +R_0 \tag{5.35}$$

The impedance matrix of an ideal circulator is, therefore,

$$\bar{Z} = \begin{bmatrix} 0 & -R_0 & R_0 \\ R_0 & 0 & -R_0 \\ -R_0 & R_0 & 0 \end{bmatrix} \tag{5.36}$$

This last matrix is skew symmetric.

Additional useful relation between the coefficients of the $\bar{Z}$ matrix are obtained provided $z_0 = 0$

$$Z_{13} = -Z_{12}^* \tag{5.37}$$

and

$$Z_{12} = \frac{-Z_{11}}{2} + R_0 \tag{5.38}$$

The general form of the $\bar{Z}$ matrix for a nonreciprocal junction is, therefore,

$$\bar{Z} = \begin{bmatrix} Z_{11} & Z_{12} & -Z_{12}^* \\ -Z_{12}^* & Z_{11} & Z_{12} \\ Z_{12} & -Z_{12}^* & Z_{11} \end{bmatrix} \tag{5.39}$$

5.4.   *ADMITTANCE MATRIX OF IDEAL 3-PORT CIRCULATOR*

For an ideal 3-port circulator, at the reference plane where $S_{12} = -1$, the matrix $(\bar{I} + \bar{S})$ given in Section 3.2 is singular. This means that an admittance matrix cannot be constructed directly at this plane. This can be seen immediately by noting that the admittance matrix is constructed by first obtaining its eigenvalues. As already indicated the latter eigenvalues are the reciprocal of the impedance ones. Since the impedance eigenvalue z_0 is zero when $S_{12} = -1$, the eigenvalue becomes infinitely large. This implies that it is impossible to construct a finite admittance matrix at this plane. However, an admittance matrix may be constructed, provided new reference terminals are chosen for which $S_{12} = +1$ instead of -1. At this new reference plane it would not be possible to construct an impedance matrix in this usual way now. In this text, it is assumed throughout, that $S_{12} = -1$ at the reference terminals of the junction. However, some boundary conditions will obviously result in $S_{12} = +1$ instead of -1. In this instance the admittance matrix and its eigenvalues are obtained in exactly the same way as in the case of the $\bar{Z}$ matrix. One such plane is the characteristic plane to be defined later in this chapter.

It is apparent from the above discussion that whether an admittance or impedance matrix can be formed is determined by the signs of the coefficients of the scattering matrix. As already mentioned, one way of altering the signs of the coefficients of the scattering matrix is to shift the reference plane of the network. Another way in which the signs of the scattering coefficients can be altered is by extracting ideal 2-port gyrators from one of the ports of the network. This has the effect of altering the signs along one row of the scattering matrix.

5.5   EQUIVALENT CIRCUIT OF JUNCTION CIRCULATOR IN TERMS OF IMPEDANCE MATRIX

An equivalent circuit can be synthesized from the impedance matrix by making use of simple matrix addition

$$
\bar{Z} =
\begin{bmatrix}
Z_{11} & Z_{12} & -Z_{12}^* \\
-Z_{12}^* & Z_{11} & Z_{12} \\
Z_{12} & -Z_{12}^* & Z_{11}
\end{bmatrix}
=
\begin{bmatrix}
Z_{11} & 0 & 0 \\
0 & 0 & 0 \\
0 & 0 & 0
\end{bmatrix}
+
\begin{bmatrix}
0 & Z_{12} & 0 \\
-Z_{12}^* & 0 & 0 \\
0 & 0 & 0
\end{bmatrix}
$$

$$
+
\begin{bmatrix}
0 & 0 & 0 \\
0 & Z_{11} & 0 \\
0 & 0 & 0
\end{bmatrix}
+
\begin{bmatrix}
0 & 0 & -Z_{12}^* \\
0 & 0 & 0 \\
Z_{12} & 0 & 0
\end{bmatrix}
$$

$$
+
\begin{bmatrix}
0 & 0 & 0 \\
0 & 0 & Z_{12} \\
0 & -Z_{12}^* & 0
\end{bmatrix}
+
\begin{bmatrix}
0 & 0 & 0 \\
0 & 0 & 0 \\
0 & 0 & Z_{11}
\end{bmatrix}
\tag{5.40}
$$

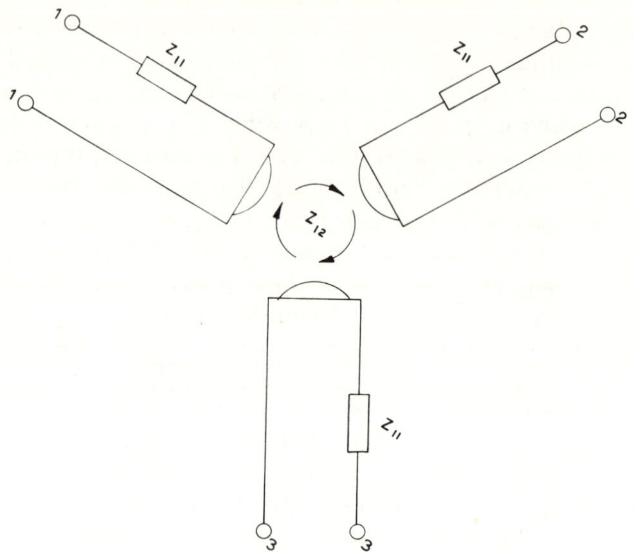

Figure 5.4. $\bar{Z}$ *matrix equivalent circuit of 3-port circulator using 2-port gyrators.*

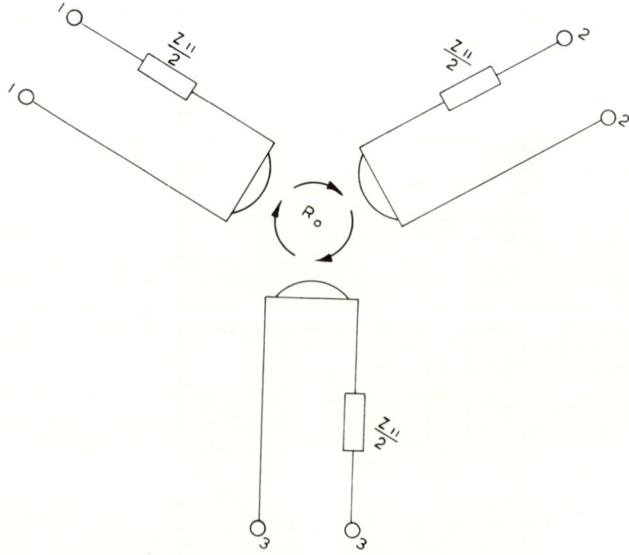

Figure 5.5. $\bar{Z}$ *matrix equivalent circuit of 3-port circulator using frequency independent 2-port gyrators.*

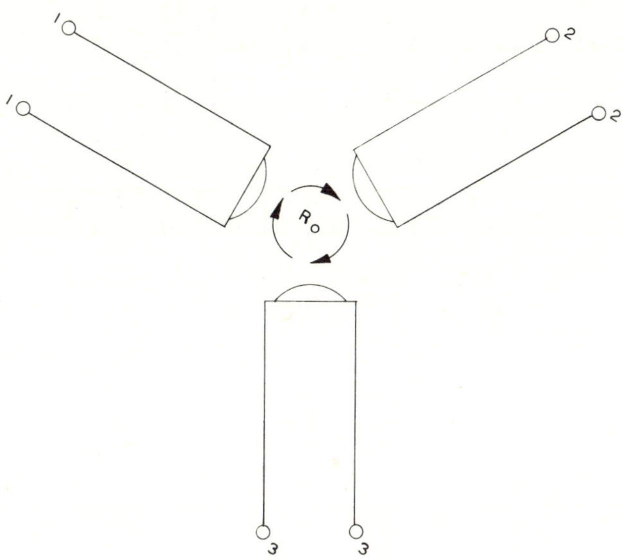

Figure 5.6. $\overline{Z}$ matrix equivalent circuit of ideal 3-port circulator using 2-port gyrators.

The first impedance matrix represents an impedance connected in series with the input port. The second term represents a 2-port gyrator connected between ports 1 and 2. The third impedance is a series network Z_{11} at port 2. The next two impedance matrices are 2-port gyrators connecting ports 3 to 1 and ports 2 to 3. The last term represents a series network at port 3. The equivalent circuit in terms of frequency dependent 2-port gyrators is shown in Figure 5.4. The equivalent circuit in terms of frequency independent gyrators is shown in Figure 5.5. The circuit of an ideal circulator is depicted in Figure 5.6.

5.6.   *EQUIVALENT CIRCUIT OF JUNCTION CIRCULATOR IN TERMS OF ADMITTANCE MATRIX*

An equivalent circuit can also be synthesized from the admittance matrix, again, by making use of simple matrix addition

$$\overline{Y} = \begin{bmatrix} Y_{11} & Y_{12} & -Y_{12}^* \\ -Y_{12}^* & Y_{11} & Y_{12} \\ Y_{12} & -Y_{12}^* & Y_{11} \end{bmatrix} = \begin{bmatrix} Y_{11} & 0 & 0 \\ 0 & 0 & 0 \\ 0 & 0 & 0 \end{bmatrix}$$

$$+ \begin{bmatrix} 0 & Y_{12} & 0 \\ -Y_{12}^* & 0 & 0 \\ 0 & 0 & 0 \end{bmatrix} + \begin{bmatrix} 0 & 0 & 0 \\ 0 & Y_{11} & 0 \\ 0 & 0 & 0 \end{bmatrix}$$

$$+ \begin{bmatrix} 0 & 0 & -Y_{12}^* \\ 0 & 0 & 0 \\ Y_{12} & 0 & 0 \end{bmatrix}$$

$$+ \begin{bmatrix} 0 & 0 & 0 \\ 0 & 0 & Y_{12} \\ 0 & -Y_{12}^* & 0 \end{bmatrix}$$

$$+ \begin{bmatrix} 0 & 0 & 0 \\ 0 & 0 & 0 \\ 0 & 0 & Y_{11} \end{bmatrix} \tag{5.41}$$

The first admittance matrix represents an admittance connected in shunt with the input port. The second admittance matrix represents a 2-port gyrator connected between ports 1 and 2. The third admittance is a shunt network Y_{11} at port 2. The next two admittance matrices are 2-port gyrators connecting ports 3 to 1 and ports 2 to 3. The last admittance matrix represents a shunt network at port 3. The equivalent circuit is shown in Figure 5.7. The equivalent circuit in terms of frequency independent gyrator is shown in Figure 5.8., while that of an ideal circulator is indicated in Figure 5.9.

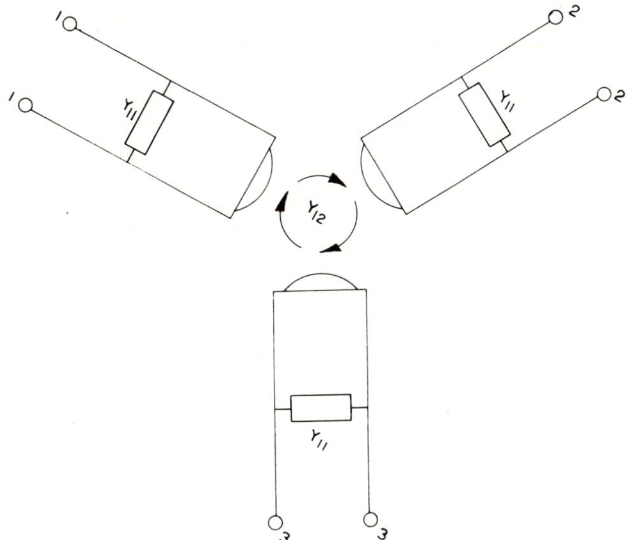

Figure 5.7. $\overline{Y}$ *matrix equivalent circuit of 3-port circulator using 2-port gyrators.*

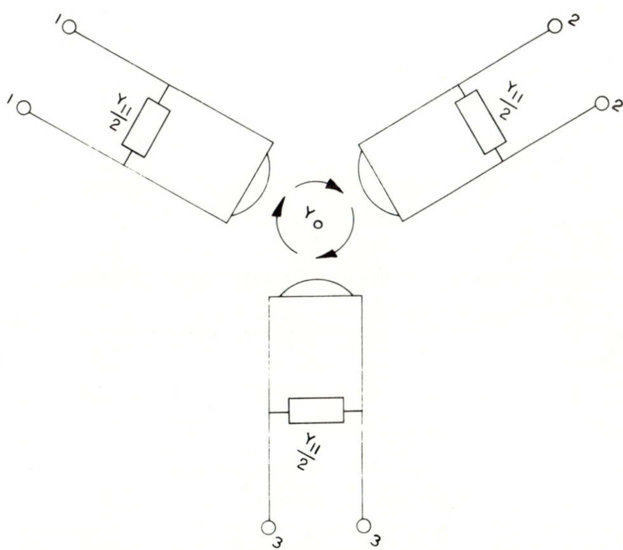

Figure 5.8. $\overline{Y}$ *matrix equivalent circuit of 3-port circulator using frequency independent 2-port gyrators.*

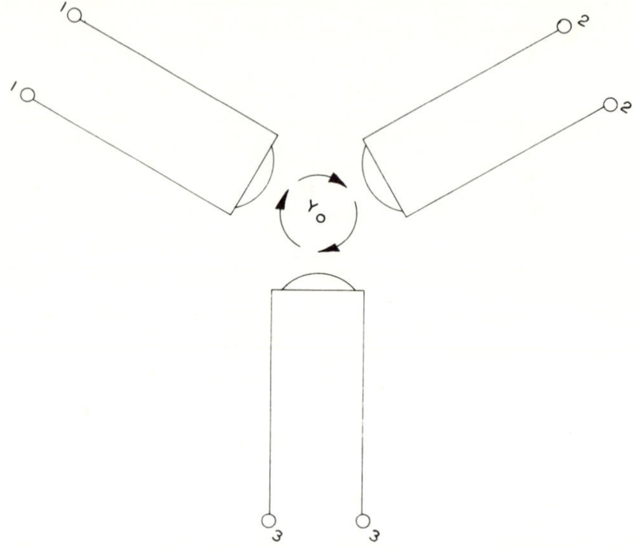

Figure 5.9. $\overline{Y}$ *matrix equivalent circuit of ideal 3-port circulator using 2-port gyrators.*

5.7.   *EQUIVALENT CIRCUIT OF RECIPROCAL 3-PORT JUNCTION*

The equivalent circuit of a reciprocal junction can be obtained by starting with the relation between the reflection coefficient S_{11} and the scattering matrix eigenvalues

$$S_{11} = \frac{s_0 + s_{+1} + s_{-1}}{3} \tag{5.42}$$

The frequency dependence of the scattering matrix eigenvalues is obtained by using the relation between the scattering and admittance eigenvalues

$$s_0 = \frac{Y_0 - y_0}{Y_0 + y_0} \tag{5.43}$$

$$s_{+1} = \frac{Y_0 - y_{+1}}{Y_0 + y_{+1}} \tag{5.44}$$

$$s_{-1} = \frac{Y_0 - y_{-1}}{Y_0 + y_{-1}} \tag{5.45}$$

For a reciprocal junction

$$s_{+1} = s_{-1} = s_1 \tag{5.46}$$

and

$$y_{+1} = y_{-1} = y_1 \tag{5.47}$$

Furthermore, the reference eigenvalue is

$$s_0 = -1 \tag{5.48}$$

which corresponds to $y_0 = \infty$.

Making use of the above relations in Eq. 5.42 gives

$$S_{11} = \frac{Y_0 - 3y_1}{3Y_0 + 3y_1} \tag{5.49}$$

The equivalent circuit, for which S_{11} is given by the last equation, is shown in Figure 5.10.

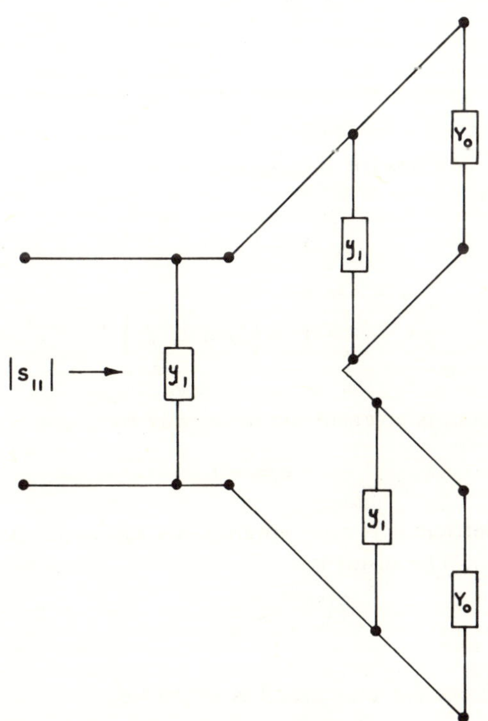

Figure 5.10.   Equivalent circuit of reciprocal 3-port junction.

5.8. 1-PORT APPROXIMATION OF 3-PORT JUNCTION CIRCULATOR

To obtain the 1-port approximation of the 3-port circulator, it is necessary to obtain the reflection coefficient S_{11} in terms of the frequency variable in the vicinity of the circulation condition. This can be done by first writing the unnormalized admittance eigenvalues in the following way:

$$y_{+1}=y_1-\frac{jY_0}{\sqrt{3}} \tag{5.50}$$

$$y_{-1}=y_1+\frac{jY_0}{\sqrt{3}} \tag{5.51}$$

where y_1 is a pure imaginary number that determines the frequency behavior of the junction. The reflection coefficient s_{+1} of an ideal circulator in the vicinity of the circulation frequency is

$$s_{+1}=\frac{Y_0-\left(y_1-\dfrac{jY_0}{\sqrt{3}}\right)}{Y_0+\left(y_1-\dfrac{jY_0}{\sqrt{3}}\right)} \tag{5.52}$$

In a similar way, one has for the eigenvalue s_{-1}

$$s_{-1}=\frac{Y_0-\left(y_1+\dfrac{jY_0}{\sqrt{3}}\right)}{Y_0+\left(y_1+\dfrac{jY_0}{\sqrt{3}}\right)} \tag{5.53}$$

The third eigenvalue is given in the usual way by

$$s_0=-1 \tag{5.54}$$

The reflection coefficient is now obtained by taking a linear combination of s_0, s_{+1}, and s_{-1}. The result is

$$S_{11}\approx\frac{-y_1}{2Y_0} \tag{5.55}$$

The input admittance for this circuit is therefore

$$Y_{in}=Y_0+y_1 \tag{5.56}$$

An approximate equivalent network for this equation is an ideal circulator available at any frequency with an admittance y_1 connected at each port. This is shown in Figure 5.11. This result indicates that the circulator input admittance is constructed in terms of the characteristic admittance of the reciprocal eigennetworks.

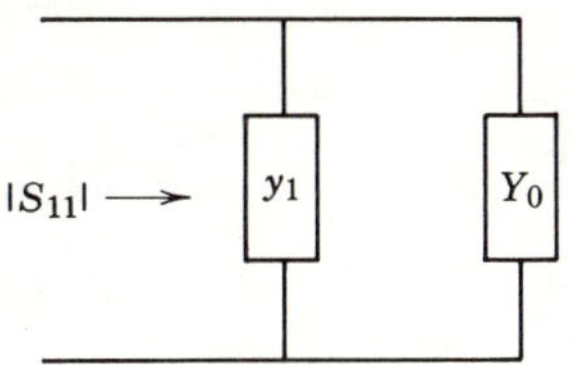

Figure 5.11.   1-Port equivalent circuit of junction circulator.

5.9.   COMPLEX GYRATOR ADMITTANCE OF 3-PORT CIRCULATOR

In this section the complex gyrator admittance is derived from the impedance matrix given by equation 5.39 by assuming the conditions of an ideal circulator at port 3.

$$V_3 = I_3 = 0 \tag{5.57}$$

and

$$\begin{bmatrix} V_1 \\ V_2 \\ 0 \end{bmatrix} = \begin{bmatrix} Z_{11} & Z_{12} & -Z_{12}^* \\ -Z_{12}^* & Z_{11} & Z_{12} \\ Z_{12} & -Z_{12}^* & Z_{11} \end{bmatrix} \begin{bmatrix} I_1 \\ I_2 \\ 0 \end{bmatrix} \tag{5.58}$$

The result is

$$I_1 = I_2 \tag{5.59}$$

$$V_2 = -V_1 \tag{5.60}$$

and

$$Z_{in} = \frac{V_1}{I_1} = Z_{11} + \frac{Z_{12}^2}{Z_{12}^*} \tag{5.61}$$

which is a general result.
Assuming $z_0 = 0$, Z_{in} becomes

$$Z_{in} = Z_{11} + Z_{12} \tag{5.62}$$

The input admittance is now given by

$$Y_{in} = \frac{1}{Z_{in}} = \frac{Z_{11}^* + Z_{12}^*}{(Z_{11} + Z_{12})(Z_{11}^* + Z_{12}^*)} \tag{5.63}$$

where

$$Z_{11} = \frac{z_{+1} + z_{-1}}{3} \tag{5.64}$$

$$Z_{12} = \frac{z_{+1}e^{j2\pi/3} + z_{-1}e^{-j2\pi/3}}{3} \tag{5.65}$$

provided $z_0 = 0$.

Z_{11}^* and Z_{12}^* are the complex conjugate equations. Assuming that Y_{in} has the form of Z_{in} in equation 5.62 gives

$$Y_{11} = \frac{3(z_{+1} + z_{-1})}{z_{+1}^2 + z_{-1}^2 - z_{+1}z_{-1}} \tag{5.66}$$

and

$$Y_{12} = \frac{3(z_{+1}e^{-j2\pi/3} + z_{-1}e^{j2\pi/3})}{z_{+1}^2 + z_{-1}^2 - z_{+1}z_{-1}} \tag{5.67}$$

In the vicinity of the circulation frequency the result is

$$Y_{11} \approx (y_{+1} + y_{-1}) \tag{5.68}$$

$$Y_{12} \approx (y_{+1}e^{j2\pi/3} + y_{-1}e^{-j2\pi/3}) \tag{5.69}$$

provided

$$\frac{z_{+1}}{z_{-1}} \approx -1 \tag{5.70}$$

The impedance eigenvalues in the above equations have been eliminated by making use of the fact that they are the reciprocal of the admittance ones. There is also a relation between Y_{11} and Y_{12} which is given by

$$Y_{12} = \frac{-Y_{11}}{2} + \frac{\sqrt{3}}{2}(y_{+1} - y_{-1}) \tag{5.71}$$

The input admittance is therefore

$$Y_{in} = \left(\frac{y_{+1} + y_{-1}}{2}\right) + j\sqrt{3}\left(\frac{y_{+1} - y_{-1}}{2}\right) \tag{5.72}$$

This equation reduces to equation (5.56) with the boundary conditions given by equations (5.50) and (5.51)

5.10   *IMMITANCE EIGENVALUES OF IDEAL CIRCULATOR AT CHARACTERISTIC PLANES*

The characteristic planes in a 3-port junction coincide with those at which a short circuit placed in one port will cause a wave at the input port to be completely reflected, none entering the third one. The reflection coefficient at port 1 is then the eigenvalue s_1. These planes need not coincide with the physical terminals of the junction. Another property of these planes is that when a wave incident on the junction at one port is totally reflected by the location of a short circuit at a characteristic plane of a second port, the electric field vanishes at all characteristic planes in the other ones. A further property of characteristic planes is that the values of the admittance in each port extrapolated to the characteristic planes in each port are related in a standard way. These properties are fully discussed in Chapter 13. It also provides a method of measuring s_1.

At the characteristic planes the eigennetworks s_1 consist of halfwave-long short-circuited transmission lines. These eigennetworks coincide with those of the counter-rotating ones of demagnetized ideal quarterwave coupled circulators. This boundary condition, therefore, establishes the input terminals of such circulators. At these terminals, the admittance matrix exists, while the impedance one does not. The boundary conditions of a circulator formed from these eigennetworks are

$$\theta_0 = \frac{\pi}{2} \tag{5.73}$$

$$\theta_1 = \frac{\pi}{2} \tag{5.74}$$

$$\theta_{\pm 1} = \mp \frac{\pi}{6} \tag{5.75}$$

$$\theta_t = \frac{\pi}{2} \tag{5.76}$$

The immittance eigenvalues are

$$y_0 = 0 \tag{5.77}$$

$$y_{+1} = j\sqrt{3}\, Y_0 \tag{5.78}$$

$$y_{-1} = -j\sqrt{3}\, Y_0 \tag{5.79}$$

$$z_0 = \infty \tag{5.80}$$

$$z_{+1} = \frac{-jR_0}{\sqrt{3}} \tag{5.81}$$

$$z_{-1} = \frac{jR_0}{\sqrt{3}} \tag{5.82}$$

where for simplicity it has been assumed that $y_t = 1$.

5.11. *ABCD PARAMETERS*

For any linear two-port network, there will be a linear relationship between input voltage and current (V_1, I_1), and output voltage and current (V_2, I_2), which may be expressed in the form

$$V_1 = AV_2 + jBI_2 \tag{5.83}$$

$$I_1 = jCV_2 + DI_2 \tag{5.84}$$

or

$$\begin{bmatrix} V_1 \\ I_1 \end{bmatrix} = \begin{bmatrix} A & jB \\ jC & D \end{bmatrix} \begin{bmatrix} V_2 \\ I_2 \end{bmatrix} \tag{5.85}$$

Where A, B, C, D are real quantities, Eqns. (5.5) and (5.6) define the ABCD matrix for a section of uniform transmission line. Fig. 5.12 lists some further ABCD matrices for some standard 2-port networks encountered in the theory of junctions. The importance of such matrices lies in the fact that the overall ABCD network of a cascade arrangement of 2-port networks can be obtained thru simple matrix multiplication of the individual matrices of each cascaded 2-port. This nomenclature is recommended in the theory of circulators. For symmetrical networks.

$$A = D \tag{5.86}$$

and for reciprocal passive networks

$$AD + BC = 1 \tag{5.87}$$

Simple Components and Their Corresponding Matices

Component	Circuit	*ABCD* Matrix
Series impedance	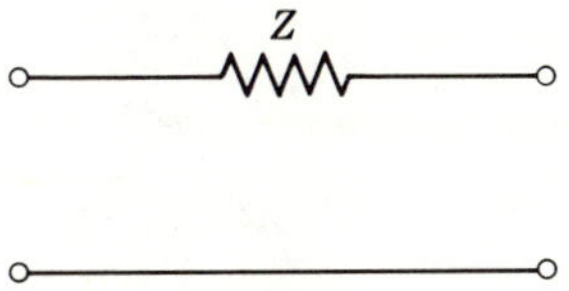 (a)	$\begin{bmatrix} 1 & Z \\ 0 & 1 \end{bmatrix}$
Shunt admittance	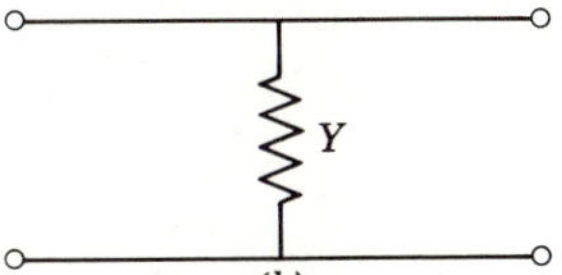 (b)	$\begin{bmatrix} 1 & 0 \\ Y & 1 \end{bmatrix}$
Section of line	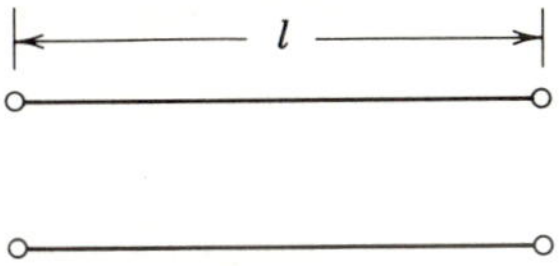 (c)	$\begin{bmatrix} \cosh \gamma l & Z_0 \sinh \gamma \\ \frac{\sinh \gamma l}{Z_0} & \cosh \gamma l \end{bmatrix}$ Z_0 = Characteristic impedance γ = Propagation function
Ideal transformer	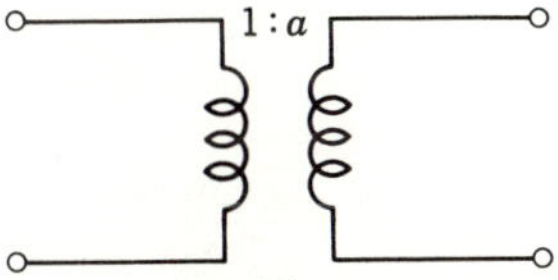 (d)	$\begin{bmatrix} \frac{1}{a} & 0 \\ 0 & a \end{bmatrix}$
Radial line	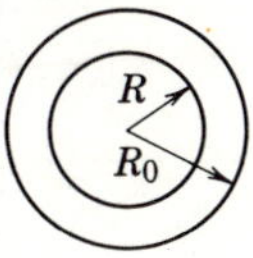 (e)	$\begin{bmatrix} A & jB \\ jC & D \end{bmatrix}^*$

Figure 5.12. *ABCD matrices of some common networks.*

$$*A = \frac{\pi(k_0 R)}{2}\left[J_n(k_0 R_0)\,Y_n'(k_0 R) - J_n'(k_0 R)\,Y_n(k_0 R_0)\right]$$

$$jB = j\eta_0 \frac{\pi(k_0 R)}{2}\left[J_n(k_0 R)\,Y_n(k_0 R_0) - J_n(k_0 R_0)\,Y_n(k_0 R)\right]$$

$$jC = \frac{j}{\eta_0}\frac{\pi(k_0 R)}{2}\left[J_n'(k_0 R)\,Y_n'(k_0 R_0) - J_n'(k_0 R_0)\,Y_n'(k_0 R)\right]$$

$$D = \frac{\pi(k_0 R)}{2}\left[J_n(k_0 R)\,Y_n'(k_0 R_0) - J_n'(k_0 R_0)\,Y_n(k_0 R)\right]$$

CHAPTER SIX

Circular Ferrite Substrate Disk Resonators

This chapter considers the conditions for resonance in circular unmagnetized (dielectric) and magnetized ferrite structures under Transverse Electric (TE) conditions. The configuration considered in this chapter is that of microstrip depicted in Figure 6.1. In the case of stripline with dielectric disks on each side of the metal disk, the bottom disk behaves as a mirror image of the top one. The analysis of the stripline resonator is, therefore, identical with that of the microstrip case. For the dielectric configuration the lowest frequency resonance consists of two field patterns, which rotate in opposite directions at the same frequency. These two counter-rotating modes then form a standing wave within the disk. Higher order modes are also possible depending on the radius of the disk resonator. These too usually occur in pairs, except for one mode for which the electric field has no variation around the periphery of the disk. It is found that the amplitude of the electric field patterns at the edge of the circular cavity are proportional to the eigenvectors discussed in Chapter 2. If the dielectric disk is replaced by a ferrite one magnetized along its axis, the two counter-rotating field patterns are no longer resonant at the same frequency. This condition satisfies the splitting of the degenerate eigenvalues of the scattering matrix discussed in Chapter 2. In addition, the standing wave pattern formed by two individual field patterns is rotated within the disk. The direction in which the field pattern is rotated is determined by the sense of the direct magnetic field.

101

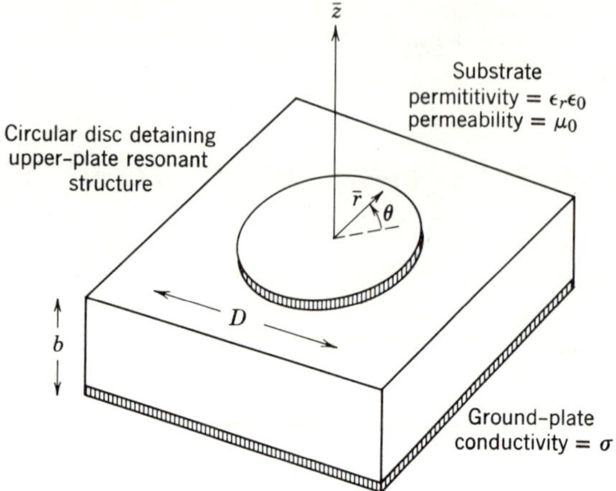

Figure 6.1.   *Microstrip circular resonator.*

6.1.   *MODE PATTERNS IN UNMAGNETIZED CIRCULAR DISK*

In the substrate Maxwell's equations with time dependence $e^{j\omega t}$ are

$$\nabla X \overline{E} = -j\omega \overline{B} \tag{6.1}$$

$$\nabla X \overline{H} = j\omega \overline{D} \tag{6.2}$$

$$\nabla \cdot \overline{D} = 0 \tag{6.3}$$

$$\nabla \cdot \overline{B} = 0 \tag{6.4}$$

where

$$\overline{B} = \mu_0 \mu \overline{H} \tag{6.5}$$

$$\overline{D} = \epsilon_0 \epsilon_r \overline{E} \tag{6.6}$$

Here μ is the relative permeability of the unmagnetized ferrite substrate, and ϵ_r is the relative dielectric constant.

In what follows it is assumed that the substrate thickness and relative dielectric constant are such that no integral half wave-long value can be accommodated in the z-direction. This means that

$$\frac{\partial}{\partial z} = 0 \tag{6.7}$$

and there is only a z-component of the electric field.

$$\bar{E} = \bar{i}_z E_z \qquad (6.8)$$

where $\bar{i}_z$ is a unit vector. Maxwell's equations in cylindrical coordinates then give for the other field components in terms of E_z

$$H_r = \frac{j}{\omega\mu_0\mu}\left[\frac{1}{r}\frac{\partial E_z}{\partial\phi}\right] \qquad (6.9)$$

$$H_\phi = \frac{-j}{\omega\mu_0\mu}\left[\frac{\partial E_z}{\partial r}\right] \qquad (6.10)$$

$$E_\phi = E_r = H_z = 0 \qquad (6.11)$$

Substituting this result in Eq. 6.2 gives the following wave equation for E_z:

$$\left[\frac{\partial^2}{\partial r^2} + \frac{1}{r}\frac{\partial}{\partial r} + \frac{1}{r^2}\frac{\partial^2}{\partial\phi^2} + k^2\right]E_z = 0 \qquad (6.12)$$

where

$$k = \omega\sqrt{\epsilon_0\epsilon_r\mu_0\mu} \ . \qquad (6.13)$$

The free space wave admittance and impedance are

$$\zeta = \frac{1}{\eta} = \sqrt{\frac{\epsilon_0\epsilon_r}{\mu_0\mu}} \qquad (6.14)$$

Since all fields must be finite at the origin, the solution to Eq. 6.12 employs Bessel functions of the first kind of order $n, J_n(kr)$

$$E_z = A_n J_n(kr)e^{jn\phi} \qquad (6.15)$$

Substituting this last equation into Eqs. 6.9 and 6.10 gives

$$H_r = -nA_n\zeta\frac{J_n(kr)}{kr}e^{jn\phi} \qquad (6.16)$$

$$H_\phi = -jA_n\zeta J_n'(kr)e^{jn\phi} \qquad (6.17)$$

and TE conditions now prevail.

At the edge of the disk it is assumed that the radial component of the surface current must vanish, and, consequently,

$$H_\phi(R) = 0 \text{ at } r = R \tag{6.18}$$

Equation 6.18 is satisfied whenever

$$J_n'(kR) = 0 \tag{6.19}$$

This last equation determines the radius of the ferrite disk for a particular mode. The solution to Eq. 6.18 is denoted by

$$(kR)_{n,j} \tag{6.20}$$

where n,j denotes the jth root of the nth order equation.

For $n = 0$, the radius of the disk for the first root is

$$(kR)_{0,1} = 3.83171 \tag{6.21}$$

and the field components are

$$E_z = k^2 J_0(kr) \tag{6.22}$$

$$H_\phi = -j\zeta k^2 J_0'(kr) \tag{6.23}$$

$$H_r = E_r = E_\phi = H_z = 0 \tag{6.24}$$

The field configuration for this mode is indicated in Figure 6.2.

For $n = 1$, the radius of the disk is

$$(kR)_{1,1} = 1.84118 \tag{6.25}$$

and the field components are

$$E_z = k^2 J_1(kr)\cos\phi \tag{6.26}$$

$$H_r = -j\zeta k \frac{J_1(kr)}{r}\sin\phi \tag{6.27}$$

$$H_\phi = -j\zeta k^2 J_1'(kr)\cos\phi \tag{6.28}$$

$$H_z = E_r = E_\phi = 0 \tag{6.29}$$

The field components for this mode are depicted in Figure 6.3.

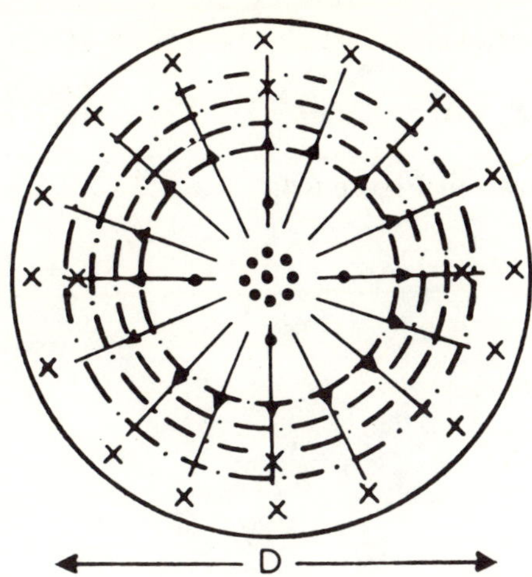

Figure 6.2. *Electromagnetic field patterns for n=0 mode (Ref. 1).*

For $n=2$, the radius of the ferrite disk is determined by

$$(kR)_{2,1} = 3.05424 \tag{6.30}$$

The radii for the $n=1$ and 2 are smaller than for the $n=0$ one. The field components for this mode are

$$E_z = k^2 J_2(kr)\cos 2\phi \tag{6.31}$$

$$H_r = -j2\zeta k \frac{J_2(kr)}{r}\sin 2\phi \tag{6.32}$$

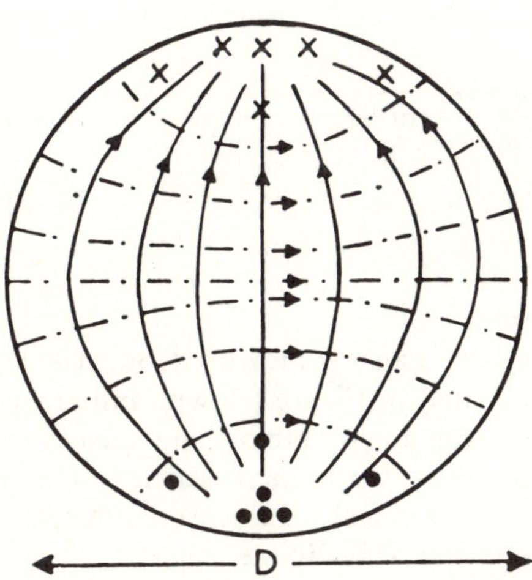

Figure 6.3. *Electromagnetic field patterns for n=1 mode (Ref. 1).*

$$H_\phi = -j\zeta k^2 J_2'(kr)\cos 2\phi \qquad (6.33)$$

$$H_z = E_r = E_\phi = 0 \qquad (6.34)$$

The field patterns for this mode is shown in Figure 6.4.

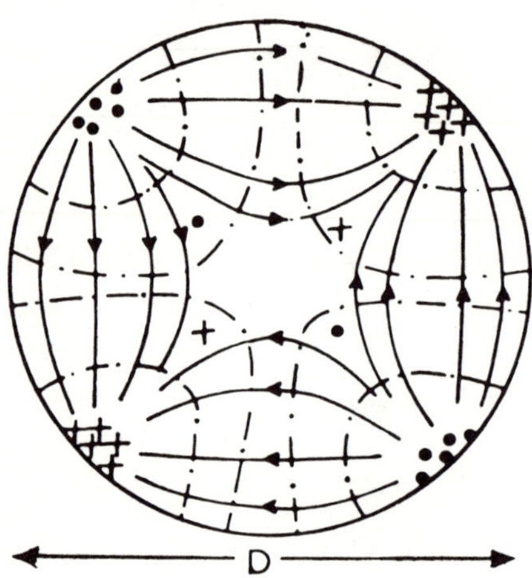

Figure 6.4.   *Electromagnetic field patterns for* $n=2$ *mode (Ref. 1).*

For $n=3$ the results are

$$(kR)_{3,1} = 4.20119 \qquad (6.35)$$

$$E_z = k^2 J_3(kr)\cos 3\phi \qquad (6.36)$$

$$H_r = -j3\zeta k\frac{J_3(kr)}{r}\sin 3\phi \qquad (6.37)$$

$$H_\phi = -j\zeta k^2 J_3'(kr)\cos 3\phi \qquad (6.38)$$

The field pattern are indicated in Figure 6.5.

The resonant frequency of these modes can be measured by placing transmission probes in the position indicated in Figure 6.6 with initially a coupling gap of about 0.001 cm. Transmission Q-factors are measured using a frequency sweeper, and the coupling gap is increased by etching until the unloaded Q-factor is measured. This technique ensures that the field perturbation due to the coupling probes is negligible.

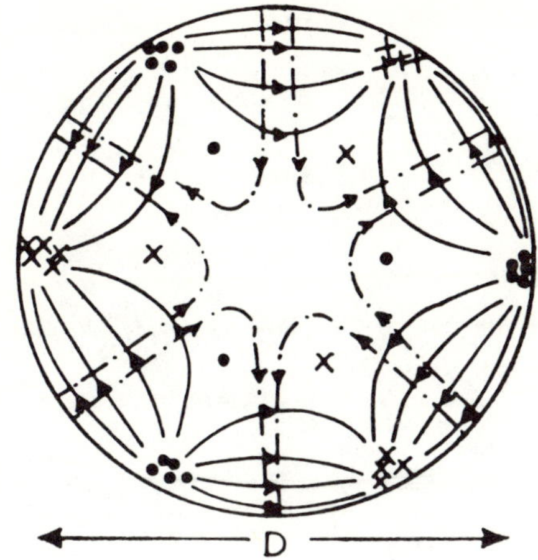

Figure 6.5.   *Electromagnetic field patterns for* $n = 3$ *mode (Ref.* 1).

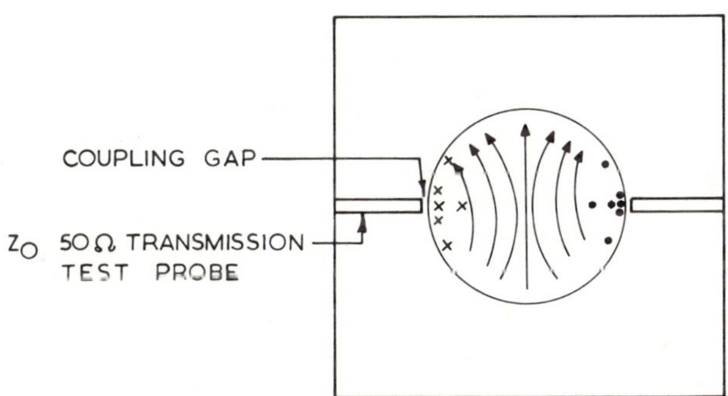

Figure 6.6.   *Test probes for transmission cavity.*

6.2.   ELECTRIC FIELDS AND EIGENVECTORS

This section relates the electric field patterns at the edge of the circular cavity discussed in the previous section to the eigenvectors derived in Chapter 2. This is done by making symmetrical connections to the circular cavity with loosely coupled probes. Figures 6.7 through 6.9 indicate an $n = 1$ cavity loosely coupled by symmetrically located probes for the cases of $m = 3$, 4, and 5. It is now recognized that the electric fields at the terminals of Figure 6.7, which are described by Eq. 6.26 are proportional to a linear combination of the eigenvectors given by Eqs. 2.78 and 2.79. Similarly, the electric fields at the four terminals of Figure 6.8 are proportional to a linear combination of Eqs. 2.81 and 2.82. Finally, the electric fields at the terminals of Figure 6.9 are proportional to a linear combination of the eigenvectors described by Eqs. 2.85 and 2.86.

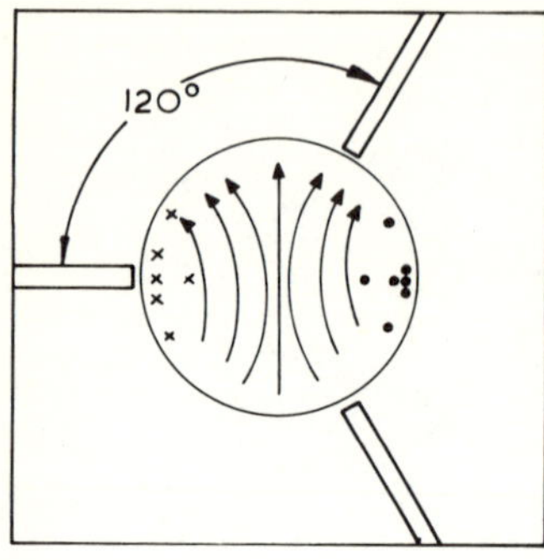

Figure 6.7.   *Test probes for* m = 3 *and* n = 1 *mode.*

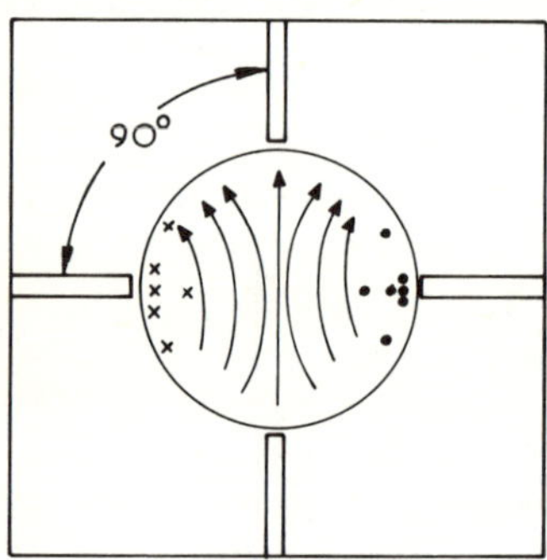

Figure 6.8.   *Test probes for* m = 4 *and* n = 1 *mode.*

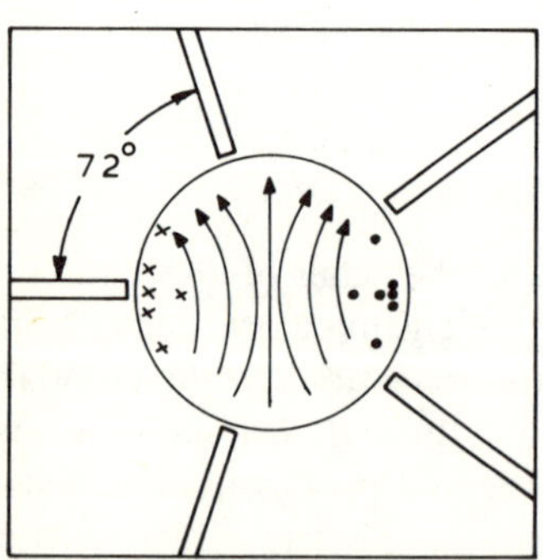

Figure 6.9.   *Test probes for* m = 5 *and* n = 1 *mode.*

A similar argument indicates that the electric field patterns of the other resonant modes are proportional to the corresponding eigenvectors introduced in Chapter 2.

6.3.   MODE PATTERNS IN MAGNETIZED CIRCULAR DISK

If the dielectric substrate is replaced by a magnetized ferrite substrate the derivation proceeds in essentially the same way as in the last section except that μ in Eq. 6.5 now takes on the tensor form derived in Chapter 1.

$$[\mu_r] = \begin{bmatrix} \mu & -jK & 0 \\ jK & \mu & 0 \\ 0 & 0 & 1 \end{bmatrix} \tag{6.39}$$

If a magnetic field is applied along the axis of the disk the two counter-rotating modes are no longer resonant at the same frequency. In addition the standing pattern formed by the two field patterns is rotated within the disk. Assuming once more that there is no variation in the z-direction the field components are given by

$$H_r = j\left[\frac{1}{r} \frac{\partial E_z}{\partial \phi} - j\frac{K}{\mu} \frac{\partial E_z}{\partial r} \right] / \omega\mu_0 \mu_e \tag{6.40}$$

$$H_\phi = -j\left[\frac{\partial E_z}{\partial r} + j\frac{K}{\mu} \frac{1}{r} \frac{\partial E_z}{\partial \phi} \right] / \omega\mu_0 \mu_e \tag{6.41}$$

where E_z again satisfies the wave equation given by Eq. 6.12

$$E_z = A_n J_n(k_e r) e^{jn\phi} \tag{6.42}$$

Combining the last three equations gives

$$H_r = -A_n \zeta_e \left[\frac{n J_n(k_e r)}{k_e r} - \frac{K}{\mu} J_n'(k_e r) \right] e^{jn\phi} \tag{6.43}$$

$$H_\phi = -jA_n \zeta_e \left[J_n'(k_e r) - \frac{Kn}{\mu} \frac{J_n(k_e r)}{k_e r} \right] e^{jn\phi} \tag{6.44}$$

                    Circular Ferrite Substrate Disk Resonators

where

$$k_e = \omega \sqrt{\epsilon_0 \epsilon_r \mu_0 \mu_e} \tag{6.45}$$

$$\zeta_e = \frac{1}{\eta_e} = \sqrt{\frac{\epsilon_0 \epsilon_r}{\mu_0 \mu_e}} \tag{6.46}$$

Applying the boundary conditions at $r = R$ given by Eq. 6.18 to H_ϕ above gives

$$J_n'(k_e R) - \frac{Kn}{\mu} \frac{J_n(k_e R)}{k_e R} = 0 \tag{6.47}$$

There are, therefore, two roots for the magnetized disk:

$$(k_e R)_{-n,j} \tag{6.48}$$

$$(k_e R)_{+n,j} \tag{6.49}$$

which define two resonant frequencies

$$\omega_{+n,j} \sqrt{\epsilon_0 \epsilon_r \mu_0 \mu_e} \, R = (k_e R)_{+n,j} \tag{6.50}$$

$$\omega_{-n,j} \sqrt{\epsilon_0 \epsilon_r \mu_0 \mu_e} \, R = (k_e R)_{-n,j} \tag{6.51}$$

The resonant frequency of the unmagnetized disk, for which K/μ is zero, lies between the two split frequencies as already determined by Eq. 6.20

$$\omega_{n,j} \sqrt{\epsilon_0 \epsilon_r \mu_0 \mu_e} \, R = (kR)_{n,j} \tag{6.52}$$

If K/μ is small the difference of $(k_e R)_{n,j}$ values of a pair of resonances is determined by

$$(\Delta k_e R)_{n,j} \approx \frac{2n(k_e R)_{n,j}}{(k_e R)_{n,j}^2 - n^2} \cdot \frac{K}{\mu} \tag{6.53}$$

This last equation states that the splitting of a pair of resonant frequencies is proportional to K/μ for K/μ small. Figure 6.10 indicates the K/μ dependence for several of these modes.

Another important result is that the field patterns in the ferrite disk are rotated as the junction is magnetized. The direction in which the field

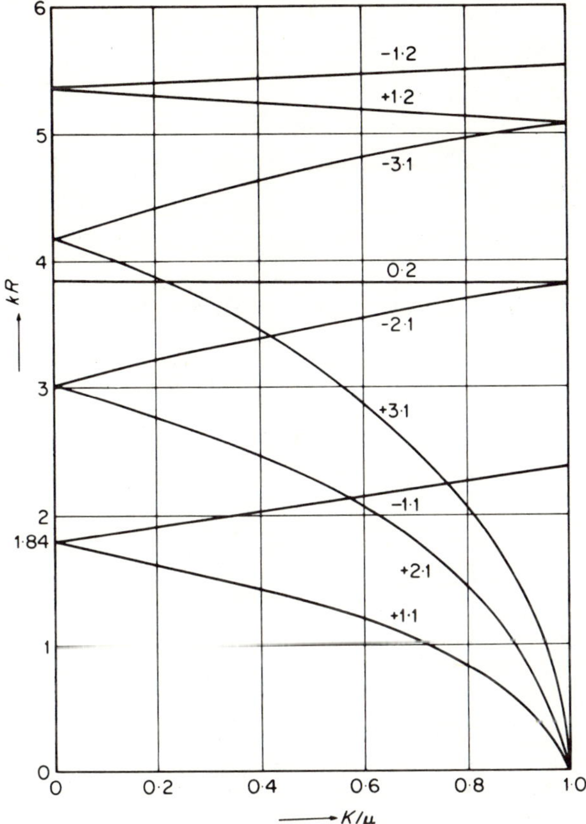

Figure 6.10.   *Mode spectrum for magnetized disk* (*Ref.* 4).

patterns are rotated is determined by the direct magnetic field. If they are rotated by 30°, one of the ports is completely decoupled, and transmission occurs between the other two ports. In this configuration the 3-port junction behaves as a circulator. This is shown in Figure 6.11. It can be seen from this diagram that the transmission coefficient S_{12} is

$$S_{12} = -1 \tag{6.54}$$

6.4.   MODE PATTERNS IN CIRCULAR RESONATORS USING FERRITE—DISCS AND DIELECTRIC RINGS

This section derives the resonance conditions in a ferrite disk resonator surrounded by a dielectric ring. The geometry considered here is shown in Figure 6.12.

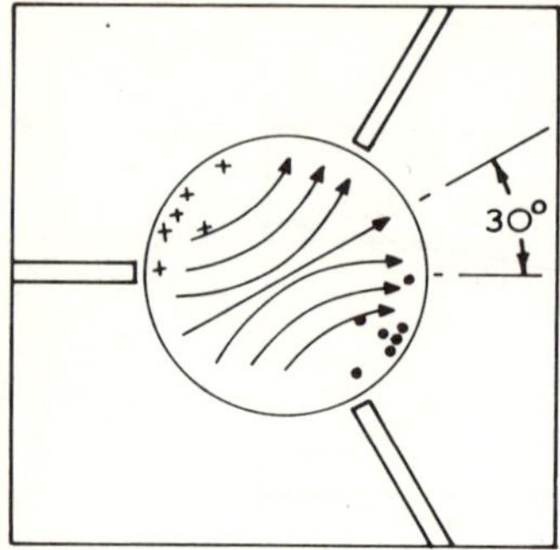

Figure 6.11.   Test probes for m = 3 and n = 1 mode rotation through 30°.

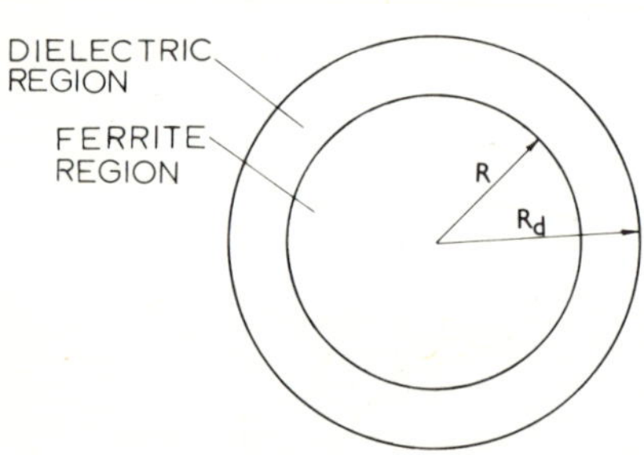

Figure 6.12.   Schematic of composite dielectric ferrite resonator.

Solutions are sought with electric fields purely in the z direction and magnetic fields purely in the x–y plane. The fields are independent of z and are taken to have time dependence exp $(j\omega t)$. A complete expansion for the fields in the ferrite is, therefore,

$$E_{zfn} = \sum_{n=-\infty}^{\infty} A_n J_n(k_e r) e^{jn\phi} \tag{6.55}$$

$$H_{\phi fn} = \frac{-j}{\eta_e} \sum_{n=-\infty}^{\infty} A_n \left[J_n'(k_e r) - \frac{Kn}{\mu} \frac{J_n(k_e r)}{k_e r} \right] e^{jn\phi} \tag{6.56}$$

$$H_{rfn} = -\frac{1}{\eta_e} \sum_{n=-\infty}^{\infty} A_n \left[\frac{n J_n(k_e r)}{k_e r} - \frac{K}{\mu} J_n'(k_e r) \right] e^{jn\phi} \tag{6.57}$$

$$E_r = E_\phi = H_z = 0 \tag{6.58}$$

where

$$k_e = \omega \sqrt{\epsilon_0 \epsilon_r \mu_0 \mu_e} \tag{6.59}$$

$$\eta_e = \sqrt{\frac{\mu_0 \mu_e}{\epsilon_0 \epsilon_r}} \tag{6.60}$$

Similarly fields outside the ferrite $(r \geqslant R)$ are of the form

$$E_{zdn} = \sum_{n=-\infty}^{\infty} A_n [B_n J_n(k_d r) + C_n Y_n(k_d r)] e^{jn\phi} \tag{6.61}$$

$$H_{\phi dn} = \frac{-j}{\eta_d} \sum_{n=-\infty}^{\infty} A_n [B_n J_n'(k_d r) + C_n Y_n'(k_d r)] e^{jn\phi} \tag{6.62}$$

$$H_{rdn} = \frac{-1}{\omega \mu_0 \mu_d} \sum_{n=-\infty}^{\infty} n A_n [B_n J_n(k_d r) + C_n Y_n(k_d r)] e^{jn\phi} \tag{6.63}$$

$$E_r = E_\phi = H_z = 0 \tag{6.64}$$

where

$$k_d = \omega \sqrt{\mu_0 \mu_d \epsilon_0 \epsilon_d} \tag{6.65}$$

$$\eta_d = \sqrt{\frac{\mu_0 \mu_d}{\epsilon_0 \epsilon_d}} \tag{6.66}$$

The constants B_n and C_n are now obtained by applying the boundary conditions at $r = R$ to the individual modes of the above series to give

$$E_{zfn} = B_n J_n(k_d R) + C_n Y_n(k_d R) \tag{6.67}$$

$$H_{\phi fn} = B_n J_n'(k_d R) + C_n Y_n'(k_d R) \tag{6.68}$$

The complete solution for E_{zdn} and $H_{\phi dn}$ is now given by

$$E_{zdn} = [A_n E_{zfn} + j B_n H_{\phi fn}] e^{jn\phi} \tag{6.69}$$

$$H_{\phi dn} = [j C_n E_{zfn} + D_n H_{\phi fn}] e^{jn\phi} \tag{6.70}$$

where $A_n, B_n, C_n,$ and D_n are listed in Chapter 5. The eigenadmittances for such a resonator has the following form:

$$y_n' = \frac{H_{\phi dn}}{E_{zdn}} = \frac{jC_n + D_n y_n}{A_n + jB_n y_n} \tag{6.71}$$

where

$$y_n = \frac{H_{\phi fn}}{E_{zfn}} \tag{6.72}$$

REFERENCES

1. J. Watkins "Circular Resonant Structures on Microstrip," *Electron. Lett.*, **5**, 524–525 (1969).
2. P. Troughton, "High Q Factor Resonators in Microstrip," *Electron. Lett.*, **4**, 520–522 (1968).
3. P. Troughton, "Measurement Techniques in Microstrip," *Electron. Lett.*, **5**, 25–26 (1969).
4. H. Bosma, "On the principle of stripline Circulation," *Proc. IEEE*, Part B *Suppl.* (21), **109**, 137–146 (1962).

CHAPTER SEVEN

Dissipation Matrix of Lossy Junctions

This chapter gives the general relations between the coefficients of the scattering matrix for lossy symmetrical junctions. This is done by constructing the individual entries of the scattering matrix directly in terms of the eigenvalues of the dissipation matrix. It also includes relations for semi-ideal circulators in which the insertion loss is not zero and either the isolation or VSWR is ideal.

The chapter starts by relating the eigenvalues of the dissipation, scattering, and admittance matrices. The eigenvalues of the dissipation matrix give the dissipation associated with each possible way of exciting the junction. Those of the scattering matrix give the reflection coefficients associated with these different excitations. The admittance eigenvalues define in each instance the eigennetworks of the junction. This leads to the definition of the entries of the dissipation matrix in terms of the loaded and unloaded Q-factors of the junction eigennetworks. In the most usual arrangement one of the eigenvalues is associated with a shunt network that appears as a short circuit at the reference terminals of the junction and is therefore always unity. The other eigenvalues are the reflection coefficients of resonant shunt networks and the presence of loss means that their eigenvalues will depart from unity. The amplitudes of the eigenvalues are in general unequal.

The theory is applied to reciprocal 3-port junctions, to 3-port junction circulators, and to semi-ideal 3-port circulators. The chapter includes the frequency variation of lossy circulators.

7.1. *EIGENVALUES OF SCATTERING AND DISSIPATION MATRICES*

For a lossy circulator the dissipation matrix $\overline{Q}$ must be positive real

$$\overline{Q} = \overline{I} - (\overline{S}^*)^T (\overline{S}) \tag{7.1}$$

where $\overline{I}$ is a unit matrix and $\overline{S}$ is the scattering matrix. In the case of a symmetrical 3-port junction, one has

$$\overline{S} = \begin{bmatrix} S_{11} & S_{12} & S_{13} \\ S_{13} & S_{11} & S_{12} \\ S_{12} & S_{13} & S_{11} \end{bmatrix} \tag{7.2}$$

The matrix $\overline{Q}$ is given by

$$\overline{Q} = \begin{bmatrix} q_{11} & -q_{12} & -q_{13} \\ -q_{13} & q_{11} & -q_{12} \\ -q_{12} & -q_{13} & q_{11} \end{bmatrix} \tag{7.3}$$

where

$$q_{11} = |-|S_{11}|^2 - |S_{12}|^2 - |S_{13}|^2 \tag{7.4}$$

$$q_{12} = S_{11}S_{12}^* + S_{12}S_{13}^* + S_{13}S_{11}^* \tag{7.5}$$

$$q_{13} = q_{12}^* \tag{7.6}$$

The necessary and sufficient condition for the dissipation matrix to be positive real is that the principle minors of the dissipation matrix be nonnegative.

$$q_{11} \geqslant 0 \tag{7.7}$$

$$\begin{vmatrix} q_{11} & -q_{12} \\ -q_{12}^* & q_{11} \end{vmatrix} \geqslant 0 \tag{7.8}$$

$$\begin{vmatrix} q_{11} & -q_{12} & -q_{12}^* \\ -q_{12}^* & q_{11} & -q_{12} \\ -q_{12} & -q_{12}^* & q_{11} \end{vmatrix} \geqslant 0 \tag{7.9}$$

A more simple definition to this last one is also obtained by requiring that the dissipation matrix eigenvalues lie between zero and unity.

If the scattering and dissipation matrices have common eigenvectors, their eigenvalues may be related by the following theorem, which has already been used in Chapter 3: If

$$\overline{Q}\,\overline{U}_n = q_n \overline{U}_n \tag{7.10}$$

then

$$f(\overline{Q}\,)\overline{U}_n = f(q_n)\overline{U}_n \tag{7.11}$$

where $\overline{U}_n$ is an eigenvector.

Using Eq. 7.1 in conjunction with Eq. 7.11 gives in the case of the 3-port junction

$$q_0 = 1 - s_0 s_0^* \tag{7.12}$$

$$q_{+1} = 1 - s_{+1} s_{+1}^* \tag{7.13}$$

$$q_{-1} = 1 - s_{-1} s_{-1}^* \tag{7.14}$$

The last three equations may be used to construct the scattering matrix in terms of the eigenvalues of the dissipation one. In a lossless junction, the amplitudes of the scattering matrix eigenvalues are unity, while if the junction is lossy their amplitudes will depart from unity.

It is seen from the above that the eigenvalues of the dissipation matrix represent the dissipation associated with each possible way of exciting the junction. These eigenvalues are real quantities that become zero when those of the scattering matrix become unity.

The entries of the dissipation matrix are

$$q_{11} = \frac{q_0 + q_{+1} + q_{-1}}{3} \tag{7.15}$$

$$q_{12} = \frac{q_0 + q_{+1}e^{j2\pi/3} + q_{-1}e^{-j2\pi/3}}{3} \tag{7.16}$$

$$q_{13} = \frac{q_0 + q_{+1}e^{-j2\pi/3} + q_{-1}e^{j2\pi/3}}{3} \tag{7.17}$$

Here, q_{11} represents the total dissipation of the junction, and q_{12} is a complex quantity that determines the allowable relation between the scattering parameters.

The entries of the dissipation and scattering matrices may be directly evaluated by relating their eigenvalues to the loaded and unloaded Q-

factors of the junction eigennetworks. This does away with the need to rely on the inequality relations given by Eqs. 7.7 through 7.9.

7.2. *EIGENVALUES OF ADMITTANCE AND DISSIPATION MATRICES*

The dissipation and scattering eigenvalues can be related to the physical variables of the junction by adding a damping term to the phase angles of the distributed networks defined in Chapter 5. Adding an imaginary part $j\alpha_{\pm1}$ to the phase angles $\theta_1+\theta_{\pm1}$ of the reflection coefficients gives

$$s_0 = e^{-j2\theta_0} \tag{7.18}$$

$$s_{+1} = e^{-j2(\theta_1+\theta_{+1}+\pi/2-j\alpha_{+1})} \tag{7.19}$$

$$s_{-1} = e^{-j2(\theta_1+\theta_{-1}+\pi/2-j\alpha_{-1})} \tag{7.20}$$

s_0 is the reflection coefficient of an open circuited transmission line of length θ_0, and $s_{\pm1}$ are the reflection coefficients of short circuited transmission lines of length $\theta_1+\theta_{\pm1}$. The angles θ_1 and θ_0 represent the electrical lengths of the demagnetized transmission lines, and $\theta_{\pm1}$ gives the change in the electrical length θ_1, when the networks are magnetized, $\alpha_{\pm1}$ is the total distributed loss of the transmission lines. The eigenvalues are indicated on the unit circle in Figure 7.1. The admittance eigenvalues are related to the

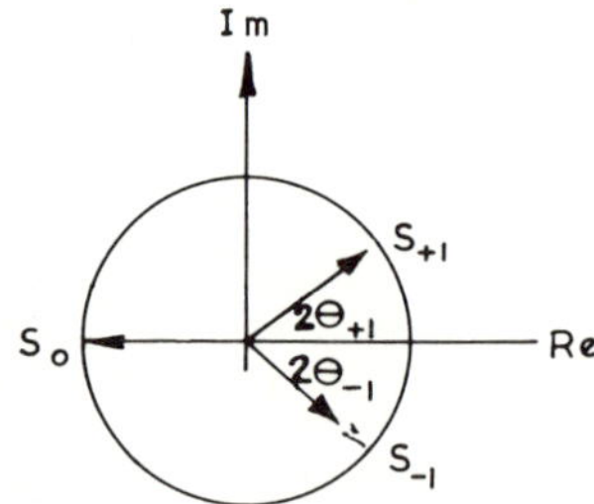

Figure 7.1   Scattering matrix eigenvalues with unequal amplitudes.

angles that the scattering matrix eigenvalues make by

$$y_0 = j\tan\theta_0 \tag{7.21}$$

$$y_{+1} = j\tan(\theta_1+\theta_{+1}+\pi/2-j\alpha_{+1}) \tag{7.22}$$

$$y_{-1} = j\tan(\theta_1+\theta_{-1}+\pi/2-j\alpha_{-1}) \tag{7.23}$$

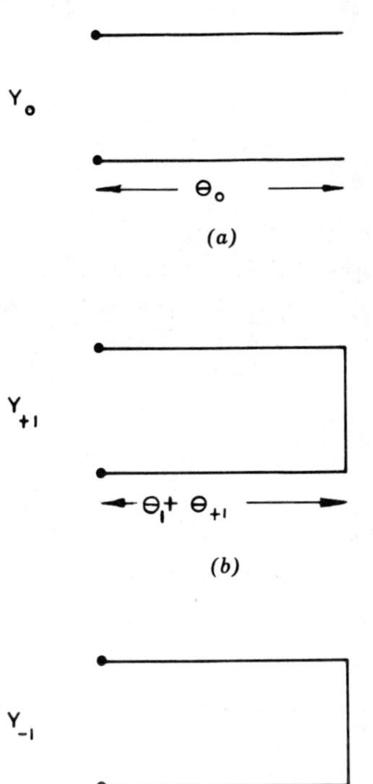

Figure 7.2   Distributed eigennetworks.

The three eigennetworks are given in Figure 7.2.

For an ideal circulator the results of Chapter 5 apply

$$\theta_0 = \frac{\pi}{2} \tag{7.24}$$

$$\theta_1 = \frac{\pi}{2} \tag{7.25}$$

$$\theta_{\pm 1} = \mp \frac{\pi}{6} \tag{7.26}$$

which satisfies the eigenvalues of the ideal circulator derived in Chapter 2.

If $2\alpha_{+1}$ is small the result for s_{+1} becomes

$$s_{+1} \approx (1 - 2\alpha_{+1})e^{-j2(\theta_1 + \theta_{+1} + \pi/2)} \tag{7.27}$$

Substituting this equation into Eq. 7.13 gives

$$q_{+1} \approx 4\alpha_{+1} \tag{7.28}$$

This result again states that the dissipation eigenvalue gives the total dissipation of the eigennetwork.

In the vicinity of the frequency at which the network is $\lambda/4$ long $(\theta_1 = \pi/2)$, it may be replaced by a shunt lumped element resonator. This allows the variables of the distributed and lumped element networks to be related. The input admittance of the distributed network in the vicinity of the frequency at which $\theta_1 = \pi/2$ is

$$y_{+1} \approx j\tan\theta_{+1} + \alpha_{+1}(1 + \tan^2\theta_{+1}) \tag{7.29}$$

The circuit considered for the lumped element resonator is illustrated in Figure 7.3. It consists of a lumped element shunt capacitance and inductance. The dielectric and magnetic losses are represented by equivalent shunt resistors.

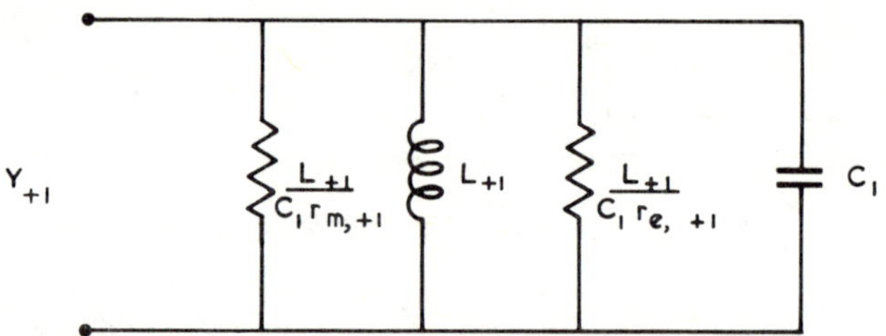

Figure 7.3   *Approximate lumped equivalent eigennetwork for y_{+1}.*

The total normalized input impedance for this circuit at the circulator center frequency ω_0 is

$$y_{+1} = \frac{Y_{+1}}{Y_0} = j2\delta_{+1}Q_{+1} + \frac{Q_{+1}}{Q_{u,+1}} \tag{7.30}$$

where

$$2\delta_{+1} \approx 2\left(\frac{\omega_0 - \omega_{+1}}{\omega_{+1}}\right) \tag{7.31}$$

Here Q_{+1} and $Q_{u,+1}$ are the loaded and unloaded Q-factors of the equivalent lumped network, ω_{+1} is the resonant frequency of the network.

Comparing Eqs. 7.29 and 7.30 gives

$$\theta_{+1} \approx 2\delta_{+1} Q_{+1} \tag{7.32}$$

$$\alpha_{+1} \approx \frac{Q_{+1}}{Q_{u,+1}} \cos^2 \theta_{+1} \tag{7.33}$$

Combining Eq. 7.28 and 7.33, one now has

$$q_{+1} \approx 4\left(\frac{Q_{+1}}{Q_{u,+1}}\right) \cos^2 \theta_{+1} \tag{7.34}$$

In a similar way one has

$$q_{-1} \approx 4\left(\frac{Q_{-1}}{Q_{u,-1}}\right) \cos^2 \theta_{-1} \tag{7.35}$$

$$q_0 = 0 \tag{7.36}$$

q_0 is zero because the frequency variation of this eigennetwork is neglected compared with that of the other two.

7.3. *SCATTERING MATRIX OF LOSSY RECIPROCAL 3-PORT JUNCTION*

For a reciprocal 3-port junction the degenerate eigenvalues are equal.

$$s_{+1} = s_{-1} = s_1 \tag{7.37}$$

$$q_{+1} = q_{-1} = q_1 \tag{7.38}$$

$$y_{+1} = y_{-1} = y_1 \tag{7.39}$$

At the frequency at which maximum power transfer through the junction occurs the angles of the eigenvalues are

$$\theta_{+1} = \theta_{-1} = 0 \tag{7.40}$$

The scattering matrix eigenvalues are, therefore,

$$s_1 = \left(1 - \frac{q_1}{2}\right) \tag{7.41}$$

$$s_0 = -1 \tag{7.42}$$

where

$$q_1 = 4\left(\frac{Q_1}{Q_u}\right) \tag{7.43}$$

The coefficients of the scattering matrix are

$$|S_{11}| = \frac{1 - q_1}{3} \tag{7.44}$$

$$|S_{12}| = |S_{13}| = \frac{2 - q_1/2}{3} \tag{7.45}$$

In terms of the original variables one has

$$|S_{11}| = \frac{1}{3} - \frac{4}{3}\left(\frac{Q_1}{Q_u}\right) \tag{7.46}$$

$$|S_{12}| = |S_{13}| = \frac{2}{3} - \frac{2}{3}\left(\frac{Q_1}{Q_u}\right) \tag{7.47}$$

7.4. SCATTERING MATRIX OF LOSSY 3-PORT CIRCULATOR

For an ideal circulator the angles of the eigenvalues are

$$\theta_{+1} = -\frac{\pi}{6} \tag{7.48}$$

$$\theta_{-1} = \frac{\pi}{6} \tag{7.49}$$

The eigenvalues of the scattering matrix are, therefore,

$$s_{+1} = \left(1 - \frac{q_{+1}}{2}\right)e^{-j\pi/3} \tag{7.50}$$

$$s_{-1} = \left(1 - \frac{q_{-1}}{2}\right)e^{j\pi/3} \tag{7.51}$$

$$s_0 = -1 \tag{7.52}$$

Using the above eigenvalues the entries of the scattering matrix are

$$|S_{11}| = |S_{13}| = \frac{q_{+1}}{6} \tag{7.53}$$

$$|S_{12}| = 1 - \frac{q_{+1}}{3} \tag{7.54}$$

provided

$$q_{+1} \approx q_{-1} \qquad (7.55)$$

In terms of the original variables the scattering coefficients become

$$|S_{11}| = |S_{13}| = \frac{1}{2}\frac{Q_1}{Q_u} \qquad (7.56)$$

$$|S_{12}| = 1 - \frac{Q_1}{Q_u} \qquad (7.57)$$

Hence $S_{11} \neq 0$, $S_{12} \neq 1$, and $S_{13} = S_{11}$.

The above scattering coefficients satisfy Eq. 7.1 as is observed by forming Eq. 7.8. Figure 7.4 gives the relation between S_{11} and $S_{12}(S_{11} = S_{13})$.

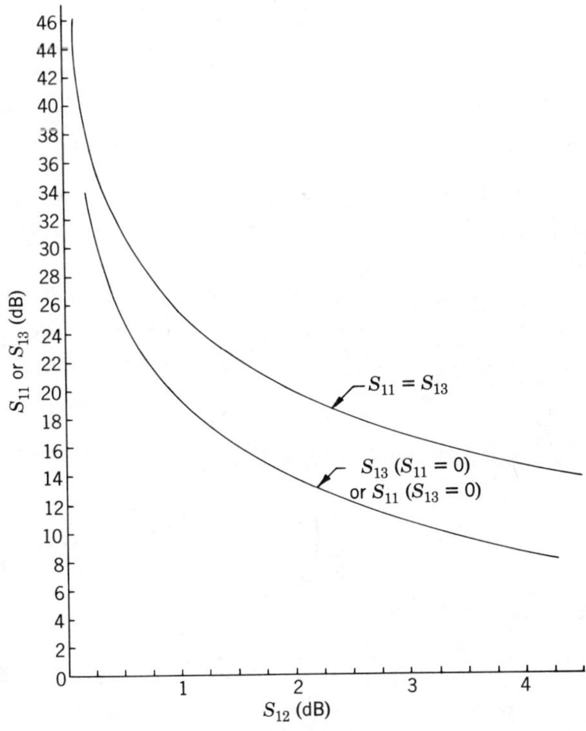

Figure 7.4   *Relation between S parameters of semi-ideal circulators (Ref. 7.13).*

7.5. SCATTERING MATRIX OF SEMI-IDEAL CIRCULATORS

Semi-ideal circulators are those for which either $S_{11}=0$, $S_{12}\neq1$, and $S_{13}\neq0$, or $S_{11}\neq0$, $S_{12}\neq1$, and $S_{13}=0$. The first situation is obtained when the angle between s_{+1} and s_{-1} is less than $|120|$. The second case is obtained when the angle between the eigenvalues is larger than $|120|$.

The scattering parameters are given in terms of their eigenvalues by

$$3S_{11} = -1 + \left(1 - \frac{q_{+1}}{2}\right)e^{-j2\theta_{+1}} + \left(1 - \frac{q_{-1}}{2}\right)e^{j2\theta_{+1}} \tag{7.58}$$

$$3S_{12} = -1 + \left(1 - \frac{q_{+1}}{2}\right)e^{-j(2\theta_{+1}-120)} + \left(1 - \frac{q_{-1}}{2}\right)e^{j(2\theta_{+1}-120)} \tag{7.59}$$

$$3S_{13} = -1 + \left(1 - \frac{q_{+1}}{2}\right)e^{-j(2\theta_{+1}+120)} + \left(1 - \frac{q_{-1}}{2}\right)e^{j(2\theta_{+1}+120)} \tag{7.60}$$

where it has been assumed that the splitting is symmetrical:

$$\theta_{-1} = -\theta_{+1} \tag{7.61}$$

In semi-ideal circulators S_{11} and S_{13} are completely determined by S_{12}, provided it is assumed that the amplitudes of s_{+1} and s_{-1} are equal. This means that the latter quantity can be obtained simply by measuring either S_{11} or S_{13}.

The first case to be considered is that in which the angle between s_{+1} and s_{-1} is such that $S_{11}=0$. This condition is obtained by setting $S_{11}=0$ in Eq. 7.58. The result is

$$-1 + 2\left[1 - 2\left(\frac{Q_1}{Q_u}\right)\cos^2\theta_{+1}\right]\cos2\theta_{+1} = 0 \tag{7.62}$$

S_{12} is now obtained by substituting Eq. 7.62 into Eq. 7.59. The result is

$$|S_{12}| = \frac{1}{2}\left(1 - \frac{\tan^2\theta_{+1}}{\sqrt{3}}\right) \tag{7.63}$$

Similarly the result for S_{13} is

$$|S_{13}| = \frac{1}{2}\left(1 + \frac{\tan^2\theta_{+1}}{\sqrt{3}}\right) \tag{7.64}$$

The relation between S_{12} and S_{13} is given graphically in Figure 7.4. Here the inequality relation, given by Eq. 7.8 with $S_{11}=0$, is

$$S_{12}S_{13}^* < 1 - |S_{12}|^2 - |S_{13}|^2 \tag{7.65}$$

which is always satisfied with the above entries for S_{12} and S_{13}.

The second case to be considered here is that for which the angle between s_{+1} and s_{-1} is such that $S_{13}=0$. This condition is obtained by setting $S_{13}=0$ in Eq. 7.60. The result is

$$-1 + 2\left[1 - 2\left(\frac{Q_1}{Q_u}\right)\cos^2\theta_{+1} \right]\cos(120 + 2\theta_{+1}) = 0. \tag{7.66}$$

S_{11} and S_{12} are now obtained by substituting the last equation into Eqs. 7.58 and 7.59. The results are

$$|S_{11}| = \frac{\left[1 + \left(\tan^2\theta_{+1}/\sqrt{3}\,\right)\right]}{1 + \sqrt{3}\,\tan^2\theta_{+1}} \tag{7.67}$$

and

$$|S_{12}| = \frac{2\tan^2\theta_{+1}/\sqrt{3}}{1 + \sqrt{3}\,\tan^2\theta_{+1}} \tag{7.68}$$

The relation between S_{11} and S_{12} is shown in Figure 7.4. The inequality equation, given by Eq. 7.8 with $S_{13}=0$, is

$$S_{11}S_{12}^* < 1 - |S_{11}|^2 - |S_{12}|^2 \tag{7.69}$$

which is always satisfied with the above entries for S_{11} and S_{12}

7.6.   FREQUENCY RESPONSE OF LOSSY CIRCULATORS

The frequency response of lossy 3-port circulators may be obtained by using the general form for s_0, s_{+1}, and s_{-1} to construct the scattering coefficients.

The scattering parameters have been evaluated in Figures 7.5 and 7.6 for the case where the damping terms for the two eigennetworks are identical. Figure 7.5 shows the frequency response of S_{11} as a function of the angle

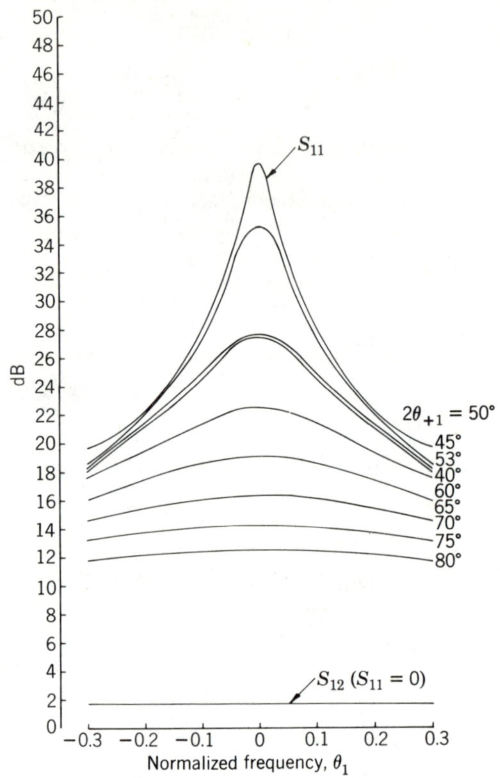

Figure 7.5   *Frequency variation of S_{11} as a function of magnetic field for $Q_1/Q_u = 0.15$.*

between s_{+1} and s_{-1} for $Q_1/Q_u = 0.15$. It also gives S_{12} for $S_{11} = 0$ and $S_{11} = S_{13}$. Figure 7.6 gives the frequency response of S_{13} as a function of the same variables and also shows S_{12} for $S_{13} = 0$ and $S_{11} = S_{13}$. The relations between S_{12} and S_{13} and between S_{12} and S_{11} are consistent with the results shown in Figure 7.4. The last two graphs also illustrate the relations between the scattering coefficients and the magnetic tuning. It is observed that S_{12} is a minimum when $S_{11} = S_{13}$.

7.7.   FREQUENCY RESPONSE OF LOSSY CIRCULATORS WITH UNEQUAL DAMPING TERMS

In the last section it has been assumed that the loss term for the two eigennetworks are identical. However, this will not be the case if the circulator is biased close to the main resonance, as in the above resonance

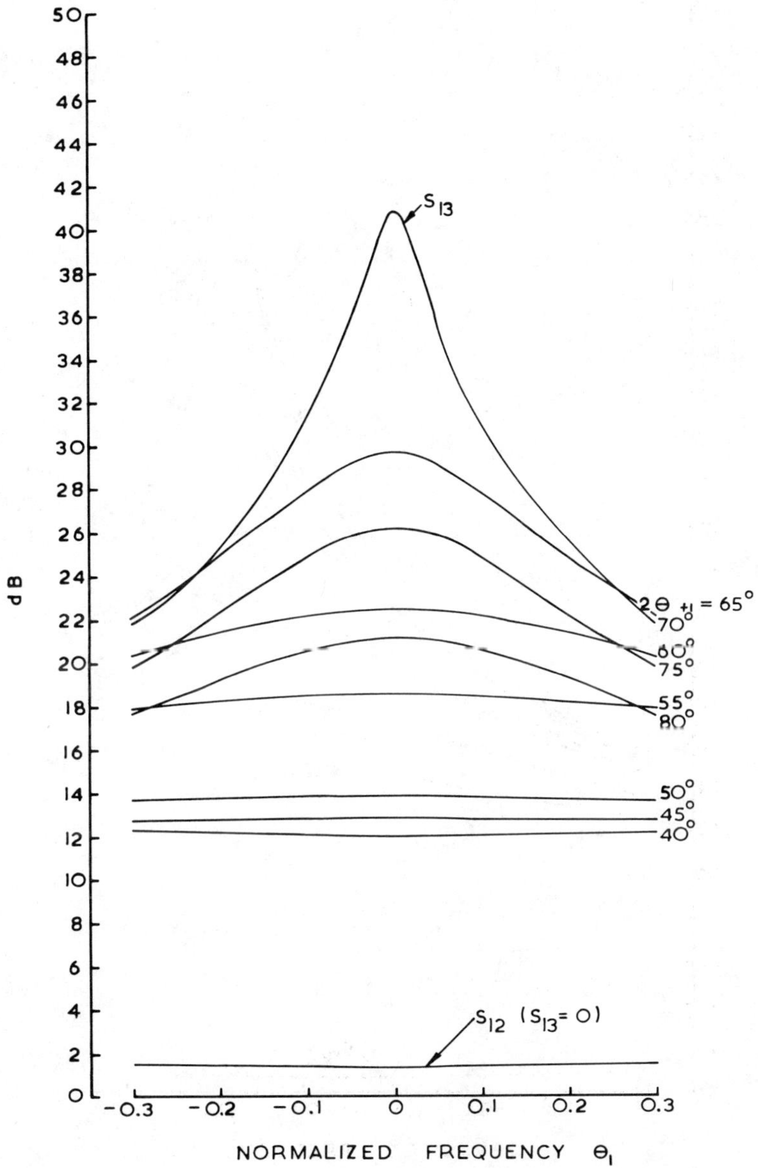

Figure 7.6 Frequency variation of S_{13} as a function of magnetic field for $Q_1/Q_u = 0.15$.

127

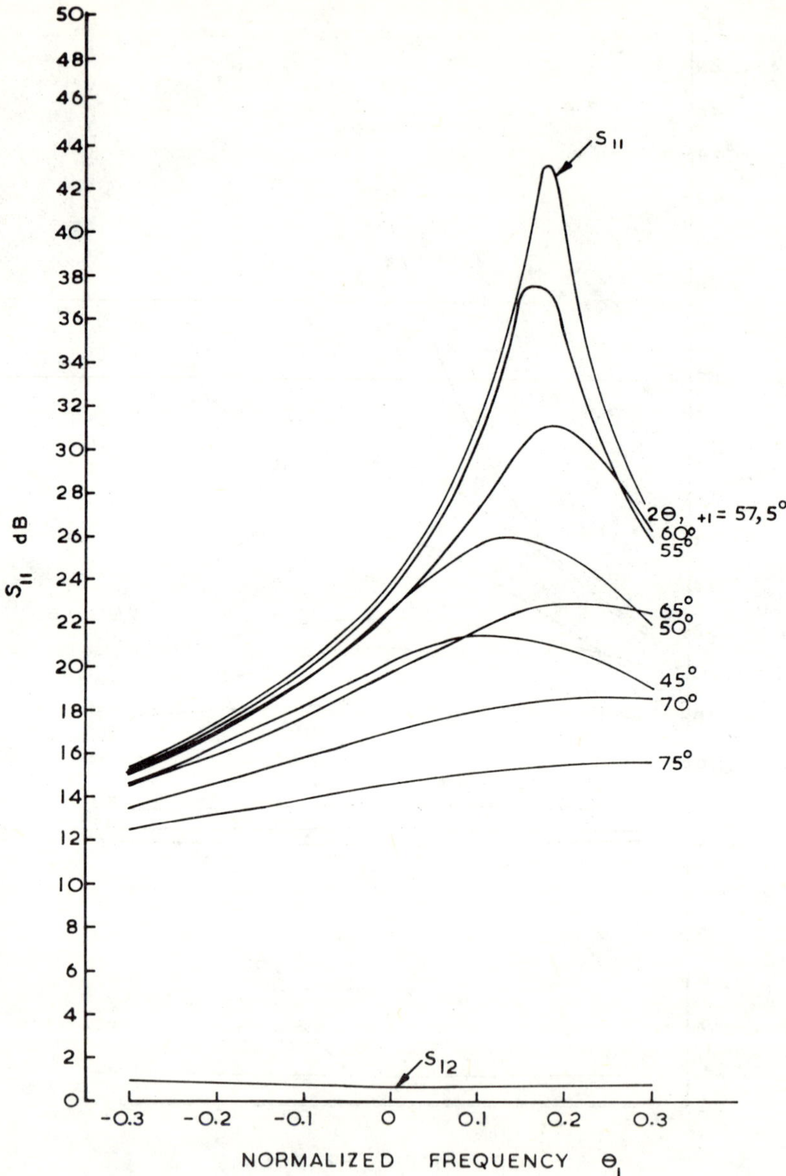

Figure 7.7 Frequency variation of S_{11} as a function of magnetic field for $Q_1/Q_{u,+1}=0.15$.

128

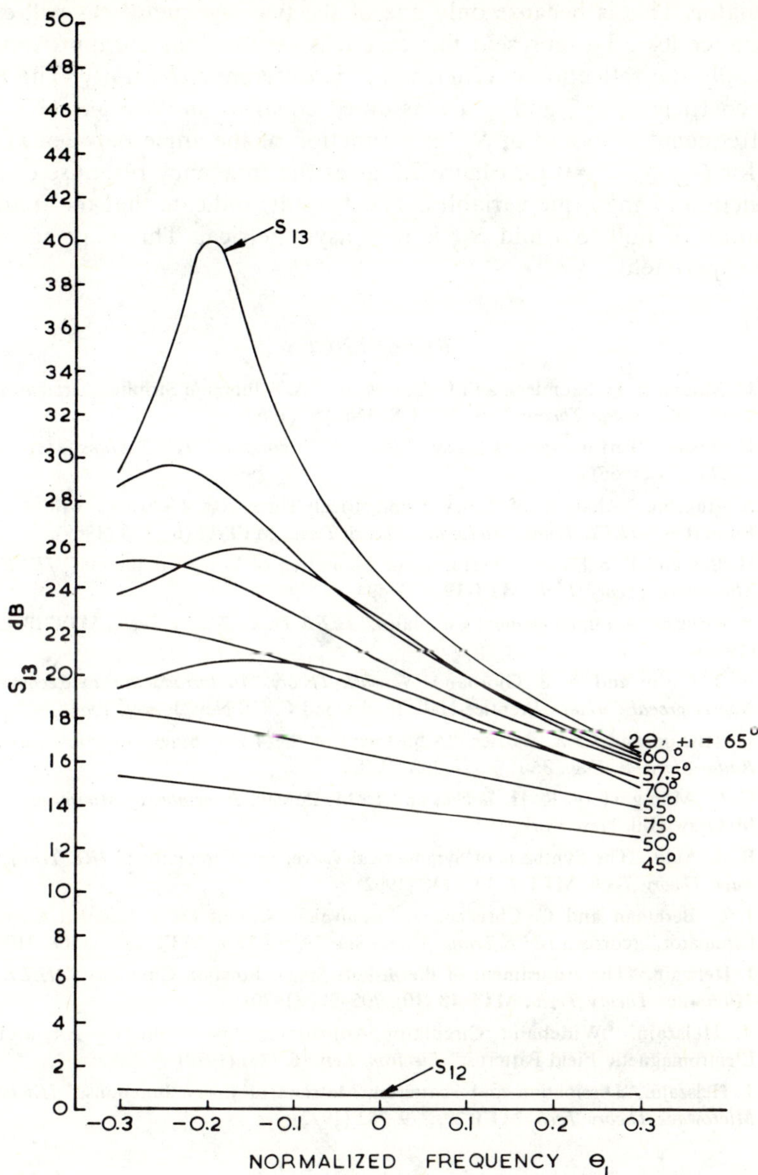

Figure 7.8   Frequency variation of S_{13} as a function of magnetic field for $Q_1/Q_{u,+1} = 0.15$.

circulator. This is because only one of the two eigennetworks will exhibit resonance loss. To represent this case it is assumed, as an approximation, that only the reflection coefficient s_{+1} is different from unity. The reflection coefficients s_{-1} and s_0 are assumed equal to unity. Figure 7.7 shows the frequency response of S_{11} as a function of the angle between s_{+1} and s_{-1} for $Q_1/Q_{u,+1} = 0.15$. Figure 7.8 gives the frequency response of S_{13} as a function of the same variables. These results indicate that the frequency variation of both S_{11} and S_{13} is now asymmetrical. This is in agreement with experiment.

REFERENCES

1. U. Milano, J. H. Saunders, and L. Davies, Jr., "A Y-Junction Stripline Circulator," *IRE Trans. Microwave Theory Tech.* **MTT-8**, 346–350 (1960).

2. H. Bosma, "Performance of Lossy H-Plane Y-Circulators," *IEEE Trans. Magn.*, **MAG-2**, 273–277 (1966).

3. S. Hagelin, "Analysis of Lossy Symmetrical Three-Port Networks with Circulator Properties", *IEEE Trans. Microwave Theory Tech.*, **MTT-17** (6), 328 (1969).

4. H. Bex and E. Schwartz, "Performance Limitation of Lossy Circulators," *IEEE Trans. Microwave Theory Tech.* **MTT-19**, 493–494.

5. Y. Konishi, "Lumped element Circulator," *IEEE Trans. Theory Tech.*, **MTT-13**, 852–864 (1965).

6. H. J. Carlin and A. B. Giordano, *Network Theory. An Introduction to Reciprocal and Nonreciprocal Circuits*, Prentice-Hall, Englewood Cliffs, New Jersey, 1964.

7. J. Helszajn and C. R. Buffler, "Adjustment of the 4-Port Single Junction Circulator," *Radio Electron. Eng.*, **35** (6), 357–360 (1968).

8. C. G. Montgomery, R. H. Dicke, and E. M. Purcell, *Principles of Microwave Circuits*, McGraw-Hill, New York, 1948.

9. B. A. Auld, "The Synthesis of Symmetrical Waveguide Circulators," *IRE Trans. Microwave Theory Tech.* **MTT-7**, 137–146 (1962).

10. J. C. Bergman and C. Christenson, "Equivalent Circuit for a Lumped Element Y-Circulator," (corres.) *IEEE Trans. Microwave Theory Tech.* **MTT-16** (5), 308–310 (1968).

11. J. Helszajn, "The Adjustment of the m-Port Single Junction Circulator," *IEEE Trans. Microwave Theory Tech.*, **MTT-18** (10), 705–711 (1970).

12. J. Helszajn, "Wideband Circulator Adjustment Using the $n = \pm 1$ and $n = 0$ Electromagnetic Field Patterns," *Electron. Lett.*, **6**, (23) (1970).

13. J. Helszajn, "Dissipation and Scattering Matrices of Lossy Junctions," *IEEE Trans. Microwave Theory Tech.* **MTT-20**, 779–782 (1972).

CHAPTER EIGHT

Adjustment of the m-Port Single Junction Circulator

One method of synthesizing symmetrical junctions consists of adjusting the eigenvalues of the scattering matrix by symmetric perturbations of the junction until the required scattering coefficients are satisfied. In the case of the m-port single-junction circulator the eigenvalues must be adjusted until they coincide with those of an ideal circulator. For an ideal m-port circulator they lie equally-spaced on a unit circle in the complex plane. To obtain this arrangement it is necessary to adjust the phases of $(m-1)$ of the eigenvalues with respect to one of them, and therefore $(m-1)$ physical perturbations of the junction are necessary. This leads to $(m-1)$ distinct scattering matrices that can be calculated once the eigenvectors of the junction are known. Since these are determined from symmetry conditions only, they remain unchanged under a symmetric perturbation of the junction. Hence, the eigenvectors already coincide with those of an ideal circulator. The symmetry considered is discussed in Chapter 2.

The synthesis procedure then consists of adjusting the phases of $(m-1)$ of the eigenvalues one at a time by symmetric perturbations of the junction until the coefficients of the scattering matrix appropriate to each required eigenvalue arrangement is obtained. The purpose of this chapter is to give a systematic way in which this can be done. In this approach it is necessary to know the initial and final locations of the eigenvalues in the complex plane. The final location of the eigenvalues are found from the scattering matrix of an ideal circulator. These lie equally spaced on a unit circle. The initial location of the eigenvalues can be obtained by forming the scattering matrix of the junction assuming that all the junction modes are cut off.

The matrix formed in this way has all of its transmissions coefficients zero and all of its reflections coefficients unity. This matrix is simply a unit matrix for which all the eigenvalues are equal. This can be seen immediately by setting the spur of the unit matrix, defined as a sum of the diagonal elements of the matrix, equal to the sum of the eigenvalues. Once these two diagrams are known a systematic adjustment of the eigenvalues is possible, which will perturb the initial set of eigenvalues to the final one. This approach also leads to $(m-1)$ distinct boundary value problems. If the electric field pattern is known the nature of the physical adjustment is strongly suggested.

8.1.   ELECTRICAL FIELD AMPLITUDES FOR 3-PORT CIRCULATOR

To compute the total electric field it is assumed as an approximation that $E_z(\phi, R)$ is proportional to a linear combination of the eigenvectors of the scattering matrix:

$$E_z(\phi, R) = a_0 + a_{+1} e^{j\phi} + a_{-1} e^{-j\phi} \tag{8.1}$$

Since the eigenvectors given by Eqs. 2.77 through 2.79 are not unique, the following general form is assumed for a_{+1} and a_{-1}

$$a_{+1} = a_1' + jb_1' \tag{8.2}$$

$$a_{-1} = a_1' - jb_1' \tag{8.3}$$

This gives

$$E_z(\phi, R) = a_0 + a_1 \cos(\tau_1 + \phi) \tag{8.4}$$

where

$$\tau_1 = \tan^{-1}\left(\frac{b_1'}{a_1'}\right) \tag{8.5}$$

$$a_1 = \sqrt{(a_1')^2 + (b_1')^2} \tag{8.6}$$

The electric field pattern given by Eq. 8.4 with $\tau_1 = 0$ consists of a linear combination of an electromagentic field pattern for which the electric field has no variation around the periphery of the ferrite disk and one for which the variation is cosinusoidal. Such field patterns have been described in Chapter 6. Figure 8.1*a* indicates the $n = \pm 1$ field patterns in a ferrite disk. The 3-port circulator can now be synthesized from this field pattern by

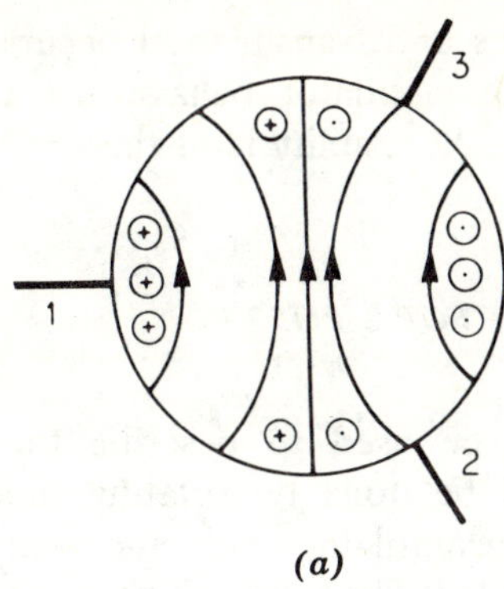

(a)

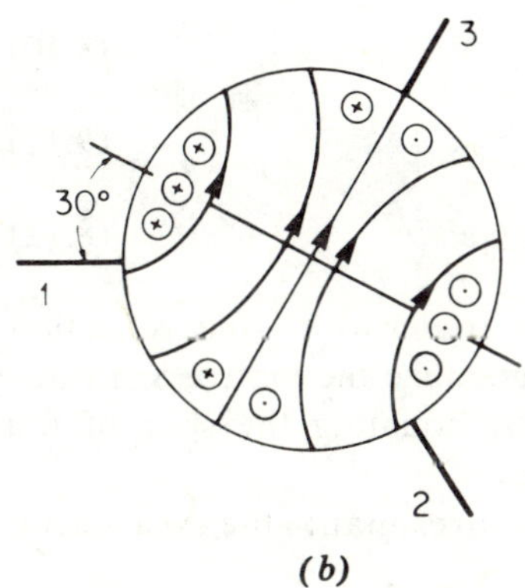

(b)

Figure 8.1.   (a) Field pattern for $n=+1$ modes for unmagnetized 3-port junction (Ref. 7). (b) Field pattern for $n=\pm 1$ modes rotated through 30° by magnetizing the junction (Ref. 7).

rotating it by an angle τ_1. This is done by splitting the degeneracy between the $n=\pm 1$ modes by magnetizing the junction.

The required rotation of the field patterns is obtained by applying the boundary conditions for a perfect circulator at the three ports. From Eq. 8.4 the electric field amplitudes at the terminals are

$$E_z(0,R)=a_0+a_1\cos\tau_1 \tag{8.7}$$

$$E_z\left(\frac{2\pi}{3},R\right)=a_0+a_1\cos\left(\tau_1+\frac{2\pi}{3}\right) \tag{8.8}$$

$$E_z\left(\frac{4\pi}{3},R\right)=a_0+a_1\cos\left(\tau_1+\frac{4\pi}{3}\right) \tag{8.9}$$

The result is $\tau_1=30°$, $a_0=0$, $a_1=2/\sqrt{3}$. Figure 8.1b shows the $n=\pm 1$ field patterns rotated through 30°. When this adjustment is satisfied port 3 is located at a null of the electric field patterns and transmission occurs between ports 1 and 2. In this condition the circulator behaves as a transmission line cavity between ports 1 and 2. The amplitude of the $n=0$ field pattern is zero in this instance.

8.2.   CIRCULATION ADJUSTMENT FOR 3-PORT CIRCULATOR

In this section the synthesis procedure will be used to describe the adjustment of the 3-port circulator. This will be done by rotating the eigenvalues in the complex plane one at a time until they coincide with those of the scattering matrix of an ideal circulator. The relations between the scattering coefficients and the eigenvalues are from Chapter 2

$$3S_{11} = s_0 + s_{+1} + s_{-1} \tag{8.10}$$

$$3S_{12} = s_0 + s_{+1}e^{j2\pi/3} + s_{-1}^{-j2\pi/3} \tag{8.11}$$

$$3S_{13} = s_0 + s_{+1}e^{-j2\pi/3} + s_{-1}e^{j2\pi/3} \tag{8.12}$$

If the adjustment procedure is to be made in terms of the reflection coefficient S_{11} only, it is not necessary to diagonalize the matrix $\bar{S}$. In this instance S_{11} can be obtained quite simply by equating the spur of the scattering matrix to the sum of the eigenvalues.

If all the electromagnetic fields are nonresonant the scattering coefficients are

$$S_{11} = -1 \tag{8.13}$$

and

$$S_{12} = S_{13} = 0 \tag{8.14}$$

The eigenvalues corresponding to these scattering coefficients are

$$s_0 = s_{+1} = s_{-1} = -1 \tag{8.15}$$

Hence all the eigenvalues lie together on the unit circle. This is shown in Figure 8.2.

The first circulation adjustment is now obtained by rotating the degenerate pair of eigenvalues s_{+1} and s_{-1} through 180° on the unit circle. This leads to the arrangement between the eigenvalues in Figure 8.3

$$s_0 = -1 \tag{8.16}$$

$$s_{+1} = s_{-1} = -s_0 \tag{8.17}$$

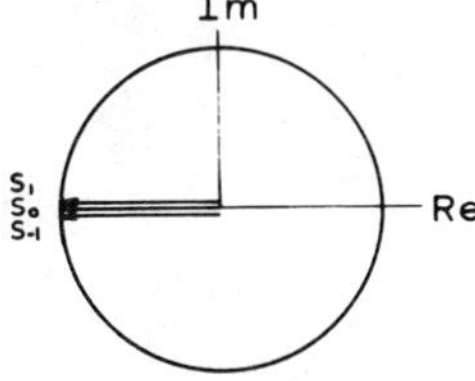

Figure 8.2.   *Scattering eigenvalues for 3-port junction having all its transmission coefficients zero and all its reflection coefficients unity.*

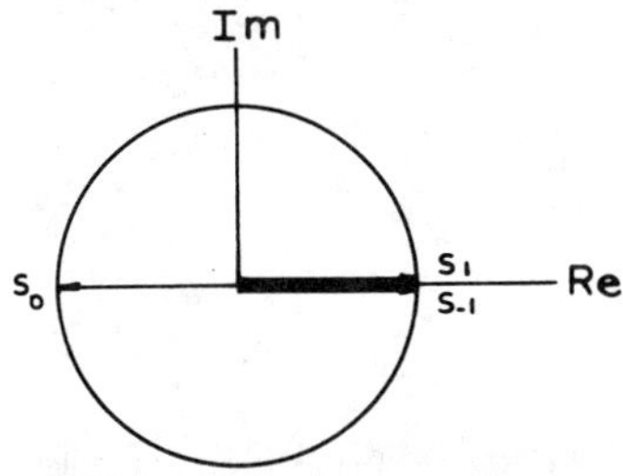

Figure 8.3.   *Scattering eigenvalues after first circulation adjustment for 3-port junction.*

Substituting the above eigenvalues into Eqs. 8.10 through 8.12 gives the following scattering coefficients:

$$S_{11} = \frac{1}{3} \tag{8.18}$$

$$S_{12} = S_{13} = \frac{-2}{3} \tag{8.19}$$

This adjustment minimizes the reflection coefficient S_{11} and establishes the $n = \pm 1$ field patterns within the junction.

This relation determines the diameter of the ferrite disk.

Substituting the above boundary conditions into Eqs. 8.7 through 8.9 gives for the electric field

$$E_z(\phi, R) = a_1 \cos\phi \tag{8.20}$$

where

$$a_1 = 1$$

The second circulation condition is now obtained by splitting the degenerate scattering matrix eigenvalues by magnetizing the junction until they coincide with those for an ideal circulator. This gives the eigenvalue arrangement in Figure 8.4

$$s_0 = -1 \tag{8.21}$$

$$s_{+1} = s_0 e^{-j2\pi/3} \tag{8.22}$$

$$s_{-1} = s_0 e^{j2\pi/3} \tag{8.23}$$

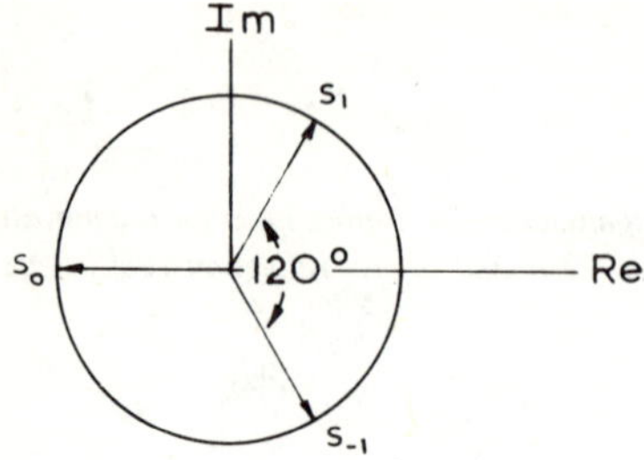

Figure 8.4.   *Scattering eigenvalues after second circulation adjustment for 3-port junction (ideal circulator).*

The coefficients of the scattering matrix now become

$$S_{11} = S_{13} = 0 \tag{8.24}$$

$$S_{12} = -1 \tag{8.25}$$

which coincides with those of an ideal circulator. Substituting the last boundary conditions into Eqs. 8.7 through 8.9 gives for the electric field

$$E_z(\phi, R) = a_1 \cos(\tau_1 + \phi) \tag{8.26}$$

where $a_1 = 2/\sqrt{3}$ and $\tau_1 = 30°$. This adjustment rotates the $n = \pm 1$ field patterns through 30° as shown in Figure 8.1*b*.

8.3.   ELECTRIC FIELD AMPLITUDES FOR 4-PORT CIRCULATOR

To compute the total electric field at the terminals of the 4-port junction one again takes a linear combination of the eigenvectors

$$E_z(\phi, R) = a_0 + a_{+1}e^{j\phi} + a_{-1}e^{-j\phi} + a_2 e^{j2\phi} \tag{8.27}$$

Using the form for a_{+1} and a_{-1} given by Eqs. 8.2 and 8.3 the total electric field becomes

$$E_z(\phi, R) = a_0 + a_1 \cos(\tau_1 + \phi) + a_2 e^{j2\phi} \tag{8.28}$$

The electric field pattern given by Eq. 8.28 with $\tau_1 = 0$ consists of a linear combination of four electromagnetic field patterns, one for which the electric field has no variation around the periphery of the ferrite disk, one for which the variation is cosinusoidal, and one additional nonresonant one. Such field patterns have been described in Chapter 6. Figure 8.5*a* gives the $n = 0$ and $n = \pm 1$ field patterns in a ferrite disk. The $n = \pm 2$ field pattern that is not resonant is not shown. The circulator can now be

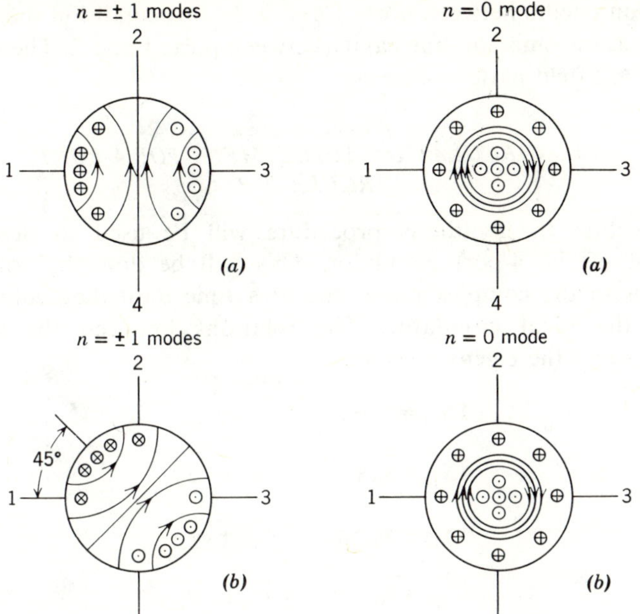

Figure 8.5. (a) Field patterns for $n=0$ and $n=\pm 1$ modes for unmagnetized 4-port junction (Ref. 7). (b) Field patterns for $n=0$ and $n=\pm 1$ modes for magnetized 4-port junction (Ref. 7).

synthesized from these modes by rotating the $n=\pm 1$ one by an angle τ_1. This is done by splitting the degeneracy between the field patterns by magnetizing the junction.

The necessary rotation of the field patterns is obtained by applying the boundary conditions for a perfect circulator at the 4 ports.

From Eq. 8.28 the electric field amplitudes at the terminals are

$$E_z(0,R) = a_0 + a_1 \cos(\tau_1) + a_2 \tag{8.29}$$

$$E_z\left(\frac{\pi}{2},R\right) = a_0 + a_1 \cos\left(\tau_1 + \frac{\pi}{2}\right) + a_2 e^{j\pi/2} \tag{8.30}$$

$$E_z(\pi,R) = a_0 + a_1 \cos(\tau_1 + \pi) + a_2 e^{j\pi} \tag{8.31}$$

$$E_z\left(\frac{3\pi}{2},R\right) = a_0 + a_1 \cos\left(\tau_1 + \frac{3\pi}{2}\right) + a_2 e^{j3\pi/2} \tag{8.32}$$

The result is $\tau_1 = 45°$, $a_0 = 1/2$, $a_1 = 2/\sqrt{3}$, $a_2 = 0$. Figure 8.5b shows that the $n=\pm 1$ field patterns are rotated through 45°. When this adjustment is satisfied ports 3 and 4 are located at a null of the electric field patterns and transmission occurs between ports 1 and 2. In this condition the circulator behaves as a transmission line cavity between ports 1 and 2. The amplitude of the $n = \pm 2$ field pattern is zero.

8.4.   CIRCULATION ADJUSTMENT FOR 4-PORT CIRCULATOR

In this section the synthesis procedure will be used to describe the adjustment of the 4-port circulator. This will be done by rotating the eigenvalues in the complex plane one at a time until they coincide with those of the ideal circulator. The relations between the scattering coefficients and the eigenvalues are

$$4S_{11} = s_0 + s_{+1} + s_{-1} + s_2 \tag{8.33}$$

$$4S_{12} = s_0 + js_{+1} - js_{-1} - s_2 \tag{8.34}$$

$$4S_{13} = s_0 - s_{+1} - s_{-1} + s_2 \tag{8.35}$$

$$4S_{14} = s_0 - js_{+1} + js_{-1} - s_2 \tag{8.36}$$

If all the electromagnetic fields are nonresonant the scattering coefficients are

$$S_{11} = -1 \tag{8.37}$$

and

$$S_{12} = S_{13} = S_{14} = 0 \tag{8.38}$$

The eigenvalues corresponding to these scattering coefficients are

$$s_0 = s_{+1} = s_{-1} = s_2 = -1 \tag{8.39}$$

Hence, they all lie together on the unit circle. This is shown in Figure 8.6.

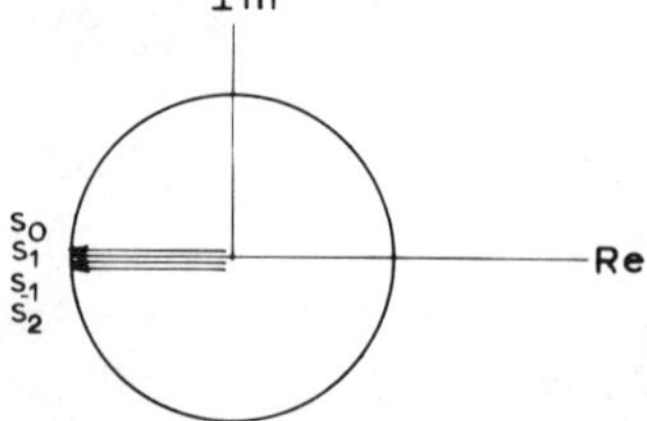

Figure 8.6.   *Scattering eigenvalues for 4-port junction having all its transmission coefficients unity.*

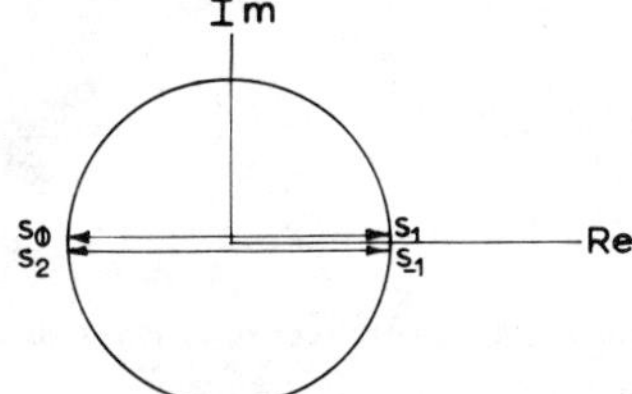

Figure 8.7.   *Scattering eigenvalues after first circulation adjustment for 4-port junction.*

The first circulation adjustment is now obtained by rotating the degenerate pair of eigenvalues s_{+1} and s_{-1} through 180° on the unit circle. This leads to the arrangement between the eigenvalues in Figure 8.7

$$s_0 = s_2 = -1 \tag{8.40}$$

$$s_{+1} = s_{-1} = -s_0 \tag{8.41}$$

Substituting the above eigenvalues into Eqs. 8.33 through 8.36 gives the following scattering coefficients

$$S_{11} = S_{12} = S_{14} = 0 \tag{8.42}$$

$$S_{13} = 1 \tag{8.43}$$

Substituting these boundary conditions into Eqs. 8.29 through 8.32 gives for the electric field

$$E_z(\phi, R) = a_1 \cos\phi \tag{8.44}$$

where $a_1 = 1.0$. This adjustment minimizes the reflection coefficient S_{11} and establishes the $n = \pm 1$ field patterns within the junction. This relation therefore determines the diameter of the ferrite disk.

The second circulation adjustment can now be made by rotating the eigenvalue s_0 through 180° in the complex plane. This gives

$$s_2 = -s_0 \tag{8.45}$$

$$s_{+1} = s_{-1} = s_0 \tag{8.46}$$

This eigenvalue arrangement is shown in Figure 8.8. The phase of this eigenvalue is varied by introducing a thin pin through the center of the junction, which rotates the s_0 eigenvalue in the complex plane but leaves the others undisturbed. The corresponding scattering coefficients are

$$S_{11} = S_{12} = S_{13} = S_{14} = \frac{1}{2} \tag{8.47}$$

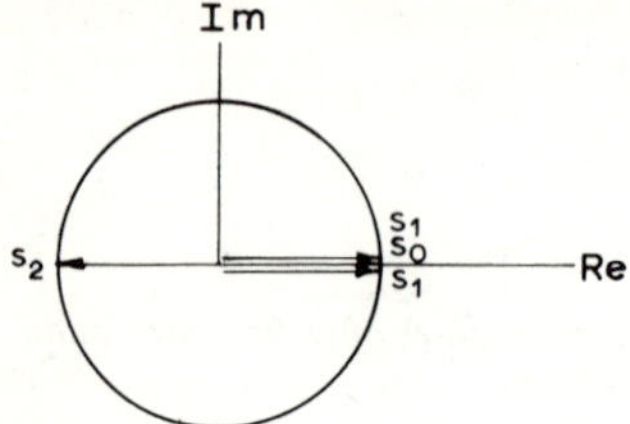

Figure 8.8. Scattering eigenvalues after second circulation adjustment for 4-port junction.

Using the above boundary conditions gives for the electric field

$$E_z(\phi, R) = a_0 + a_1 \cos\phi \qquad (8.48)$$

where $a_0 = 1/2$ and $a_1 = 1/2$. This adjustment establishes the $n = 0$ field patterns depicted in Figure 8.5*a* within the junction.

The third circulation condition is now immediately obtained by splitting the degeneracy of the s_{+1} and s_{-1} eigenvalues by 180° with the magnetic field. This gives the eigenvalue arrangement of the ideal circulator shown in Figure 8.9:

$$s_2 = -1 \qquad (8.49)$$

$$s_0 = -s_2 \qquad (8.50)$$

$$s_{+1} = s_2 e^{-j2\pi/4} \qquad (8.51)$$

$$s_{-1} = s_2 e^{j2\pi/4} \qquad (8.52)$$

If the splitting is not symmetrical, the eigenvalues will not be in the position required for circulation. A small adjustment of the ferrite disk and metal pin may then be necessary. However, an essential feature of the circulator adjustment is that the splitting is very small and essentially symmetrical to the first order. It is also noted that the s_0 eigenvalue cannot

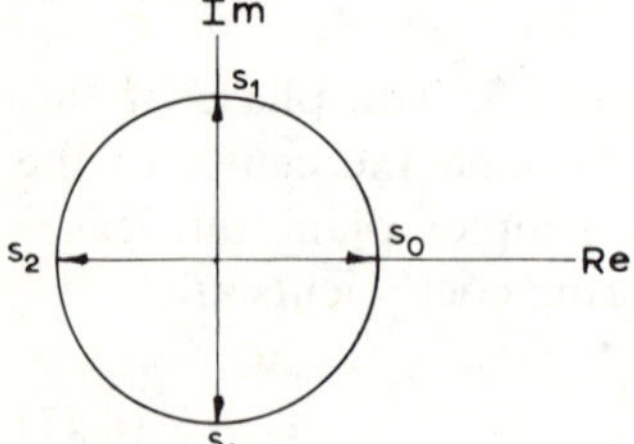

Figure 8.9. Scattering eigenvalues after third circulation adjustment for 4-port junction (ideal circulator).

be split by the magnetic field. The final scattering coefficients are

$$S_{11} = S_{13} = S_{14} = 0 \tag{8.53}$$

$$S_{12} = -1 \tag{8.54}$$

Using the above boundary conditions the electric field is

$$E_z(\phi, R) = a_0 + a_1 \cos(\tau_1 + \phi) \tag{8.55}$$

where $a_0 = 1/2$, $a_1 = 1/2$, and $\tau_1 = 45°$. This adjustment, therefore, rotates the $n = \pm 1$ field patterns through $45°$ as shown in Figure 8.5b.

8.5.   ELECTRIC FIELD AMPLITUDES FOR 5-PORT CIRCULATOR

The total electric field at the terminals of the 5-port junction in terms of the junction eigenvectors is

$$E_z(\phi, R) = a_0 + a_{+1} e^{j\phi} + a_{-1} e^{-j\phi} + a_{+2} e^{j2\phi} + a_{-2} e^{-j2\phi} \tag{8.56}$$

Using the form for a_{+n} and a_{-n} given by Eqs. 8.2 and 8.3 gives

$$E_z(\phi, R) = a_0 + a_1 \cos(\tau_1 + \phi) + a_2 \cos(\tau_2 + 2\phi) \tag{8.57}$$

The electric field pattern given by Eq. 8.57 with $\tau_1 = \tau_2 = 0$ consists of a linear combination of electromagnetic field patterns for which the electric fields have cosinusoidal variations around the periphery of the ferrite disk. Such field patterns have been described in Chapter 6.

Figure 8.10a shows the $n = \pm 1$ and $n = \pm 2$ field patterns of a ferrite disk. A suitable nonresonant field pattern (not shown) is obtained with $n = 0$. A 5-port circulator can now be synthesized from these field patterns by rotating the two field patterns by the angles τ_1 and τ_2. This is done by splitting the degeneracy of each of the field patterns by magnetizing the junction.

The rotation of the two field patterns is obtained by applying the boundary conditions for a perfect circulator at the 5-ports:

$$E_z(0, R) = a_0 + a_1 \cos\tau_1 + a_2 \cos\tau_2 \tag{8.58}$$

$$E_z\left(\frac{2\pi}{5}, R\right) = a_0 + a_1 \cos\left(\tau_1 + \frac{2\pi}{5}\right) + a_2 \cos\left(\tau_2 + \frac{4\pi}{5}\right) \tag{8.59}$$

$$E_z\left(\frac{4\pi}{5}, R\right) = a_0 + a_1 \cos\left(\tau_1 + \frac{4\pi}{5}\right) + a_2 \cos\left(\tau_2 + \frac{8\pi}{5}\right) \tag{8.60}$$

$$E_z\left(\frac{6\pi}{5},R\right)=a_0+a_1\cos\left(\tau_1+\frac{6\pi}{5}\right)+a_2\cos\left(\tau_2+\frac{12\pi}{5}\right) \tag{8.61}$$

$$E_z\left(\frac{8\pi}{5},R\right)=a_0+a_1\cos\left(\tau_1+\frac{8\pi}{5}\right)+a_2\cos\left(\tau_2+\frac{16\pi}{5}\right) \tag{8.62}$$

The result is $\tau_1=54°$, $\tau_2=18°$, $a_0=0$, $a_1=0.4707$, and $a_2=0.7607$. Figure 8.10*b* shows the field patterns in Figure 8.10*a* rotated through 54° and 18°, respectively. If the magnitude of the E_z fields for the two patterns are equal at the 5-ports, circulation occurs between ports 1 and 2 and all the other ports are isolated.

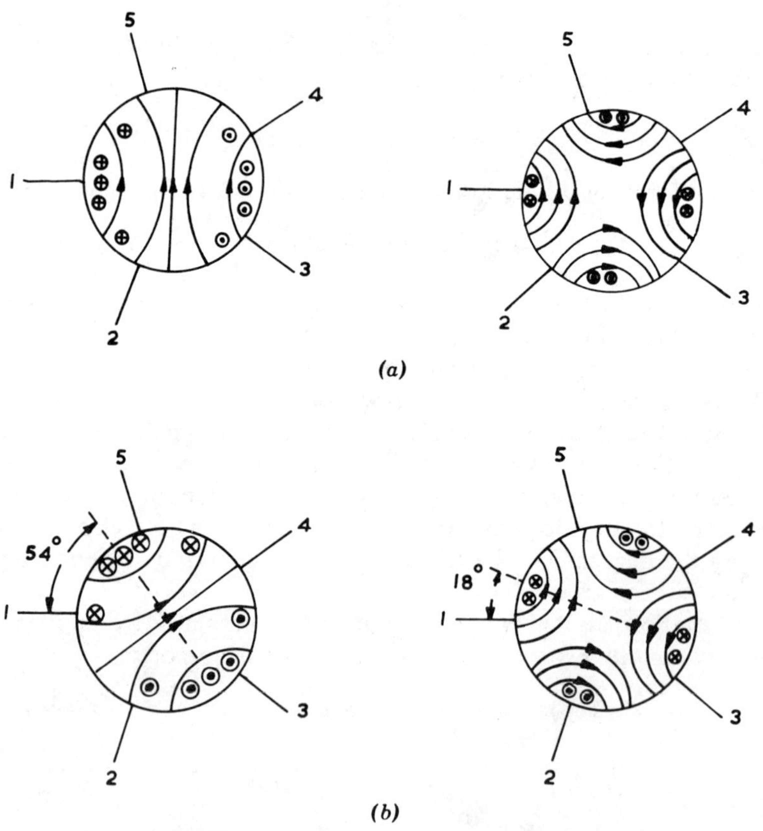

Figure 8.10.   (a) *Field patterns for* $n=\pm1$ *and* $n=\pm2$ *modes for unmagnetized junction* (*Ref.* 8). (*b*) *Field patterns for rotated* $n=\pm1$ *and* $n=\pm2$ *modes for magnetized junction* (*Ref.* 8).

8.6.   *CIRCULATION ADJUSTMENT FOR 5-PORT CIRCULATOR*

In this section the synthesis procedure will be used to describe the adjustment of the 5-port circulator. The physical perturbation described in this section applies to the stripline junction only.

The relation between the scattering coefficients and the eigenvalues are in this instance given by

$$5S_{11} = s_0 + s_{+1} + s_{-1} + s_{+2} + s_{-2} \tag{8.63}$$

$$5S_{12} = s_0 + s_{+1}e^{+j2\pi/5} + s_{-1}e^{-j2\pi/5} + s_{+2}e^{+j4\pi/5} + s_{-2}e^{-j4\pi/5} \tag{8.64}$$

$$5S_{13} = s_0 + s_{+1}e^{+j4\pi/5} + s_{-1}e^{-j4\pi/5} + s_{+2}e^{+j8\pi/5} + s_{-2}e^{-j8\pi/5} \tag{8.65}$$

$$5S_{14} = s_0 + s_{+1}e^{+j6\pi/5} + s_{-1}e^{-j6\pi/5} + s_{+2}e^{+j12\pi/5} + s_{-2}e^{-j12\pi/5} \tag{8.66}$$

$$5S_{15} = s_0 + s_{+1}e^{+j8\pi/5} + s_{-1}e^{-j8\pi/5} + s_{+2}e^{+j16\pi/5} + s_{-2}e^{-j16\pi/5} \tag{8.67}$$

If all the electromagnetic fields are nonresonant the scattering coefficients are

$$S_{11} = -1 \tag{8.68}$$

and

$$S_{12} = S_{13} = S_{14} = S_{15} = 0 \tag{8.69}$$

The eigenvalues corresponding to these scattering coefficients are

$$s_0 = s_{+1} = s_{-1} = s_{+2} = s_{-2} = -1 \tag{8.70}$$

Hence all the eigenvalues lie together on the unit circle. This is shown in Figure 8.11.

The first circulation adjustment is now obtained by rotating the degenerate pair of eigenvalues s_{+2} and s_{-2} through 180° on the unit circle.

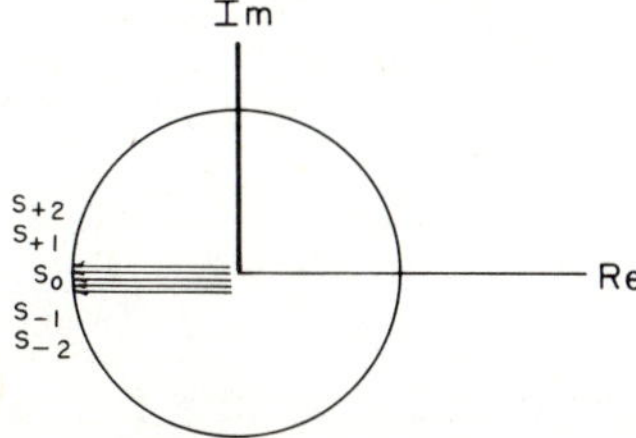

Figure 8.11.   *Scattering matrix eigenvalues for 5-port junction having all its transmission coefficients zero and all its reflection coefficients unity (Ref. 8).*

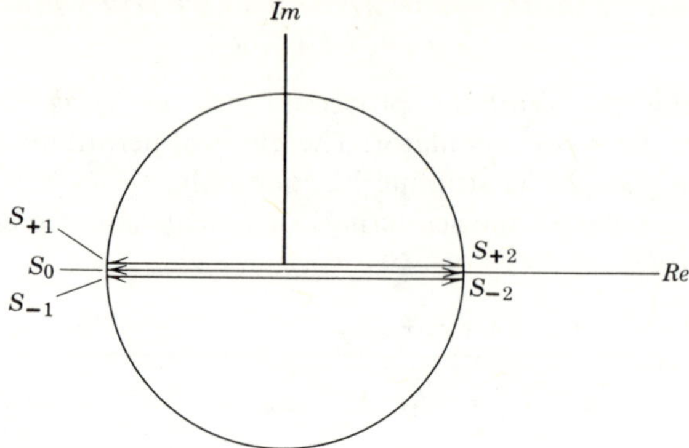

Figure 8.12.   *Scattering matrix eigenvalues after first circulation adjustment* (*Ref.* 8).

This leads to the arrangement between the eigenvalues in Figure 8.12

$$s_0 = s_{+1} = s_{-1} \tag{8.71}$$

and

$$s_0 = -s_{+2} = -s_{-2} \tag{8.72}$$

Substituting the above eigenvalues into Eqs. 8.63 through 8.67 gives the following scattering coefficients:

$$S_{11} = \frac{-1}{5} \tag{8.73}$$

$$S_{12} = \frac{-3.2360}{5} \tag{8.74}$$

$$S_{13} = \frac{+1.2360}{5} \tag{8.75}$$

$$S_{14} = \frac{+1.2360}{5} \tag{8.76}$$

$$S_{15} = \frac{-3.2360}{5} \tag{8.77}$$

Introducing the above boundary conditions into Eqs. 8.58 through 8.62 gives for the electric field

$$E_z(\phi, R) = a_2 \cos 2\phi \tag{8.78}$$

where $a_2 = 0.80$. The physical arrangement is the diameter of the ferrite disk appropriate to the $n = \pm 2$ field patterns, shown in Figure 8.10*a*.

The second circulation adjustment is now obtained by rotating the degenerate pair of eigenvalues s_{+1} and s_{-1} through 180° on the unit circle. This gives the eigenvalue diagram in Figure 8.13,

$$s_0 = -s_{+2} = -s_{-2} = -s_{+1} = -s_{-1} \tag{8.79}$$

This eigenvalue arrangement is obtained when the scattering coefficients take on the following values:

$$S_{11} = \frac{3}{5} \tag{8.80}$$

$$S_{12} = -\frac{2}{5} \tag{8.81}$$

$$S_{13} = -\frac{2}{5} \tag{8.82}$$

$$S_{14} = -\frac{2}{5} \tag{8.83}$$

$$S_{15} = -\frac{2}{5} \tag{8.84}$$

Using the above boundary conditions for the electric field gives

$$E_z(\phi, R) = a_1 \cos\phi + a_2 \cos 2\phi \tag{8.85}$$

where $a_1 = 0.80$ and $a_2 = 0.80$. The second perturbation consists of tuning the $n = \pm 1$ field patterns to the same frequency as the $n = \pm 2$ ones. One possible physical perturbation is an inductive post at the center of the

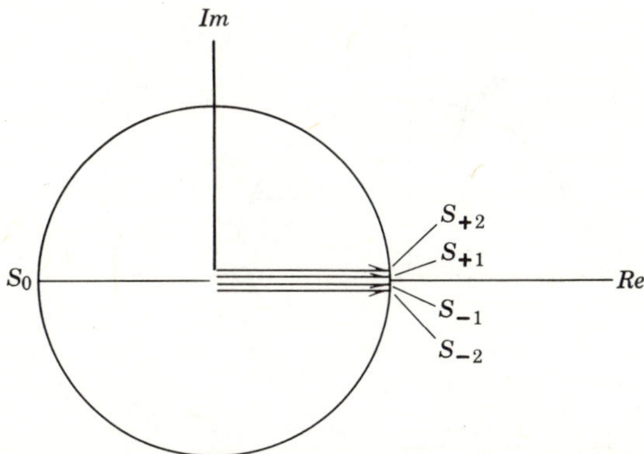

Figure 8.13.   *Scattering matrix eigenvalues after second circulation adjustment (Ref. 8).*

ferrite disk. This physical variable is obtained by noting that the $n = \pm 1$ field configuration has most of its magnetic energy at the center of the disk; whereas, the $n = \pm 2$ field pattern has no energy there. In order not to tune the $n = 0$ mode, which has a maximum electric field at the center of the junction with this adjustment, it is necessary to short circuit this mode with a thin metal post.

The third circulation adjustment is now obtained by splitting the degeneracy between the s_{+1} and s_{-1} eigenvalues without affecting the degeneracy between s_{+2} and s_{-2}

$$s_{+2} = s_{-2} = -s_0 \tag{8.86}$$

$$s_{+1} = s_0 e^{-j2\pi/5} \tag{8.87}$$

$$s_{-1} = s_0 e^{+j2\pi/5} \tag{8.88}$$

This eigenvalue arrangement is depicted in Figure 8.14. It is obtained when

$$S_{11} = \frac{0.3820}{5} \tag{8.89}$$

$$S_{12} = \frac{-4.6180}{5} \tag{8.90}$$

$$S_{13} = \frac{-1}{5} \tag{8.91}$$

$$S_{14} = \frac{1.2360}{5} \tag{8.92}$$

$$S_{15} = \frac{-1}{5} \tag{8.93}$$

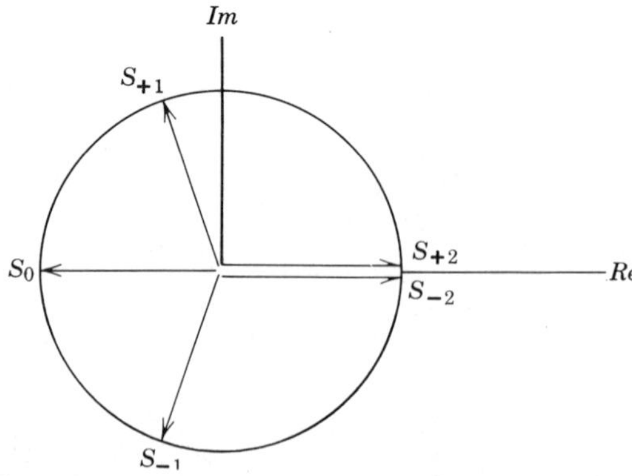

Figure 8.14.   *Scattering matrix eigenvalues after third circulation adjustment (Ref. 8).*

Substituting the above boundary conditions into Eqs. 8.58 through 8.62 gives for the electric field distribution

$$E_z(\phi, R) = a_1 \cos(\phi + \tau_1) + a_2 \cos 2\phi \tag{8.94}$$

where $a_1 = 0.4707$, $a_2 = 0.80$, and $\tau_1 = 54°$. The circulation adjustment rotates the $n = \pm 1$ field patterns by $54°$ as shown in Figure 8.10*b*.

The third physical perturbation is obtained by splitting the degeneracy between the $n = \pm 1$ field patterns. Here, a magnetic field at the center of the disk should split the degeneracy of the $n = \pm 1$ field patterns without affecting the degeneracy between the $n = \pm 2$ field patterns which have little magnetic energy there.

The final eigenvalue adjustment, in Figure 8.15 is now obtained by removing the degeneracy between s_{+2} and s_{-2}:

$$s_{+1} = s_0 e^{-j2\pi/5} \tag{8.95}$$

$$s_{-1} = s_0 e^{+j2\pi/5} \tag{8.96}$$

$$s_{+2} = s_0 e^{-j4\pi/5} \tag{8.97}$$

$$s_{-2} = s_0 e^{+j4\pi/5} \tag{8.98}$$

This eigenvalue arrangement is obtained when

$$S_{11} = S_{13} = S_{14} = S_{15} = 0 \tag{8.99}$$

$$S_{12} = -1 \tag{8.100}$$

which corresponds to an ideal circulator.

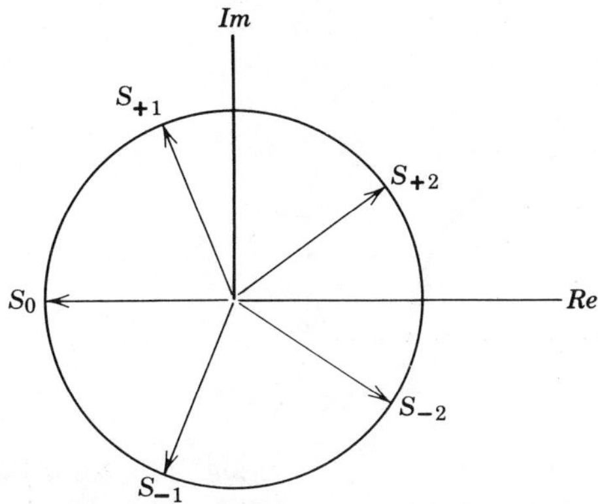

Figure 8.15.   Scattering eigenvalues after fourth circulation adjustment (ideal circulator) (Ref. 8).

These boundary conditions give for the electric field

$$E_z(\phi, R) = a_1 \cos(\phi + \tau_1) + a_2 \cos(2\phi + \tau_2) \tag{8.101}$$

where $a_1 = 0.4707$, $a_2 = 0.7607$, $\tau_1 = 54°$, and $\tau_2 = 18°$. This adjustment rotates the field patterns of the $n = \pm 2$ modes by $18°$ as shown in Figure 8.10*b*. The fourth and last physical perturbation is a ring magnet at the edge of the ferrite disk where most of the magnetic energy of the $n = \pm 2$ field patterns resides.

8.7. *WIDEBAND ADJUSTMENT OF 3-PORT CIRCULATOR USING $n=0$ AND $n = \pm 1$ MODES*

As the frequency is varied, the eigenvalues rotate in the complex plane at different speeds, and the phase relation between them no longer corresponds to an ideal circulator. This is primarily due to the fact that the reference eigenvalue is associated with a nonresonant field pattern and the other two are associated with resonant ones.

The purpose of this section is to describe an adjustment procedure in which the reference eigenvalue is associated with a resonant field pattern also. If the Q-factors of all three field patterns can be made equal, then the three eigenvalues will all rotate at the same speed around the unit circle. The bandwidth of the circulator will then be very wide. This adjustment requires one additional independent variable in the form of a thin metal pin through the center of the ferrite disk. This additional variable rotates the reference eigenvalue by $180°$ in the complex plane, thereby establishing an additional resonant field pattern. In this way all three eigenvalues are associated with resonant modes, instead of just two of them.

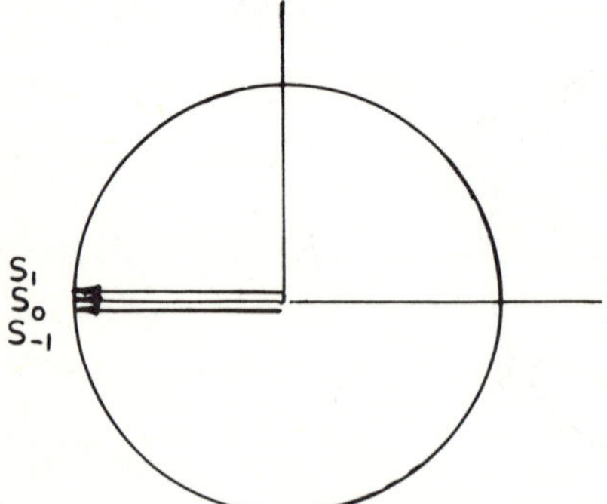

Figure 8.16. *Initial location of eigenvalues for wide band junction (Ref. 5).*

The wideband adjustment principle described now starts once more with the eigenvalue arrangement of the nonresonant junction shown in Fig. 8.16. The first perturbation of the junction proceeds in the usual way by rotating the degenerate eigenvalues s_{+1} and s_{-1} through $180°$ in the complex plane by adjusting the diameter of the ferrite disk and establishes the $n = \pm 1$ field patterns within it. This gives the eigenvalue relation in Figure 8.17,

$$s_0 = -1 \tag{8.102}$$

$$s_{+1} = s_{-1} = -s_0 \tag{8.103}$$

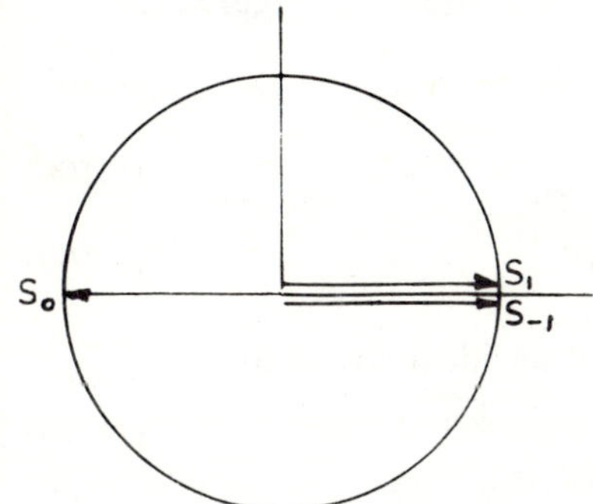

Figure 8.17.   *Location of eigenvalues after first circulation adjustment for wide band junction (Ref. 5).*

which coincides with the minimum value of the reflection coefficient

$$S_{11} = \frac{1}{3}$$

The additional broadband circulation adjustment is obtained by rotating the s_0 eigenvalue through $180°$. This is done by introducing a thin metal pin through the center of the junction which establishes the $n = 0$ field pattern within the junction. This eigenvalue arrangement is shown in Figure 8.18

$$s_0 = s_{+1} = s_{-1} \tag{8.105}$$

and is obtained when the scattering coefficient becomes

$$S_{11} = 1 \tag{8.106}$$

The final eigenvalue relation is now obtained by splitting the degeneracy between s_{+1} and s_{-1} in the usual way by magnetizing the junction until

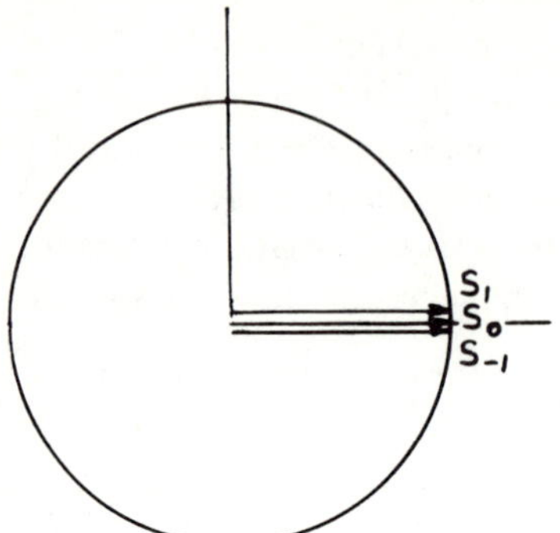

Figure 8.18. *Location of eigenvalues after second circulation adjustment for wide band junction (Ref. 5).*

they coincide with those of an ideal circulator as shown in Figure 8.19

$$s_0 = 1 \tag{8.107}$$

$$s_{+1} = s_0 e^{j120°} \tag{8.108}$$

$$s_{-1} = s_0 e^{-j120°} \tag{8.109}$$

This again gives the reflection coefficient of an ideal circulator

$$S_{11} = 0 \tag{8.110}$$

The three eigenvalues shown in Figure 8.19 are all associated with resonant field patterns. Figure 8.20*a* gives the $n = \pm 1$ and $n = 0$ modes in an unmagnetized disk. Figure 8.20*b* shows the $n = \pm 1$ and $n = 0$ modes in a magnetized disk in which the former modes are rotated by 60°. If the amplitudes of the electric fields of the two modes are the same at the edge of the disk they will cancel at port 3 and transmission will take place between ports 1 and 2.

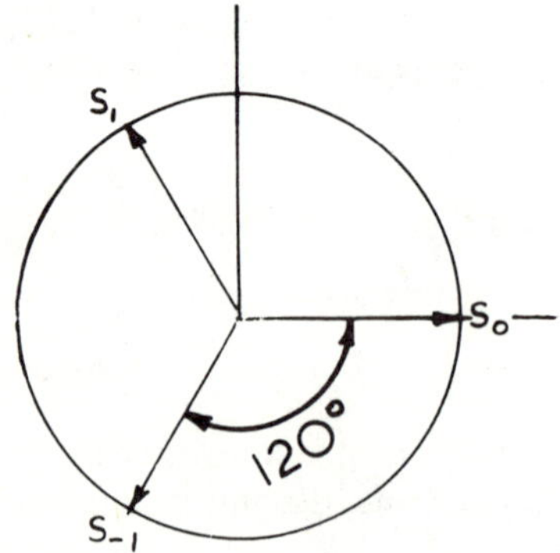

Figure 8.19. *Location of eigenvalues after third circulation adjustment for wide band junction (ideal circulator) (Ref. 5).*

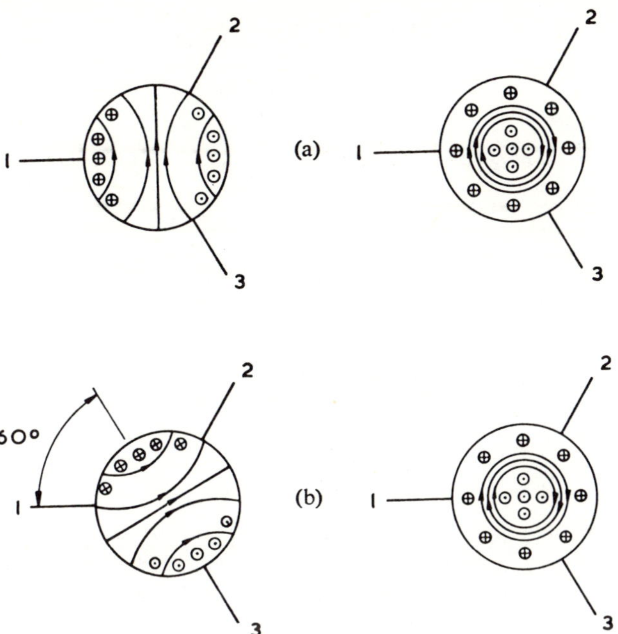

Figure 8.20.   (a) $n = \pm 1$ and $n = 0$ *field patterns in unmagnetized disk* (*Ref.* 6). (b) $n = \pm 1$ and $n = 0$ *field patterns in magnetized three resonant mode circulator* (*Ref.* 6).

REFERENCES

1. C. G. Montgomery, R. H. Dicke, and E. M. Purcell, *Principles of Microwave Circuits*, McGraw-Hill, New York, 1948.

2. B. A. Auld, "The Synthesis of Symmetrical Waveguide Circulators," *IRE Trans. Microwave Theory Tech.*, **MTT-7**, 238–246 (1959).

3. U. Milano, J. H. Saunders, and L. Davis, Jr., "A Y-Junction Stripline Circulator," *IRE Trans. Microwave Theory Tech.*, **MTT-8**, 346–351 (1960).

4. J. Helszajn and C. R. Buffler, "Adjustment of the 4-port Junction Circulator," *Radio Electron. Eng.*, **35**, 357–360 (1968).

5. J. Helszajn, "Wideband Circulator Adjustment using the $n = \pm 1$ and $n = 0$ Electromagnetic Field Patterns," *Electron. Lett.*, **6** (23) 729–731 (1970).

6. J. Helszajn, "Three-resonant-mode Adjustment of the Waveguide Circulator," *Radio Electron. Eng.*, **42** (5) pp 213–216 (1972).

7. C. E. Fay and R. L. Comstock, "Operation of the Ferrite Junction Circulator," *IEEE Trans. Microwave Theory Tech.*, **MTT-13**, 15–27 (1965).

8. J. Helszajn, "Adjustment of the *m*-port Single Junction Circulator," *IEEE Trans. Microwave Theory Tech.*, **MTT-20** 705–711 1970.

CHAPTER NINE

Lumped Element Circulator

At Ultra High Frequencies (UHF) distributed networks are normally replaced by lumped element ones because their dimensions become excessively long. Such devices are particularly simple to analyze and so the theory of circulators will start with this structure.

The lumped, nonreciprocal element in the circulator consists of a ferrite disk with three coils wound on it so that the magnetic fields of the coils are oriented at 120° with respect to each other. At UHF frequencies the geometry discussed here is usually biased above the main resonance. This symmetrical but nonreciprocal element can be used to form a circulator by connecting capacitors either in series or shunt with the load and source impedances as shown in Figures 9.1a and 9.1b.

One configuration of the shunt arrangement consists of a ferrite disk at the junction of a mesh arrangement of three short sections of short-circuited striplines at 120° that are insulated from each other. This geometry is shown in Figure 9.2. If the short-circuited striplines are electrically short the energy within the disk geometry is essentially magnetic. The junction degenerate resonances are then obtained by connecting shunt capacitances at the three terminals. It is this construction that will be studied in detail in this chapter. The three short-circuited transmission lines can be made self-resonant by making their electrical length $\lambda/4$ long.

One important property of this geometry is that the center frequency can be tuned simply by varying the external capacitance of the device.

This chapter includes design equations and experimental results for a number of UHF circulators. Also included are design equations for matching networks yielding circulators with Chebyshev characteristics. Figure 9.3 shows a planar construction of a lumped element circulator using interdigital junction capacitors.

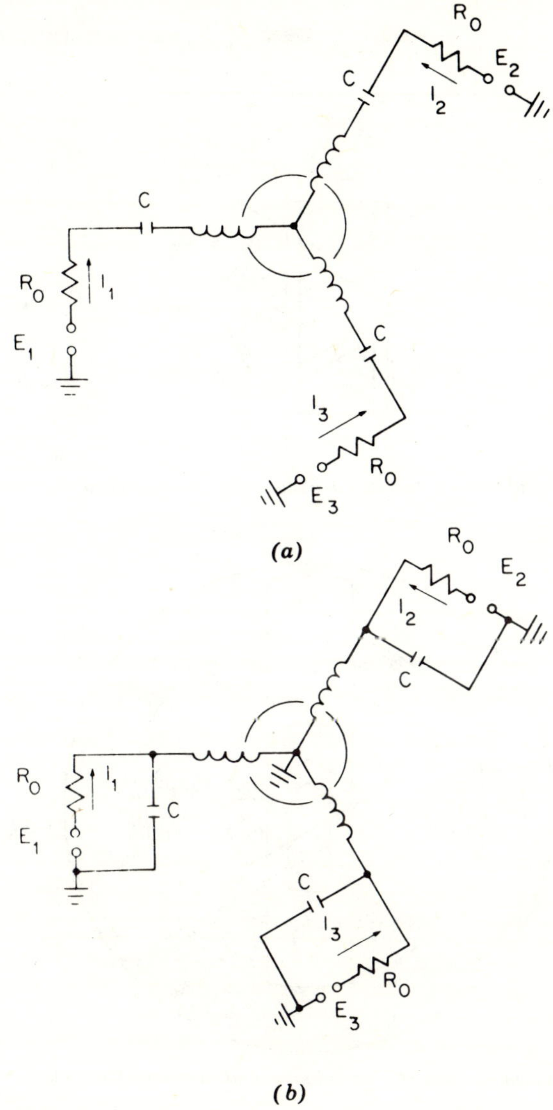

(a)

(b)

Figure 9.1. (*a*) *Schematic of lumped element circulator. Shunt network.* (*b*) *Schematic of lumped element circulator. Series network.*

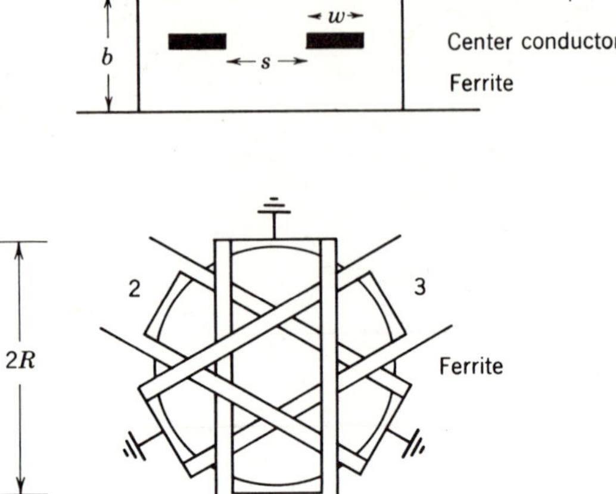

Figure 9.2.   Geometry of lumped constant circulator using short circuited striplines (Ref. 2).

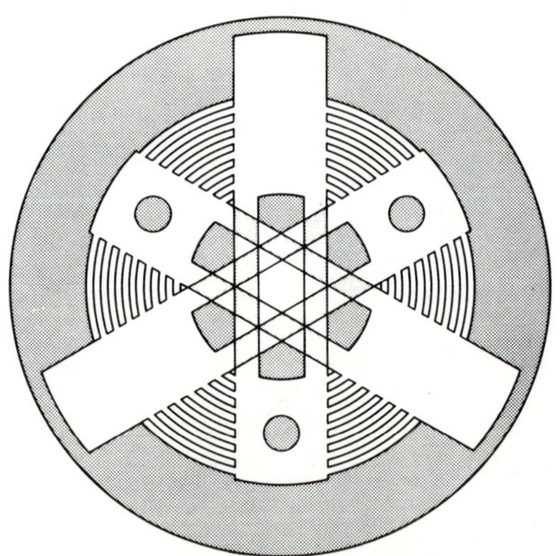

Figure 9.3.   Lumped element circulator using interdigital capacitors (Ref. 11).

9.1.   *LUMPED ELEMENT CIRCULATOR USING SHUNT NETWORK*

The schematic arrangement of the lumped element circulator for the shunt connection is given in Figure 9.1*b*. If the short-circuited transmission lines are electrically short the energy within the disk geometry is essentially magnetic. The voltage current relations at the terminals of the mesh arrangement are then given by

$$\begin{bmatrix} V_1 \\ V_2 \\ V_3 \end{bmatrix} = j\omega L_0 \begin{bmatrix} \mu_i & \dfrac{-\mu_i}{2} & \dfrac{-\mu_i}{2} \\ \dfrac{-\mu_i}{2} & \mu_i & \dfrac{-\mu_i}{2} \\ \dfrac{-\mu_i}{2} & \dfrac{-\mu_i}{2} & \mu_i \end{bmatrix} \begin{bmatrix} I_1 \\ I_2 \\ I_3 \end{bmatrix} \tag{9.1}$$

Here L_0 is the inductance of the constituent mesh, and μ_i is the permeability which applies above the main resonance.

The impedance eigenvalues for this network are

$$z_0 = 0 \tag{9.2}$$

$$z_{+1} = \frac{j3\omega L_0 \mu_i}{2} \tag{9.3}$$

$$z_{-1} = \frac{j3\omega L_0 \mu_i}{2} \tag{9.4}$$

The impedance eigenvalues consist of one nondegenerate eigenvalue and a pair of degenerate ones.

The admittance eigenvalues for this network are the reciprocal of those of the impedances.

$$y_0 = \infty \tag{9.5}$$

$$y_{+1} = \frac{2}{j3\omega L_0 \mu_i} \tag{9.6}$$

$$y_{-1} = \frac{2}{j3\omega L_0 \mu_i} \tag{9.7}$$

The above admittance eigenvalues can be made to coincide with those of the first circulation adjustment by adding shunt capacitances at each terminal:

$$y_0 = \infty \tag{9.8}$$

$$y_{+1} = j\omega C + \frac{2}{j3\omega L_0 \mu_i} = 0 \tag{9.9}$$

$$y_{-1} = j\omega C + \frac{2}{j3\omega L_0 \mu_i} = 0 \tag{9.10}$$

This adjustment establishes the isotropic junction resonances. The second circulation condition is now obtained by splitting the degenerate admittance eigenvalues until they coincide with those of an ideal circulator:

$$y_0 = \infty \tag{9.11}$$

$$y_{+1} = j\omega C + \frac{2}{j3\omega L_0(\mu - K)} = -j\frac{Y_0}{\sqrt{3}} \tag{9.12}$$

$$y_{-1} = j\omega C + \frac{2}{j3\omega L_0(\mu + K)} = j\frac{Y_0}{\sqrt{3}} \tag{9.13}$$

The eigen-networks for this situation can be put down by inspection. The two circulation conditions are therefore

$$Y_0 = \frac{\sqrt{3}}{\omega L} \cdot \frac{K}{\mu} \tag{9.14}$$

and

$$\omega_0^2 LC = 1 \tag{9.15}$$

where

$$L = \frac{3}{2}\mu_e L_0 \tag{9.16}$$

L_0 is the inductance of the constituent resonator defined in Chapter 12.

9.2. EQUIVALENT CIRCUIT OF LUMPED ELEMENT CIRCULATOR

The equivalent circuit of the lumped element circulator near its circulation condition may be obtained by forming the input admittance of the circuit given by

$$Y_{in} = \left(\frac{y_{+1} + y_{-1}}{2}\right) + j\sqrt{3}\left(\frac{y_{+1} - y_{-1}}{2}\right) \tag{9.17}$$

which applies with $s_0 = -1$,

The eigen-admittances for the lumped element circulator are given from the previous section by

$$y_{+1} = j\left(\omega C - \frac{1}{\omega L}\right) - \frac{j}{\omega L} \cdot \frac{K}{\mu} \tag{9.18}$$

$$y_{-1} = j\left(\omega C - \frac{1}{\omega L}\right) + \frac{j}{\omega L} \cdot \frac{K}{\mu} \tag{9.19}$$

The result is

$$Y_{in} = \frac{\sqrt{3}}{\omega L} \cdot \frac{K}{\mu} + j\left(\omega C - \frac{1}{\omega L}\right) \tag{9.20}$$

From the above equation an ideal circulator is obtained provided

$$Y_0 = \frac{\sqrt{3}}{\omega L} \cdot \frac{K}{\mu} \tag{9.21}$$

and

$$\omega_0^2 LC = 1 \tag{9.22}$$

which are the conditions obtained in the previous section. The general form for the admittance of the network is therefore

$$Y_{in} = Y_0 + j\omega_0 C\left(\frac{\omega}{\omega_0} - \frac{\omega_0}{\omega}\right) \tag{9.23}$$

An approximate equivalent network for this equation is an ideal circulator available at any frequency with an admittance y_1 connected at each port. This is shown in Fig. 9.4.

9.3.   BANDWIDTH OF LUMPED ELEMENT CIRCULATOR

An expression for the Q-factor in terms of the junction specification derived in Section 12.4 is

$$Q_L = \frac{B'}{Y_0} = \frac{r-1}{2\delta_{max}\sqrt{r}} \tag{9.24}$$

where r is the VSWR and $2\delta_{max}$ is the full bandwidth $(\omega_2 - \omega_1)/\omega_0$.

The physical variables are now obtained by forming the susceptance slope parameter B' in the usual way

$$B' = \frac{\omega_0}{2}\left.\frac{\partial Y_{in}}{\partial \omega}\right|_{\omega=\omega_0} = \frac{1}{\omega_0 L} = \omega_0 C \tag{9.25}$$

Combining this equation with Eq. 9.14 gives Q_L in terms of the magnetic variable.

$$Q_L = \frac{B'}{Y_0} = \frac{1}{\sqrt{3}}\frac{\mu}{K} \tag{9.26}$$

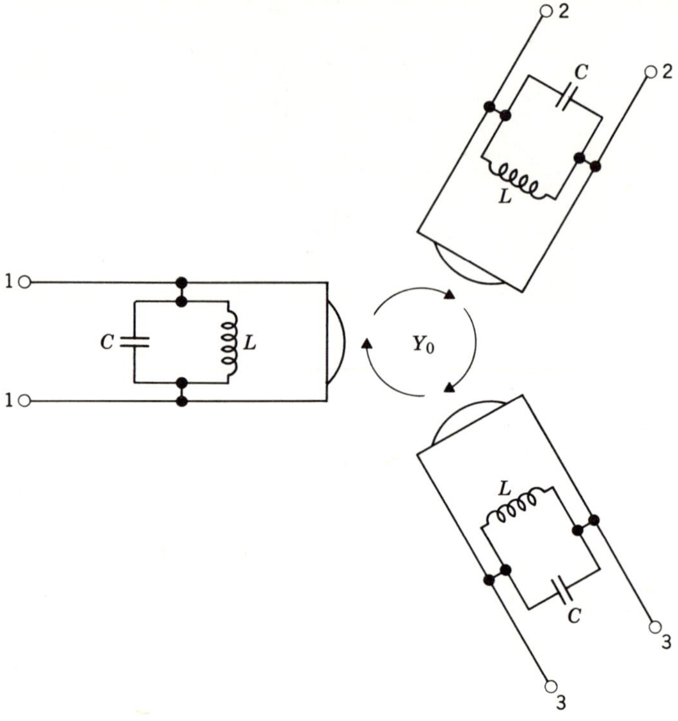

Figure 9.4.   Equivalent circuit of lumped element circulator using frequency independent 2-port gyrators.

The difference between the two resonant frequencies defined by Eqs. 9.12 and 9.13 is related to K/μ by

$$\frac{\omega_{+1} - \omega_{-1}}{\omega_0} = \frac{K}{\mu} \tag{9.27}$$

The loaded Q-factor of the junction is, therefore, also given in terms of the split frequencies ω_+ and ω_- by

$$\frac{1}{\sqrt{3}\, Q_L} = \frac{\omega_{+1} - \omega_{-1}}{\omega_0} \tag{9.28}$$

which is a general result.

9.4.   *INDUCTANCE OF CONSTITUENT RESONATOR FOR LUMPED ELEMENT CIRCULATOR*

In this section the theoretical inductance is calculated by forming the input-impedance of the short section of short-circuited rectangular stripline shown in Figure 9.2. Because of the mesh arrangement used, the characteristic impedance of the stripline is half of the even-mode characteristic impedance of two coupled strip transmission lines. It is possible to calculate this impedance in terms of the stripline dimensions by first determining the static capacitance per unit length of the striplines. The characteristic impedance is then calculated from the capacitance on the assumption that the fields on the lines are pure transverse electromagnetic (TEM).

Figure 9.5 shows S/b versus $L_0/2R$ for parametric values of W/b, for the case where $t/b=0.025$. For t/b small, the inductance is not strongly dependent on it. The mesh arrangement requires the t/b should be kept small; otherwise, it is not possible to interweave it. In the above illustrations R is the ferrite radius and L_0 is the inductance of the constituent resonantor defined in Chapter 12.

9.5.   *MAGNETIC VARIABLES OF LUMPED ELEMENT CIRCULATOR*

At UHF frequencies the lumped element circulator is usually biased above the main resonance to prevent low field loss as discussed in Chapter 19 Provided $\sigma>1$ and $p>1$ one has

$$\frac{K}{\mu}\approx\frac{1}{\sigma}\cdot\frac{p/\sigma}{1+p/\sigma} \tag{9.29}$$

and

$$\mu_e\approx1+\frac{p}{\sigma} \tag{9.30}$$

The original variables used here are defined in Chapter 19. From the above equations it is seen that for $\sigma>1$, and $p>1$, K/μ is proportional to the center frequency, and μ_e is independent of it. Therefore, the impedance relation given by Eq. 9.14 is independent of frequency. This implies that the center frequency of the device given by Eq. 9.15 can be tuned by altering the lumped capacitances at the terminals of the junction. Another property of the device is that since the bandwidth is proportional to K/μ as shown by Eqs. 9.25 and 9.26 the former is proportional to the center frequency.

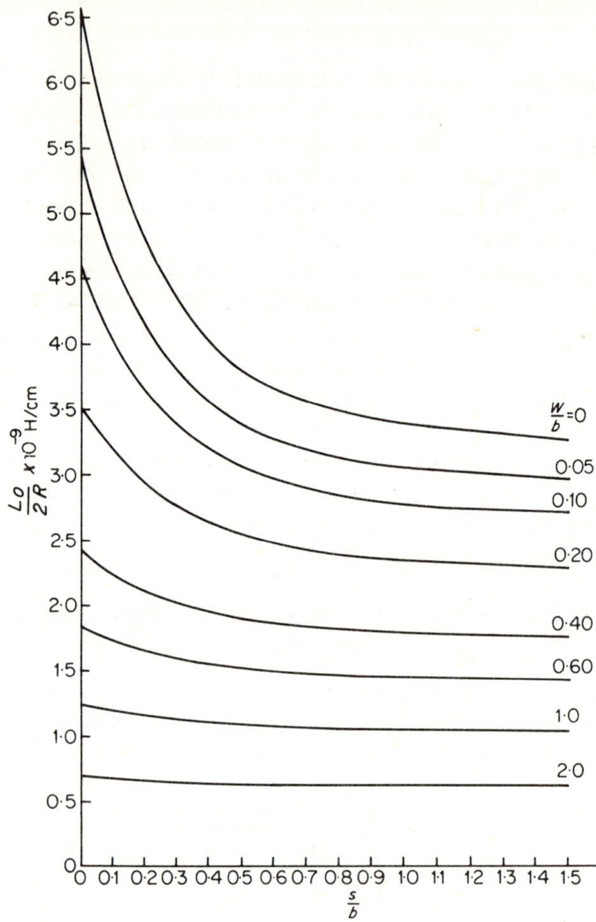

Figure 9.5. Junction inductance of stripline lumped element circulator for $t/b = .025$ (Ref. 12).

Figure 9.6 gives the experimental response of a circulator designed at 240 MHz. Figure 9.7 shows the same circulator retuned to 480 MHz by changing the terminal capacitances only. The experimental bandwidth at 480 MHz is now 5%, which is approximately twice the value at 240 MHz. This is in good agreement with the fact that the bandwidth is proportional to the center frequency of the device.

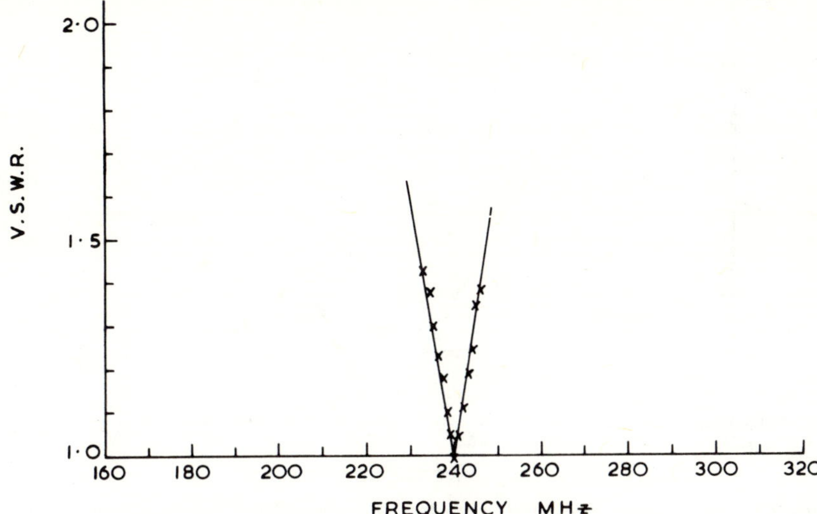

Figure 9.6. Performance of 240 MHz circulator with $n=1$ (Ref. 23).

9.6. $n=2$ NETWORK FOR LUMPED ELEMENT CIRCULATOR

The bandwidth of the junction can be improved by connecting matching networks at the three ports. One possible network consists of the series resonant circuit in Figure 9.8. The overall input admittance of such an $n=2$ network in terms of the normalized $ABCD$ matrix is

$$y_{in} = \frac{Y_{in}}{Y_0} = \frac{jC + Dy_L}{A + jBy_L} \qquad (9.31)$$

The reflection coefficient Γ is

$$\Gamma = \frac{1 - y_{in}}{1 + y_{in}} \qquad (9.32)$$

For the series network used here the $ABCD$ matrix is

$$\begin{bmatrix} A & jB \\ jC & D \end{bmatrix} = \begin{bmatrix} 1 & jz_s \\ 0 & 1 \end{bmatrix} \qquad (9.33)$$

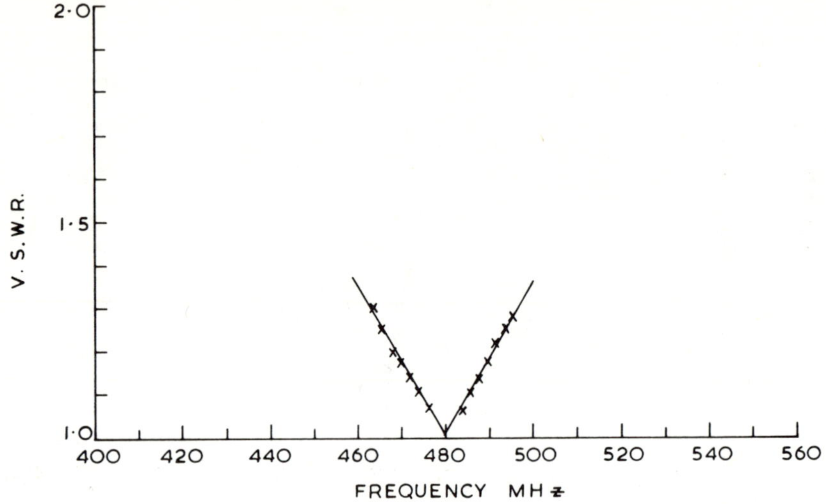

Figure 9.7. *Performance of* 480 *MHz circulator with* $n = 1$ *(Ref.* 23).

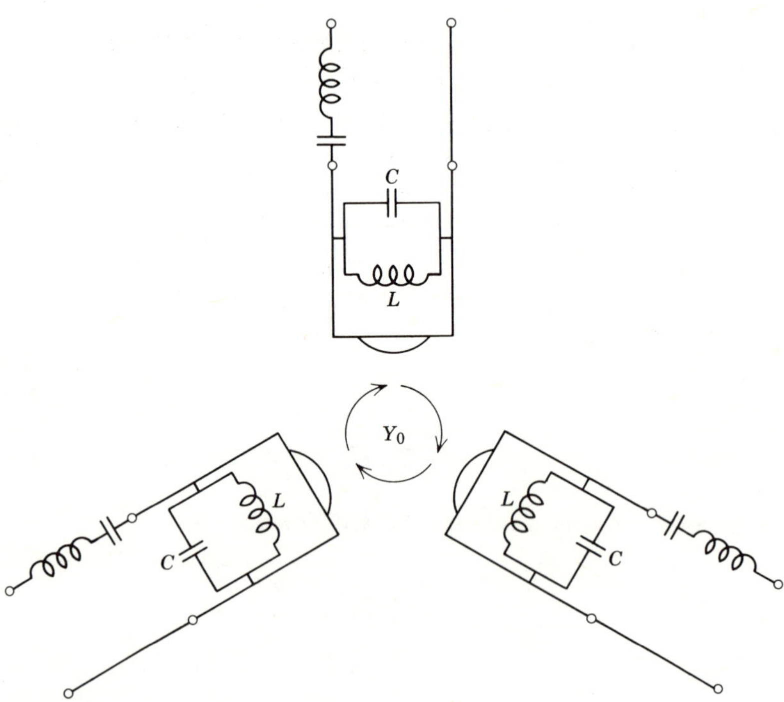

Figure 9.8. *Schematic of* $n = 2$ *circulator.*

where $A, B, C,$ and D are pure real numbers. For the lumped element circulator

$$y_L = \frac{Y_L}{Y_0} = (g_L + j2\delta Q_L) \tag{9.34}$$

$$z_s = \frac{Z_s}{R_0} = 2\delta Q_s \tag{9.35}$$

where

$$Q_L = \frac{\omega C}{Y_0} \tag{9.36}$$

$$Q_s = \frac{\omega L_s}{R_0} \tag{9.37}$$

and

$$\delta = (\omega - \omega_0)/\omega_0$$

The reflection coefficient can now be adjusted to have a Chebyshev response. The result for the circuit elements is

$$g_L = \frac{Q_s}{Q_L} \tag{9.38}$$

$$r = \frac{1}{g_L} \tag{9.39}$$

and the bandwidth relation is

$$Q_L = \frac{\sqrt{2}\,(r-1)^{1/2}}{2\delta_{\max}} . \tag{9.40}$$

which may be compared with Eq. 9.24 which applies to $n=1$.

A property of such networks is that for n even the generator admittance Y_0 and the gyrator conductance G_L are related by Eq. 9.39.

The series network is, therefore, given by

$$L_s = \frac{R_0^2 C}{r} \tag{9.41}$$

and

$$C_s = \frac{1}{\omega_0^2 L_s} \tag{9.42}$$

Figure 9.9 indicates the response obtained with such a network at 480 MHz for the basic response in Figure 9.7. The bandwidth for $r=1.22$ is

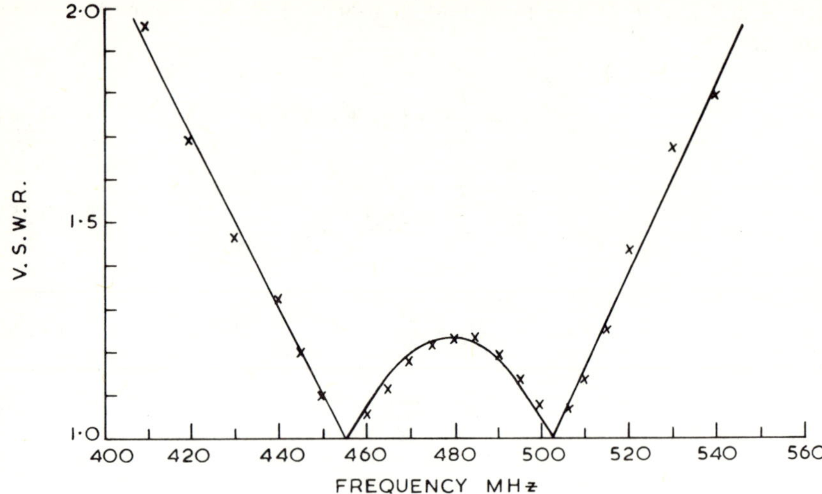

Figure 9.9.   *Performance of* 480 *MHz circulator with* $n = 2$.

improved from 5 to about 18%. The first minimum in the response occurs
at

$$2\delta_{min} = \frac{2\delta_{max}}{\sqrt{2}}$$

(9.43)

where $2\delta_{min}$ is the normalized bandwidth between the first zeros in the
frequency response.

9.7. $n = 3$ *NETWORK FOR LUMPED ELEMENT CIRCULATOR*

A wideband network with $n = 3$ is also sometimes used as a matching
network. This arrangement consists of one series resonant circuit and one
parallel resonant circuit connected as shown in Figure 9.10. The analysis
proceeds by first finding the overall *ABCD* matrix and substituting it into
Eq. 9.31. The subsequent procedure then follows that described in Chapter
14. However, the results given here for the $n = 3$ network rely on a lowpass
to highpass transformation. They are

$$L_s = \left(\frac{2\vartheta^2}{\vartheta^2 + 3/4} \right) R_0^2 C$$

(9.44)

$$C_s = \frac{1}{\omega^2 L_s}$$

(9.45)

$$L_p = \frac{1}{\omega^2 C} \tag{9.46}$$

$$C_p = C \tag{9.47}$$

where

$$\vartheta = \left(\frac{4}{\epsilon^2}\right)^{1/4} - \left(\frac{4}{\epsilon^2}\right)^{-1/4} \tag{9.48}$$

$$\epsilon^2 = \frac{(r-1)^2}{4r} \tag{9.49}$$

In this arrangement, the terminal impedance of the circulator is equal to that of the generator.

The bandwidth is increased by a factor of 4.25 for $r=1.22$ and by a factor of 8.42 for $r=1.065$.

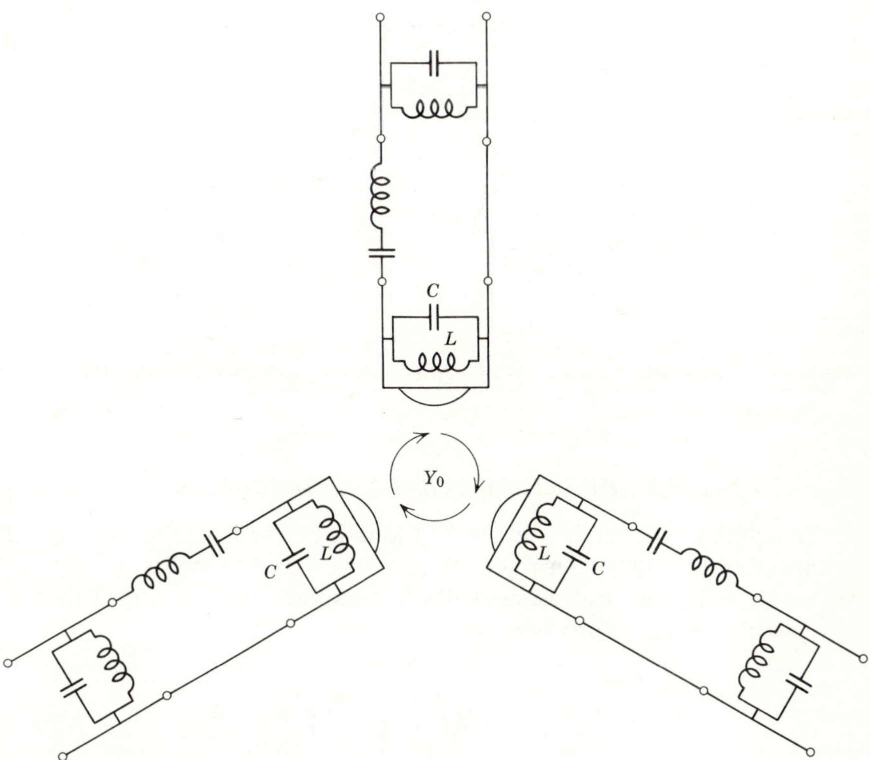

Figure 9.10.   Schematic of $n=3$ circulator.

9.8.   *VARACTOR TUNED LUMPED ELEMENT CIRCULATOR*

It has been seen in Section 9.5 that it is possible to tune the center frequency of the lumped element circulator simply by varying the terminal capacitance. This can be done provided the device is biased sufficiently above the main resonance. This property of the device leads to a voltage tuned circulator if the lumped capacitors are replaced by voltage tuned varactors. The equivalent circuit for this arrangement is indicated in Fig. 9.11.

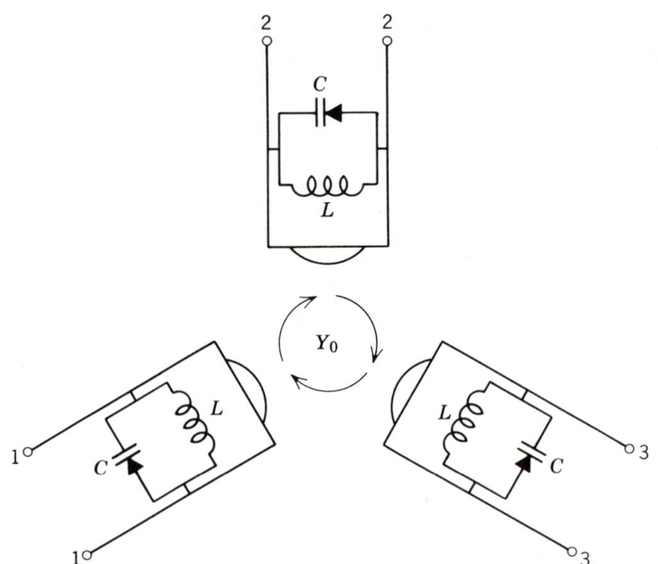

Figure 9.11.   *Equivalent circuit of varactor tuned lumped element circulator (Ref. 15).*

9.9.   *QUARTERWAVE MODE CIRCULATOR*

If the length of the mesh is made $\lambda/4$ long it becomes unnecessary to add a capacitance at each terminal of the junction. The mesh is now self-resonant. With this arrangement it will be shown in Chapter 12 that the two circulation equations become

$$R_{oe} = \sqrt{3} \sqrt{\frac{\epsilon_r}{\mu_e}} \left(\frac{\pi R_0}{6}\right) \cdot \frac{K}{\mu} \tag{9.50}$$

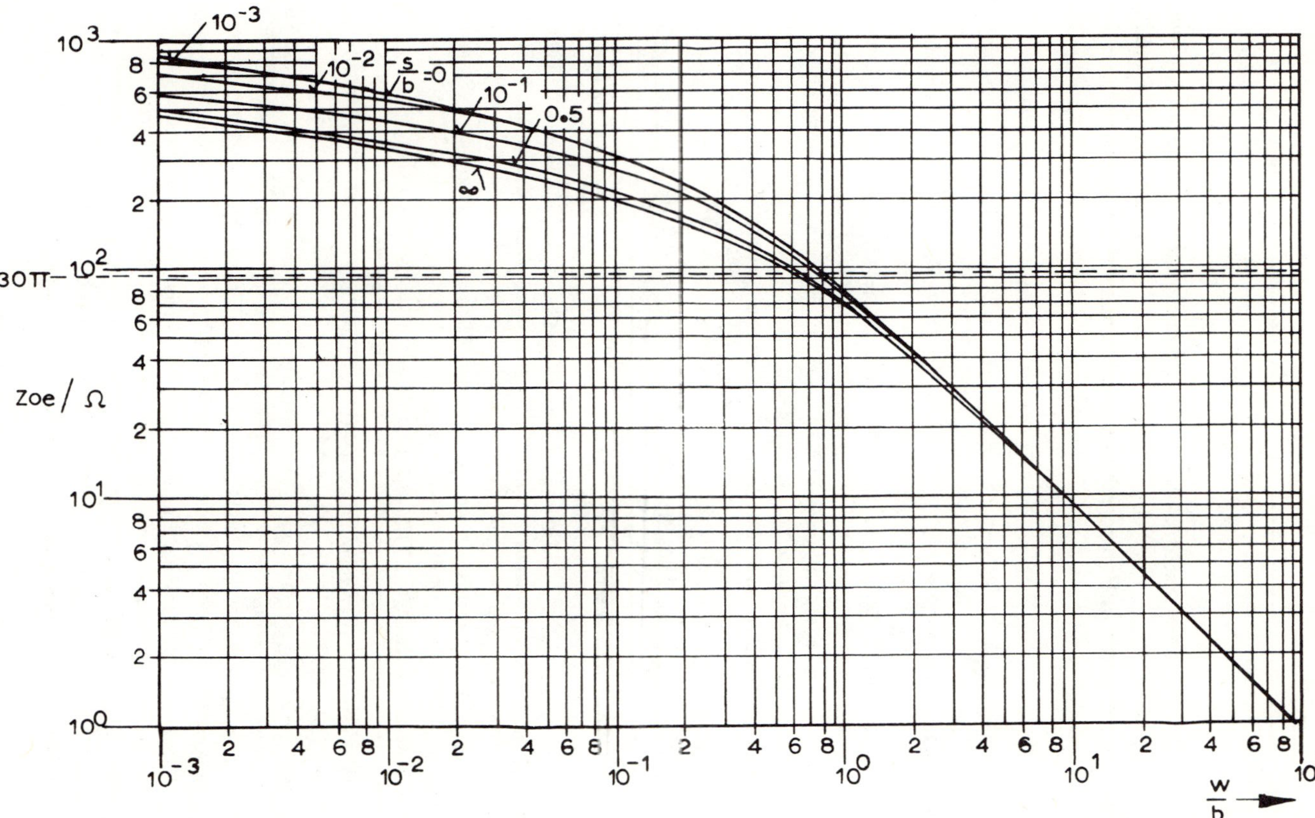

Figure 9.12.   Even mode characteristic impedance of two coupled lines (Ref. 22).

and

$$2R = \frac{\lambda_0}{4} / \sqrt{\epsilon_r \mu_e} \qquad (9.51)$$

where R_{oe} is the free space even mode characteristic impedance of two coupled striplines, and ϵ_r is the relative dielectric constant of the ferrite material. R_{oe} is given in Figure 9.12 as a function of W/b and s/b.

REFERENCES

1. V. E. Dunn and R. W. Roberts, "Miniature VHF and UHF Circulators Use Lumped Element Design," *Microwaves*, 46–47 (December 1963).

2. Y. Konishi, "Lumped Element Y Circulator," *IEEE Trans. Microwave Theory Tech.*, **MTT-13**, (6) 852–864 (1965).

3. V. E. Dunn and R. W. Roberts, "New Design Techniques for Miniature VHF Circulators," presented at the G-MTT Symposium, Clearwater, Fla., 1965.

4. J. Deutch and B. Wieser, "Resonance Isolator and Y Circulator with Lumped Elements at VHF," *IEEE Trans. Magn.*, **MAG-2**, 278–282 (1966).

5. B. Chiron, P. Girard and P. M. Prache, "Theory and Practice of Lumped Parameter Circulators," IEE International Conference on the Microwave Behaviour of Ferrimagnetics and Plasmas, (1965).

6. G. McChesney and V. Dunn "Broadband Lumped Element UHF Circulator," *IEEE Trans. Microwave Theory Tech.*, **MTT-15**, (3) 198–199 (1967).

7. J. Helszajn, "A Simplified Theory of the Three Port Junction Circulator," *Radio Electron. Eng.*, **33**, 283–288 (1967).

8. J. C. Bergman and C. Christenson, "Equivalent Circuit for a Lumped-Element Y Circulator," (Correspondence), *IEEE Trans. Microwave Theory Tech.*, **MTT 16** (5), 308–310 (1968).

9. E. Fliegler, "A 350 MHz Broadband Lumped Element Circulator as a Protective Isolator," *IEEE Trans. on Microwave Theory Tech.*, **MTT-17** (5), 275–276 (1969).

10. S. Okamura and T. Pagai, "VHF and UHF-Band Stacked Junction Circulators," *IEEE Trans. Microwave Theory Tech.*, **MTT-17** (12), 1151–1152 (1969).

11. R. H. Knerr, "A Thin Film Lumped Element Circulator," *IEEE Trans. Microwave Theory Tech.*, **MTT. 17**, (12) 1152–1154 (1969).

12. J. Helszajn and M. McDermott, " The Inductance of a Lumped Constant Circulator," *IEEE Trans. Microwave Theory Tech.*, **MTT. 18** (1), 50–52 (1970).

13. Y. Konishi and N. Hoshino, "Design of a New Broad-Band Isolator," *IEEE Trans. Microwave Theory Tech.*, **MTT-19** (3), 260–269 (1971).

14. Y. Nahito and N. Tanaka, "Broad-Banding and Changing Operation Frequency of Circulator," *IEEE Trans. Microwave Theory Tech.*, **MTT-19** (4), 367–372 (1971).

15. J. Helszajn, D. Walker, and F. M. Aitken, "Varactor-Tuned Lumped-Element Circulators," *IEEE Trans. Microwave Theory Tech.*, **MTT-19** (10) 825–826 (1971).

16. A. S. Webster "Remotely Tuned Circulator," *Electron. Lett.*, **6**, 377–379 (1970).

17. R. H. Knerr and C. E. Barnes, "A compact Broad-Band Thin Film Lumped Element L-Band Circulator," *IEEE Trans. Microwave Theory Tech.*, **MTT-18** (12), 1100–1108 (1970).

18. E. Pivit, "Circulators from Lumped Elements," *Telefunken*, **38** (2) 206–213 (1965).

19. R. W. Roberts and J. E. Lahey, "Ferrites for VHF Frequencies" *IEEE Trans. Veh. Commun.*, **VC-14**, 117–121 (1965).

20. Y. Konishi, "General Theory of Coaxial Y-Junction Circulator and design Consideration on its UHF High Power Application," *Tech. J. Japan Broadcast. Corp.*, **20** (2) 19 (1968).

21. Y. Konishi, T. Murayama, and N. Hoshino, "VHF Power Circulator with Slow-Wave Circuits," *Electron. Commun. Japan*, **51A** (1968).

22. T. G. Bryant and J. A. Weiss, "Parameters of microstrip transmission lines and of coupled pairs of microstrip lines," *IEEE Trans. Microwave Theory and Techniques*, vol. MTT-16, pp. 1021–1027, December 1968.

23. J. Helszajn and F. M. Aitken, "U. H. F. Techniques for Lumped Constant Circulators," *Electronic Engineering*, pp. 53–59, Nov. 1973.

CHAPTER TEN

The Stripline Junction Circulator

The most general form of the 3-port junction circulator consists of a symmetrical distribution of ferrite material at the junction of three transmission lines. The stripline geometry is shown in Figure 10.1. It consists of two ferrite disks, separated by a circular center conductor fed by symmetrical transmission lines. The ferrite material is magnetized perpendicularly to the plane of the disks by a static magnetic field. An important property of the device already discussed, is that a perfect circulator is obtained when the junction is matched. For a 3-port junction this requires two independent variables. Under certain simplifying conditions the adjustment of the stripline circulator can be described in terms of a standing wave of the electric field pattern within the disk due to the interference of a pair of field patterns rotating in opposite directions. When the junction is unmagnetized, the resonant frequencies of the two field patterns are the same. When the junction is magnetized, the degeneracy is removed, and the standing wave pattern within the disk is rotated. One circulation condition is obtained by operating between the two split frequencies. This condition essentially determines the diameter of the ferrite geometry. The second circulation condition is met by adjusting the splitting, in the case of the $n = \pm 1$ modes, until the standing wave pattern is rotated through 30°. From symmetry, port 3 is then situated at a null of the standing wave pattern and is therefore isolated. The junction then behaves as a transmission line resonant cavity between ports 1 and 2. This adjustment is determined by the magnitude of the direct magnetic field.

In this chapter the electromagnetic boundary problem will be given. This allows the dependence of the disk diameter and magnetic field on frequency and magnetization to be determined. The electromagnetic theory of the stripline circulator is derived in terms of the impedance matrix of the junction. The boundary conditions are satisfied by comparing this impedance matrix to that of an ideal circulator. This provides two simple relations for the gyrator resistance and operating frequency of the junction.

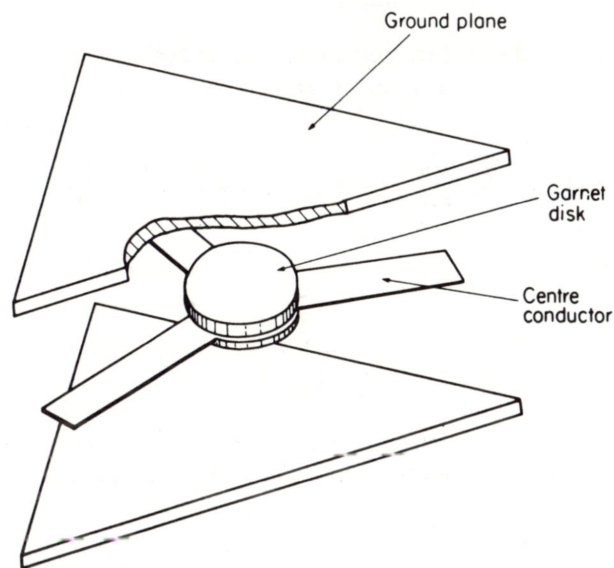

Figure 10.1.   *Stripline circulator.*

10.1.   *IMPEDANCE MATRIX OF STRIPLINE CIRCULATOR*

Figure 10.1 gives the configuration of the device being considered. In the usual way it is assumed that it is possible to neglect the variation with z of the field inside the ferrite. Then the electric field will have one component E_z that satisfies the homogeneous Helmholtz equation in cylindrical coordinates given by Eq. 6.12

$$\left[\frac{\partial^2}{\partial r^2} + \frac{1}{r}\frac{\partial}{\partial r} + \frac{1}{r^2}\frac{\partial^2}{\partial \phi^2} + k_e^2 \right] E_z = 0 \tag{10.1}$$

with time dependence $e^{j\omega t}$ and

$$k_e = \omega \sqrt{\epsilon_r \epsilon_0 \mu_e \mu_0} \tag{10.2}$$

Here, ϵ_r is the relative dielectric constant of the ferrite, μ_e is the effective permeability.

The general solution of Eq. 10.1 in cylindrical coordinates for a magnetized ferrite has the form given in Chapter 6

$$E_z = \sum_{n=-\infty}^{\infty} A_n J_n(k_e r) e^{jn\phi} \tag{10.3}$$

where $J_n(k_e r)$ is a Bessel function of the nth order.

Maxwell's equation then gives for the ϕ component of $\overline{H}_n$

$$H_\phi = \frac{-j}{\eta_e} \sum_{n=-\infty}^{\infty} A_n \left[J_n'(k_e r) - \frac{K}{\mu} \frac{n J_n(k_e r)}{k_e r} \right] e^{jn\phi} \tag{10.4}$$

where

$$\eta_e = \sqrt{\frac{\mu_0 \mu_e}{\epsilon_0 \epsilon_r}} \tag{10.5}$$

Following Bosma it is assumed that H_ϕ is a constant over the width of the strip lines and zero elsewhere, and take for the boundary conditions at $r = R$

$$-\psi < \phi < \psi, H_\phi = H_1$$

$$-120° - \psi < \phi < \psi - 120°, H_\phi = H_2$$

$$120° - \psi < \phi < \psi + 120°, H_\phi = H_3 \tag{10.6}$$

$$\text{elsewhere, } H_\phi = 0$$

where

$$\sin \psi = \frac{W}{2R} \tag{10.7}$$

and $W/b = 1.45$ for a 50Ω line. Here, W is the width of the stripline feeding the junction and b is the ground plane spacing. It is assumed that the center conductor thickness t is zero. The schematic of the configuration studied is shown in Figure 10.2. The amplitude constant A_n in Eq. 10.3 is now found by expanding H_ϕ given by Eq. 10.6 into a Fourier series with

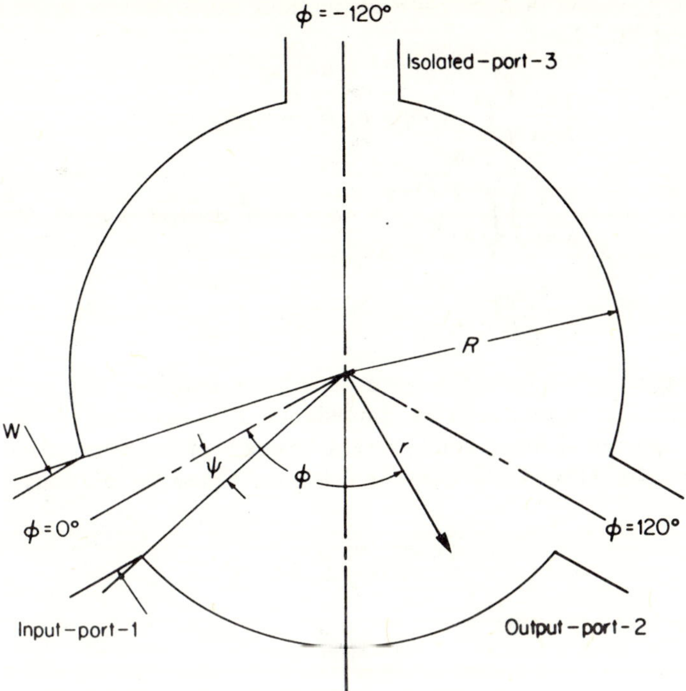

Figure 10.2.   *Schematic of stripline circulator* (*Ref.* 3).

respect to ϕ

$$H_\phi = \sum_{n=-\infty}^{\infty} b_n e^{jn\phi} \tag{10.8}$$

where

$$b_n = \frac{\sin n\psi}{\pi n}\left[H_1 + H_2 e^{j2\pi n/3} + H_3 e^{-j2\pi n/3}\right] \tag{10.9}$$

Comparing Eqs. 10.4 and 10.8 at $r = R$ gives for the amplitude constant A_n

$$A_n = \frac{-\eta_e \sin n\psi}{j\pi n}\frac{\left[H_1 + H_2 e^{j2\pi n/3} + H_3 e^{-j2\pi n/3}\right]}{\left[J_n'(k_e R) - \left(\frac{K}{\mu}\right)\frac{n J_n(k_e R)}{k_e R}\right]} \tag{10.10}$$

The average values of E_z at $r = R$ at the three ports are now given with the help of Eq. 10.3 by

$$E_1 = \left(\frac{1}{2\psi}\right)\int_{-\psi}^{\psi} E_z \, d\phi = \eta_{11}H_1 + \eta_{12}H_2 + \eta_{21}H_3 \qquad (10.11)$$

$$E_2 = \left(\frac{1}{2\psi}\right)\int_{-2\pi/3-\psi}^{-2\pi/3+\psi} E_z \, d\phi = \eta_{21}H_1 + \eta_{11}H_2 + \eta_{12}H_3 \qquad (10.12)$$

$$E_3 = \left(\frac{1}{2\psi}\right)\int_{2\pi/3-\psi}^{2\pi/3+\psi} E_z \, d\phi = \eta_{12}H_1 + \eta_{21}H_2 + \eta_{11}H_3 \qquad (10.13)$$

This last equation defines the wave impedance matrix of the junction. In order to obtain its characteristic impedance, it is necessary to develop a relation between voltage and current instead of between electric and magnetic field. This can be done by using the following relations:

$$V = \int \overline{E} \cdot d\overline{l} \qquad (10.14)$$

$$I = \oint \overline{H} \cdot d\overline{l} \qquad (10.15)$$

An approximate result for a stripline configuration having a center conductor width W and a ground plane spacing b, which applies for $W \gg b$, is

$$V = \frac{Eb}{2} \qquad (10.16)$$

$$I = 2HW \qquad (10.17)$$

The characteristic impedance matrix is now obtained by rewriting Eqs. 10.11 through 10.13 in terms of V and I with the help of the last two equations

$$\overline{R} = \begin{bmatrix} R_{11} & R_{12} & -R_{12}^* \\ -R_{12}^* & R_{11} & R_{12} \\ R_{12} & -R_{12}^* & R_{11} \end{bmatrix} \qquad (10.18)$$

The above impedance matrix makes use of the relation between R_{12} and R_{21}

$$R_{21} = -R_{12}^* \qquad (10.19)$$

where

$$R_{11} = \sum_{n=-\infty}^{\infty} \frac{r_n}{3} \qquad (10.20)$$

$$R_{12} = \sum_{n=-\infty}^{\infty} \frac{r_n e^{j2\pi n/3}}{3} \qquad (10.21)$$

$$R_{21} = \sum_{n=-\infty}^{\infty} \frac{r_n e^{-j2\pi n/3}}{3} \qquad (10.22)$$

and

$$r_n = \frac{-3R_e \sin^2(n\psi) J_n(k_e R)}{jn^2\pi\psi \left[J_n'(k_e R) - \left(\frac{K}{\mu}\right) \frac{nJ_n(k_e R)}{k_e R} \right]} \qquad (10.23)$$

where

$$R_e = \sqrt{\frac{\mu_e}{\epsilon_r}} \left(30\pi \frac{b}{W}\right) \qquad (10.24)$$

This last equation is an approximate one for the characteristic impedance for a strip line transmission line when $W \gg b$. A more exact expression is

$$R_e = \sqrt{\frac{\mu_e}{\epsilon_r}} \left(30\pi \ln\left[\frac{W+b}{W+t}\right]\right) \qquad (10.25)$$

In what follows the two circulation conditions will be derived by comparing the impedance matrix obtained above with that of an ideal circulator. In the above, the eigenvalue r_0 is infinite at the origin while all the other eigenvalues are zero there. Hence the eigennetwork for the $n=0$ mode is an open circuited transmission line, and the others are short-circuited ones. These are shown in Figure 10.3.

The impedance matrix of an ideal circulator has the following form:

$$\bar{R} = \begin{bmatrix} 0 & -R_0 & R_0 \\ R_0 & 0 & -R_0 \\ -R_0 & R_0 & 0 \end{bmatrix} \qquad (10.26)$$

With $r_0 = 0$ one also has

$$R_{12} = \frac{-R_{11}}{2} + R_0 \qquad (10.27)$$

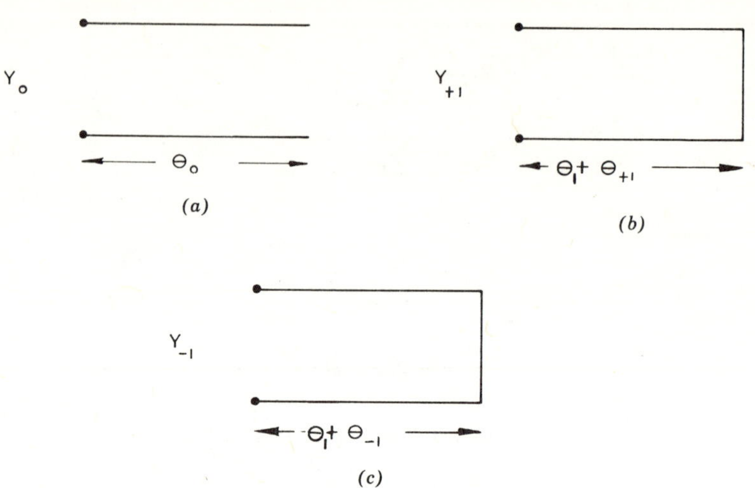

Figure 10.3.   *Eigennetworks of stripline circulators (Ref. 15).*

Here R_0 is the characteristic impedance of the transmission lines feeding
the junction

$$R_0 = \sqrt{\frac{\mu_d}{\epsilon_d}} \left(30\pi \ln\left[\frac{W + b_d}{W + t} \right] \right) \tag{10.28}$$

where μ_d and ϵ_d are the relative permeability and dielectric constants in the
stripline feeding the junction, b_d is the ground plane spacing of the stripline
that may be different from that of the junction $b \cdot W = W_d$ since W is
common to both the stripline and the junction.

10.2.   CIRCULATION CONDITIONS

Comparing the impedance matrix equation for the stripline circulator
given by Eq. 10.18 with that given for an ideal circulator by Eq. 10.26
immediately gives the two circulation equations:

$$R_{11} = 0 \tag{10.29}$$

$$R_{12} = -R_0 \tag{10.30}$$

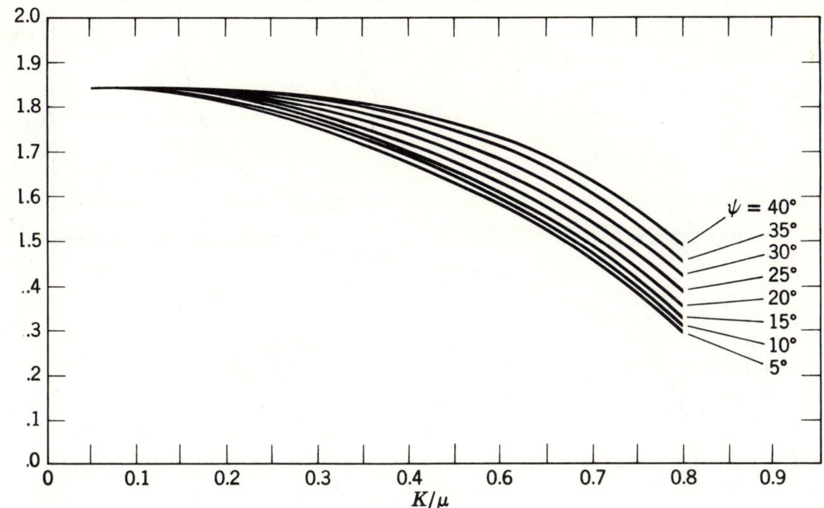

Figure 10.4.   *kR versus* K/μ *for* $n = \pm 1$ *modes with* $p = 11$.

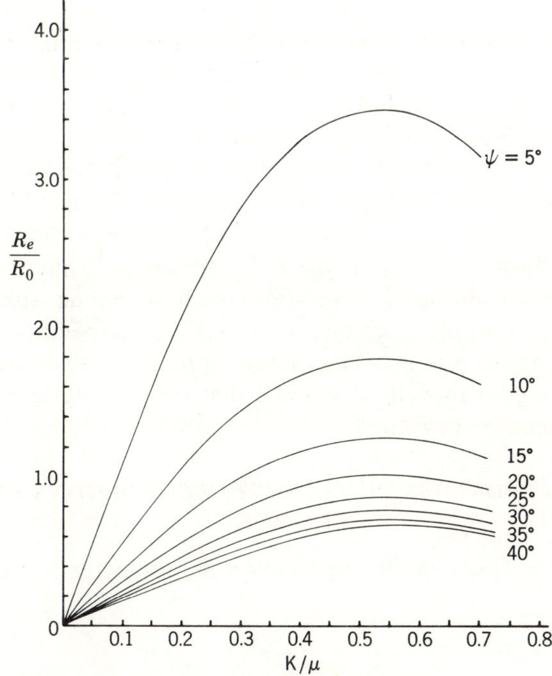

Figure 10.5. R_e/R_0 *versus* K/μ *for* $n = \pm 1$ *modes with* $p = 11$.

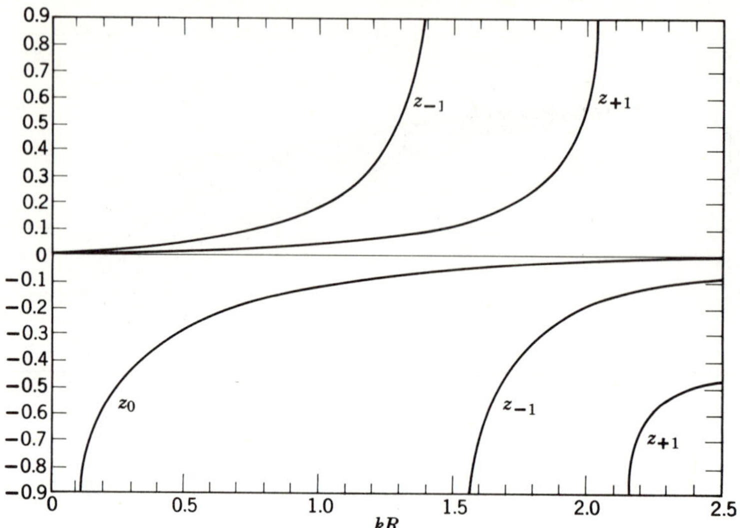

Figure 10.6.   *Impedance eigenvalues of stripline circulator.*

Equation 10.29 determines the ferrite diameter for appropriate pairs of field patterns. Equation 10.30 can then be used to calculate the gyrator resistance.

Figure 10.4 shows $k_e R$ versus K/μ obtained by satisfying Eq. 10.29 in the vicinity of the $n = \pm 1$ modes by retaining the first 11 terms in the series. Figure 10.5 gives R_e/R_0 versus K/μ obtained by straightforward calculation using Eq. 10.30. Figure 10.6 illustrates the impedance eigenvalues of the $n = 0$ and $n = \pm 1$ modes. This illustration indicates that r_0 is approximately 0 in the vicinity of the $r_{\pm 1}$ resonances, which is consistent with the approximation used here. For small magnetic splitting it is, therefore, possible to apply a selection rule whereby circulation is taken to occur in the vicinity of a pair of degenerate impedance eigenvalues. In the vicinity of those eigenvalues it is assumed that the amplitude of the others are small and so can be neglected.

10.3.   *APPROXIMATE CIRCULATION CONDITIONS FOR $n = \pm 1$ MODES*

For small magnetic splitting the approximate solutions for R_{11} and R_{12} are, therefore,

$$R_{11} \approx \frac{r_n + r_{-n}}{3} \tag{10.31}$$

and

$$R_{12} \approx \frac{r_n e^{j2\pi n/3} + r_{-n} e^{-j2\pi n/3}}{3} \tag{10.32}$$

For $n = \pm 1$ the impedance eigenvalues are

$$r_{+1} = 3 R_e \sin\psi J_1(k_e R) \left(-j\pi \left[J_1'(k_e R) - \frac{K}{\mu} \frac{J_1(k_e R)}{k_e R} \right] \right)^{-1} \tag{10.33}$$

$$r_{-1} = 3 R_e \sin\psi J_1(k_e R) \left(-j\pi \left[J_1'(k_e R) + \frac{K}{\mu} \frac{J_1(k_e R)}{k_e R} \right] \right)^{-1}. \tag{10.34}$$

where it has been assumed that $\sin\psi \approx \psi$.

The first circulation condition is now obtained when Eq. 10.29 is satisfied. This requires that

$$J_1'(k_e R) = 0 \tag{10.35}$$

The first root is given by

$$(k_e R)_{1,1} = 1.84 \tag{10.36}$$

Figure 6.10 gives a theoretical mode chart for some of the lower order modes. Figure 10.7 shows an experimental mode chart obtained on a loosely coupled circulator for which $\psi \approx 0$.

The gyrator resistance can now be calculated from Eq. 10.30 with the boundary condition given by Eq. 10.35. Replacing R by $-R$ gives

$$(R_0)_{1,1} = \frac{\sqrt{3}\,(1.84) R_e \sin\psi}{\pi}\, \frac{\mu}{K} \tag{10.37}$$

Equations 10.36 and 10.37 are the two circulation conditions to be satisfied. The first states that $(k_e R)_{1,1}$ lies somewhere between $(k_e R)_{+1,1}$ and $(k_e R)_{-1,1}$ for the normal modes of the uncoupled case. One condition for circulation is, therefore, that the operating frequency lies between the resonant frequencies of the $n = \pm 1$ modes. These modes are depicted in Fig. 10.8a when the disks are excited at port 1 and the ferrites are not magnetized. The second circulation condition states that the input impedance at the input and output ports must be matched to the impedance at the terminals. Hence, the second circulation condition is obtained by

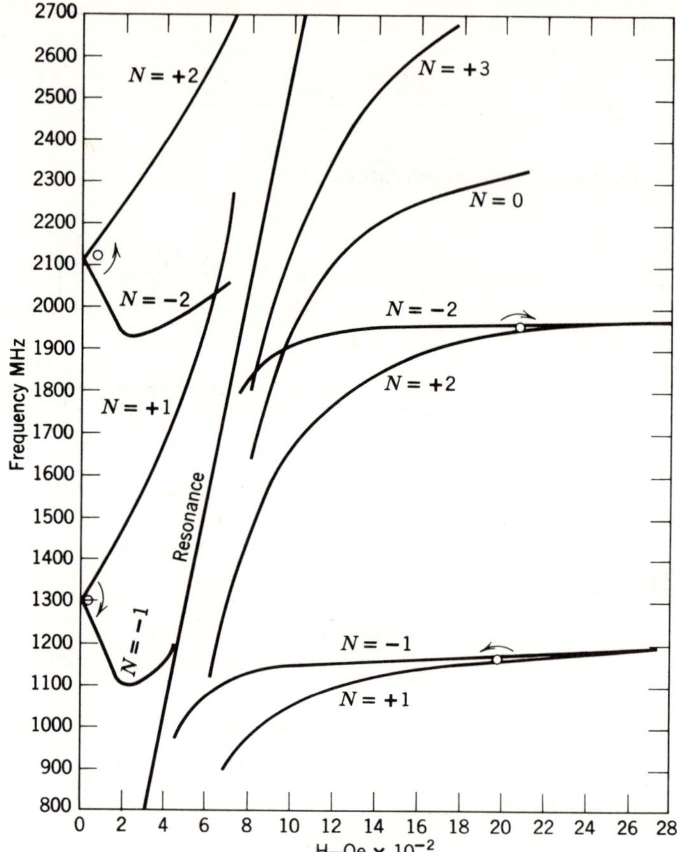

Figure 10.7.   *Mode chart of experimental circulator* (*Ref.* 3).

adjusting the splitting K/μ between the normal modes to satisfy the impedance equation. This adjustment rotates the field pattern of the $n = \pm 1$ modes through 30° by removing the degeneracy between the two normal modes. When this requirement is satisfied, port 3 is situated at a null of the electric field pattern, and transmission occurs between ports 1 and 2. This is shown in Fig. 10.8*b*. The distribution of the electric field around the periphery of a loosely coupled junction is given in Fig. 10.9. It is in good agreement with the assumption that such a circulator can be synthesized with the $n = \pm 1$ modes only.

As will be seen later on the splitting K/μ is determined by the susceptance slope parameter of the junction which dictates the bandwidth of the device.

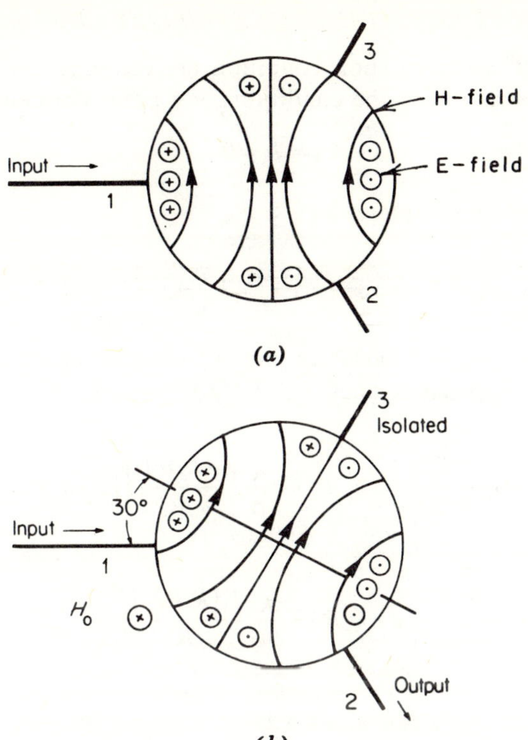

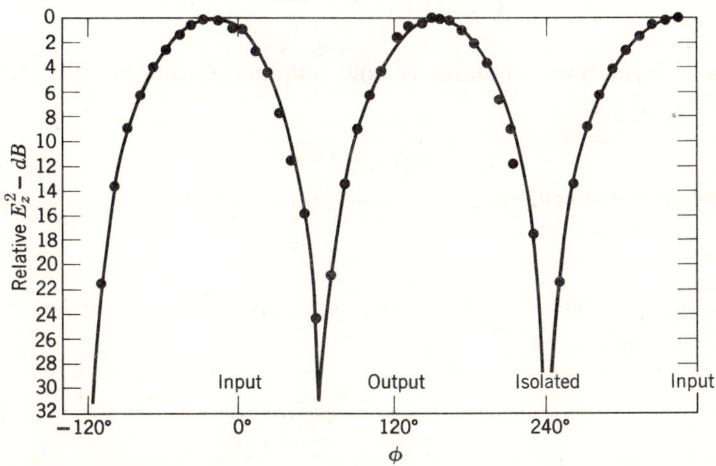

Figure 10.8. (a) *Field patterns for* $n = \pm 1$ *modes* (*Ref.* 6). (b) *3-Port circulator synthesized by* $n = \pm 1$ *modes* (*Ref.* 6).

Figure 10.9. *Measured electric field at the periphery of a loosely coupled circulator* (*Ref.* 6).

10.4. *INPUT ADMITTANCE OF STRIPLINE CIRCULATOR*

In the vicinity of the circulation condition the complex input admittance can be formed by assuming the condition for a perfect circulator at port 3.

$$V_3 = I_3 = 0 \tag{10.38}$$

The result is given in Chapter 5 by

$$Y_{in} = \left(\frac{y_{+1} + y_{-1}}{2}\right) + j\sqrt{3}\left(\frac{y_{+1} - y_{-1}}{2}\right) \tag{10.39}$$

In terms of the original variables the admittance eigenvalues are the reciprocal of the impedance ones

$$y_{+1} = \frac{-j\pi Y_e}{3\sin\psi}\left[J_1'(k_e R) - \frac{K}{\mu}\frac{J_1(k_e R)}{k_e R}\right][J_1(k_e R)]^{-1} \tag{10.40}$$

$$y_{-1} = \frac{-j\pi Y_e}{3\sin\psi}\left[J_1'(k_e R) + \frac{K}{\mu}\frac{J_1(k_e R)}{k_e R}\right][J_1(k_e R)]^{-1} \tag{10.41}$$

Substituting the above eigenvalues with R replaced by $-R$ to satisfy the positive coordinate system defined in Chapter 5 into Equation 10.39 gives

$$Y_{in} = \frac{\pi Y_e}{\sqrt{3}\,(k_e R)\sin\psi}\frac{K}{\mu} - \frac{j\pi Y_e}{3\sin\psi}\left[\frac{J_1'(k_e R)}{J_1(k_e R)}\right] \tag{10.42}$$

The first circulation condition is now obtained by setting the imaginary part of Y_{in} to zero.

$$J_1'(k_e R) = 0 \tag{10.43}$$

The first root is given by

$$(k_e R)_{1,1} = 1.84 \tag{10.44}$$

The second circulation condition is satisfied by setting the real part of Y_{in} to Y_0. The result is

$$Y_0 = \frac{\pi Y_e}{\sqrt{3}\,(1.84)\sin\psi}\cdot\frac{K}{\mu} \tag{10.45}$$

Equations 10.44 and 10.45 are identical with equations 10.36 and 10.37 obtained with the help of the impedance matrix. The equivalent circuit of the junction circulator can be constructed using the procedure developed in Chapter 5. Figure 10.10 shows the equivalent circulator of the stripline circulator in terms of frequency independent 2-port gyrators.

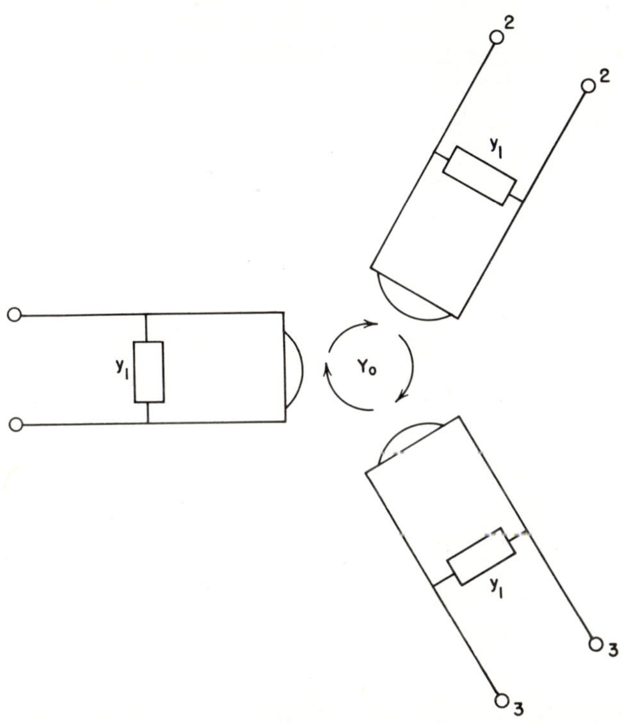

Figure 10.10.   *Equivalent circuit of circulator in terms of frequency independent ideal gyrators.*

10.5.   *BANDWIDTH OF STRIP LINE CIRCULATOR*

The bandwidth of the junction can be obtained by forming the susceptance slope parameter from the admittance function given by Eq. 10.42. The susceptance slope parameter is defined by

$$B' = \frac{\omega_0}{2} \left. \frac{\partial B}{\partial \omega} \right|_{\omega=\omega_0} \tag{10.46}$$

where B is the shunt susceptance of the network. In terms of the original parameters the result is

$$B' = \frac{\pi Y_e}{3\sin\psi}\left[\frac{(k_e R)_{1,1}^2 - 1}{2(k_e R)_{1,1}}\right] \tag{10.47}$$

It is observed from the above equation that the susceptance slope parameter is essentially independent of the magnetic splitting. It is shown in Fig. 10.11 as a function of K/μ.

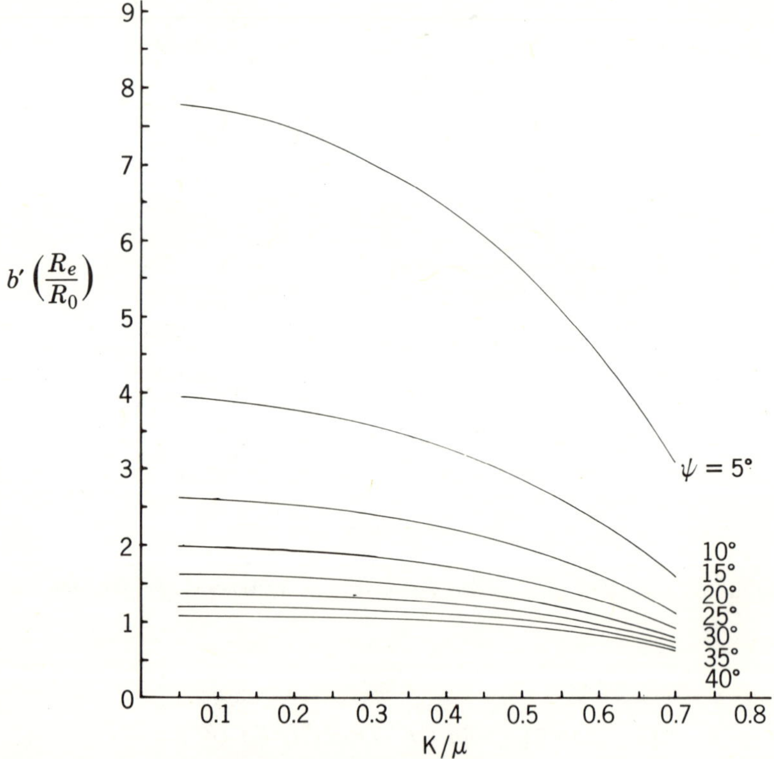

Figure 10.11. *Susceptance slope parameter with $p = 11$.*

The loaded Q-factor is now given by

$$Q_L = \frac{B'}{G} \tag{10.48}$$

where G is the shunt conductance of the network.

In terms of the original variables the result is

$$Q_L = \frac{1}{\sqrt{3}} \left[\frac{(k_e R)^2_{1,1} - 1}{2} \right] \cdot \frac{\mu}{K} \qquad (10.49)$$

The last equation may also be written in terms of the two split frequencies by making use of Eq. 6.53

$$\frac{1}{\sqrt{3}\, Q_L} = \frac{\omega_{+1,1} - \omega_{-1,1}}{\omega_{1,1}} \qquad (10.50)$$

Equation 10.50 has the same form as Eq. 9.28 and the discussion in Section 9.4 applies to it also.

10.6.   *CIRCULATION CONDITION USING $n = \pm 2$ MODES*

It is possible to construct 3-port circulators with field patterns other than the $n = \pm 1$ ones. Figure 10.12a gives the $n = \pm 2$ field patterns when the disks are excited at port 1 and the ferrites are not magnetized. Figure 10.12b shows the same field patterns rotated through 15°. Here, circulation is in the opposite direction to that obtained with the $n = \pm 1$ field patterns.

The first circulation condition is here obtained with

$$J'_2(k_e R) = 0 \qquad (10.51)$$

The first root is given by

$$(k_e R)_{2,1} = 3.05 \qquad (10.52)$$

The second circulation condition is

$$(R_0)_{2,1} = \frac{-\sqrt{3}\,(3.05)\,R_e \sin\psi}{\pi} \cdot \frac{\mu}{K} \qquad (10.53)$$

The fact that $(R_0)_{2,1}$ is negative means that the direction of circulation in this case is in the opposite direction from the mode which uses the $n = \pm 1$ field patterns.

10.7.   *EIGENNETWORKS OF STRIPLINE CIRCULATOR IN TERMS OF IDEAL TRANSFORMERS*

One way to account for the coupling of the stripline transmission lines to the ferrite disks is to introduce ideal transformers. One set of eigennet-

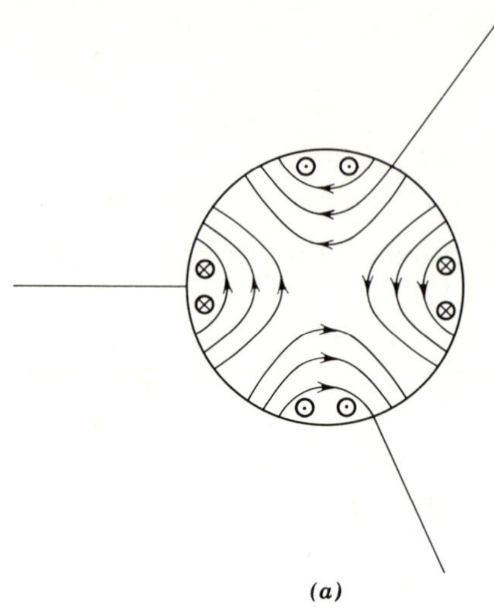

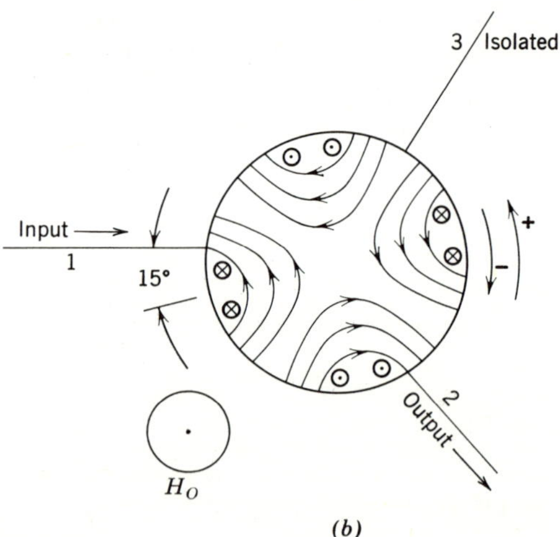

Figure 10.12.   (*a*) *Field patterns for* $n = \pm 2$ *modes* (*Ref.* 6). (*b*) 3-*Port circulator synthesized by* $n = \pm 2$ *modes* (*Ref.* 6).

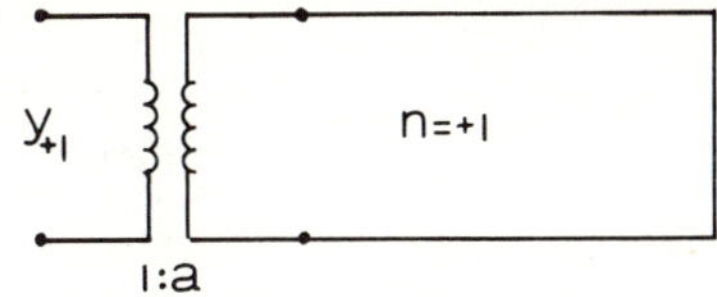

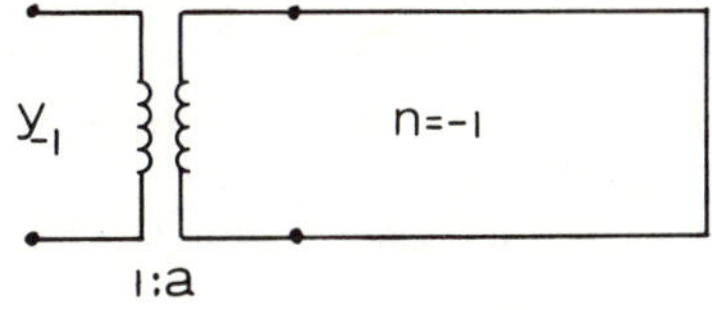

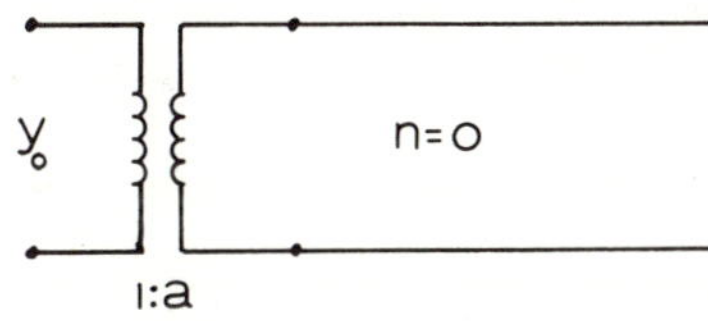

Figure 10.13.   *Eigennetworks of stripline circulators using ideal transformers.*

works which satisfy the admittance eigenvalues is shown in Figure 10.13.
The eigenadmittances for these eigennetworks have the following form:

$$y_n = \frac{jC + DH_{\phi n}/E_{zn}}{A + jBH_{\phi n}/E_{zn}} \tag{10.54}$$

where $ABCD$ are the entries of the ideal transformer, and $H_{\phi n}/E_{zn}$ is the
wave admittance at the radius R of the ferrite disk.
   The result is

$$y_0 = \frac{ja^2 \zeta_e J_0'(k_e R)}{J_0(k_e R)} \tag{10.55}$$

$$y_{+1} = -ja^2 \zeta_e \left[J_1'(k_e R) - \frac{K}{\mu} \frac{J_1(k_e R)}{k_e R} \right] [J_1(k_e R)]^{-1} \tag{10.56}$$

$$y_{-1} = -ja^2 \zeta_e \left[J_1'(k_e R) + \frac{K}{\mu} \frac{J_1(k_e R)}{k_e R} \right] [J_1(k_e R)]^{-1} \tag{10.57}$$

where

$$a^2 = \frac{3\sin\psi}{\pi} \tag{10.58}$$

REFERENCES

1. U. Milano, J. H. Saunders, and L. Davis, Jr., "A Y-Junction Strip-Line Circulator," *IRE Trans. Microwave Theory Tech.*, **MTT-8**, 346–351 (1960).

2. B. A. Auld, "The Synthesis of Symmetrical Waveguide Circulators," *IRE Trans. Microwave Theory Tech.*, **MTT-7**, 238–246 (1959).

3. H. Bosma, "On the Principle of Stripline Circulation," *Proc. IEE Suppl.* **109**, Part B (21), 137–146 (1962).

4. J. B. Davies and P. Cohen, "Theoretical Design of Symmetrical Junction Stripline Circulators," *IEEE Trans. Microwave Theory Tech.* **MTT-11**, 506–512 (1963).

5. H. Bosma, "On Stripline Y-Circulation at UHF," *IEEE Trans. Microwave Theory Tech.*, **MTT-12**, 61–72 (1964).

6. C. E. Fay and R. L. Comstock, "Operation of the Ferrite Junction Circulator," *IEEE Trans. Microwave Theory Tech.* **MTT-13**, 15–27 (1965).

7. V. G. Feoktistov, "Design of a Strip Y-circulator," *Radio Eng. Electron. Phys.*, **13**, (7) 1111–1115, (1968).

8. K. Whiting, "Design Data for UHF Circulators," *IEEE Trans. Microwave Theory Tech.* **MTT-15**, 195–198 (1967).

9. J. Helszajn, "Frequency and Bandwidth of H Plane TEM Junction Circulator," *Proc. IEE*, **117** (7), 1235–1238 (1970).

10. J. K. Richardson, "An Approximate Method of Calculating Z_0 of a Symmetrical Stripline," *IEEE Trans. Microwave Theory Tech.* **MTT-15** 130–131 (1967).

11. J. Helszajn, "Ferrite Ring Stripline Junction Circulator," *Radio Electron. Eng.*, **32** (1), 55–60 (1966).

12. E. N. Skomal, "Theory of Operation of a 3-Port Y-Junction Ferrite Circulator," *IEEE Trans. Microwave Theory Tech.*, **MTT-11**, 117–123 (1963).

13. C. V. Beuhler and A. F. Eikenberg, "A VHF High Power Y-Circulator," *IRE Trans. Microwave Theory Tech.*, **MTT-9**, 569–570 (1961).

14. J. Edrich and R. G. West, "Low-Loss Cryogenic L-Band Circulator," *IEEE Trans. Microwave Theory Tech.*, **MTT-10** (10), 743–745 (1970).

15. Y. Konishi, "A High Power UHF Circulator," *IEEE Trans. Microwave Theory Tech.* **MTT-15** (12), 700–708 (1967).

16. J. Helszajn, "An H-Plane High Power TEM Ferrite Circulator," *Radio Electron. Eng.*, **33**, 257–262 (1967).

17. C. R. Buffler and J. Helszajn, "The Use of Composite Junctions in the Design of High Power Circulator," *IEEE Int. Microwave Symp.*, May 20–22, 1968.

18. J. Edrich and R. G. West, "Very Low Loss L-Band Circulator with Gallium Substituted YIG," *IEEE Trans. Magn.*, **MAG-5**, 481–482 (1969).

19. K. Shirahata and D. Takeromi, "New Design Technique for Stable Wide Band Tunnel Diode Amplifier," *IEEE Trans. Microwave Theory Tech.* **MTT-18** (1), 52–54 (1970).

20. S. J. Salay and H. J. Peppiatt, "Input Impedance Behavior of Stripline Circulator," *IEEE Trans. Microwave Theory Tech.* **MTT-19** (1), 109–110 (1971).

21. B. Maher, D. Mitchell, and A. Reeves, "A 150 Mc/s Circulator," *IEEE Trans. Microwave Theory Tech.* **MTT-14** (1), 42–43 (1966).

22. L. Freiberg, "Lightweight Y-Junction Strip-Line Circulator," *IRE Trans. Microwave Theory Tech.*, **MTT-8**, 672 (1960).

23. D. Masse, "A High Power S-Band Ferrite Circulator," *Onde Elec.*, **44**, 150–154 (1960).

24. S. Yoshida, "Strip-Line Y Circulator," *Proc. IRE*, **48**, 1337–1338 (1960).

25. S. Yoshida, "J-Band Strip-Line Y Circulator," *Proc. IRE*, **48**, 1664 (1960).

26. G. V. Buehler and A. F. Eikenberg, "Stripline Y-Circulators for the 100 to 400 Mc Region," *Proc. IRE*, **49**, 518–519 (1961).

27. J. Clark, "Perturbation Techniques for Miniaturized Coaxial Y-Junction Circulators," *J. Appl. Phys. Suppl.*, **32**, 323–324 (1961).

28. J. Clark and J. Brown, "Miniaturized Temperature Stable Coaxial Y-Junction Circulators," *IRE Trans. Microwave Theory Tech.*, **MTT-9**, 267–269 (1961).

29. A. Clavin, "Higher Order Mode Resonances in Strip-Line Y-Junction Circulators," *IRE Trans. Microwave Theory Tech.*, **MTT-9**, 575, (1961).

30. J. Clark and G. R. Harrison, "Miniaturized Coaxial Ferrite Devices," *Microwave J.*, **5**, 108–118 (1962).

31. E. Dessert, "The Study and Production of Y-Junction Circulators in the Frequency Range 450–1000 Mc/s," *Acta Electron.*, **8**, 175 (1964).

32. O. P. Gandhi and C. Dattatryan, "A L-Band Broad Bandwidth Y-Circulator," *Proc. IEEE*, **111**, 1285–1286 (1964).

33. J. Sokolov, "Below Resonance Operation of the Stripline Y-Circulator," *IEEE Trans. Microwave Theory Tech.* **MTT-12**, 568–569 (1964).

34. B. Hershenov, "X-Band Microstrip Circulator," *Proc. IEEE Lett.*, **54**, 2022–2023 (1966).

35. J. K. Ackers, "Contributions to the Theory Stripline Junction Circulator," *Microwave J.*, **10**, 57–59 (1967).

36. B. Hershenov, "All-Garnet-Substrate Microstrip Circulator" *Proc. IEEE Lett.*, **55**, 696–697 (1967).

37. F. Bosch and Z. Moursi, "On the Determination of Low Field Working Point of Stripline Circulators," *Arch. Elek. Ubertragung*, **22**, 605–607 (1968).

38. D. Masse, "Broadband Microstrip Junction Circulators," *Proc. IEEE*, **56**, 352–353 (1968).

39. V. E. Dunn and A. J. Domenico, "Recent Advances in Microstrip Circulators," *IEEE Trans. Microwave Theory Tech.*, **MTT-16**, 1060–1061, (1968).

40. B. Hershenov and R. L. Ernst, "Miniature Microstrip Circulators using High-Dielectric-Constant Substrates," *RCA Rev.* **30**, 541–543 (1969).

41. R. C. Kumar, "S-Band Three Port Stripline Circulator," *J. Inst. Telecommun. Eng. New Delhi*, **13**, 354–358 (1967).

42. Y. S. Wu and F. J. Rosenbaum, "Wide Band Operation of Microstrip Circulators, *IEEE Mtl. Microwave Conference*, 1973.

CHAPTER ELEVEN

Waveguide Junction Circulator

The three configurations which the 3-port waveguide junction circulator takes are the *H*- or *E*-plane configurations and the turnstile one. The *H*-plane waveguide circulator relies for its operation on either a full-height ferrite post with no variation of the electromagnetic field along the ferrite axis or on ferrite disks against the wide dimensions of the junction with a standing wave of the electromagnetic field along its axis. The *E*-plane circulator consists of ferrite disks against the narrow walls of the waveguide at the junction of three rectangular waveguides instead of the wide one as is the case of the *H*-plane device. The turnstile structure has only historic significance but is still useful in that it incorporates the principles of the junction circulator in a rather simple way.

H-plane waveguide circulators using partial height ferrite structures are those most often used commercially. They have now been described with 1/4, 1/2, 3/4, and 5/4 wavelength axial dimensions. The higher order ones being particularly attractive at millimeter waves. Higher order full height configurations are also of course possible and have been described also.

Commercial devices usually incorporate quarter wave admittance transformers to enhance the bandwidth of the basic configurations. For circulators with 20 dB isolation the bandwidth of the simple structures is in the 2–5% range while for the quarterwave-coupled, values between 10–40% are more usual. Such admittance transformers consist of reduced height waveguide sections in the junction region. In the case of the full ferrite posts it is also possible to employ dielectric rings of suitable radius and dielectric constant. Figure 11.1 illustrates some typical geometries. This chapter deals in detail with the operating frequency of the partial height

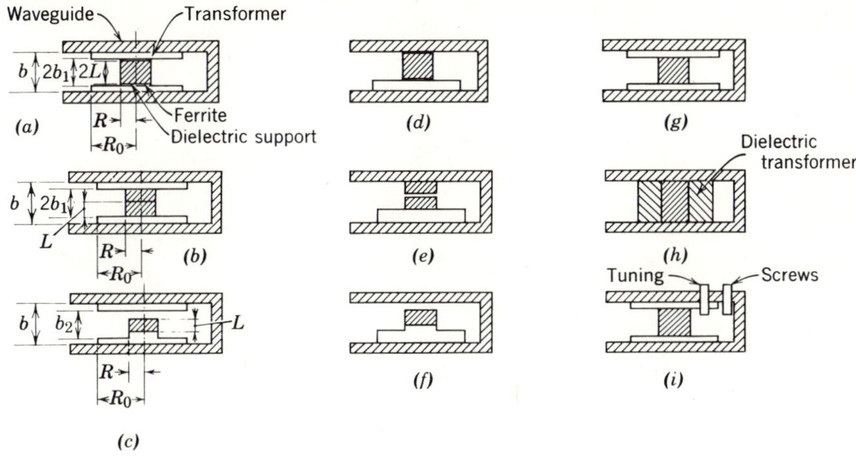

Figure 11.1. (*a*) *Cross section of waveguide circulator using* $\lambda/2$ *resonator with symmetrical radial transformers.* (*b*) *Cross section of waveguide circulator using coupled* $\lambda/4$ *resonators with symmetrical radial transformers.* (*c*) *Cross section of waveguide circulator using single* $\lambda/4$ *resonator with symmetrical radial transformers.* (*d*) *Cross section of waveguide circulator using* $\lambda/2$ *resonator with asymmetrical radial transformers.* (*e*) *Cross section of waveguide circulator using coupled* $\lambda/4$ *resonators with asymmetrical radial transformers.* (*f*) *Cross section of waveguide circulator using single* $\lambda/4$ *resonator with asymmetrical radial transformers.* (*g*) *Cross section of waveguide circulator using ferrite post with symmetrical radial transformer.* (*h*) *Cross section of waveguide circulator using ferrite post with dielectric quarter wave transformer.* (*i*) *Cross section of waveguide circulator showing fine capacitive tuning.*

ferrite configuration and with the theory of the simple full height ferrite postconstruction.

11.1.   *THE TURNSTILE JUNCTION CIRCULATOR*

The original waveguide junction circulator is the turnstile configuration first described by Tor Schaug Patterson. This device is shown in Figure 11.2. It consists of a circular waveguide containing a longitudinally magnetized ferrite section at the junction of three rectangular waveguides. It relies for its operation on the well-known Faraday rotation principle along a magnetized ferrite loaded circular waveguide. To adjust this and other circulators requires a 120° phase difference between the reflection coefficients of the three different ways it is possible to simultaneously excite the three rectangular waveguides that will give identical reflection coefficients at each port. One of these excitations corresponds to in phase fields at the three ports, while the other two require counter-rotating field patterns. These latter field patterns propagate along the magnetized ferrite section with different propagation constants. The former field pattern does

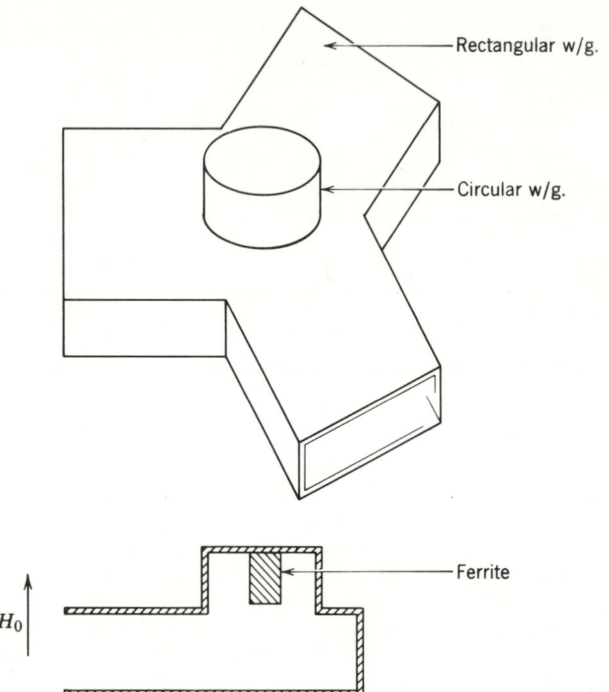

Figure 11.2.   *Schematic of turnstile junction circulator.*

not couple to the circular waveguide but is reflected at the junction of the device. Excitation of a single port establishes all three field patterns within the junction. The proper phase angles of the three reflection coefficients are established by first adjusting the length of the demagnetized ferrite section so that the angles between the in-phase and counter-rotating reflection coefficients are 180°. The phase angles of the counter rotating reflection coefficients are then separated by 120° by magnetizing the ferrite region, thereby producing the ideal circulation phase angles defined in Chapter 5.

11.2   *M-PORT H-PLANE WAVEGUIDE CIRCULATOR*

This section deals with the theory of the *m*-port waveguide circulator depicted in Figure 11.3.

The boundary conditions in the case of this circulator are best stated in terms of the coefficients of the scattering matrix. These latter parameters are known once the eigenvalues of the $\bar{S}$ matrix are obtained. The problem

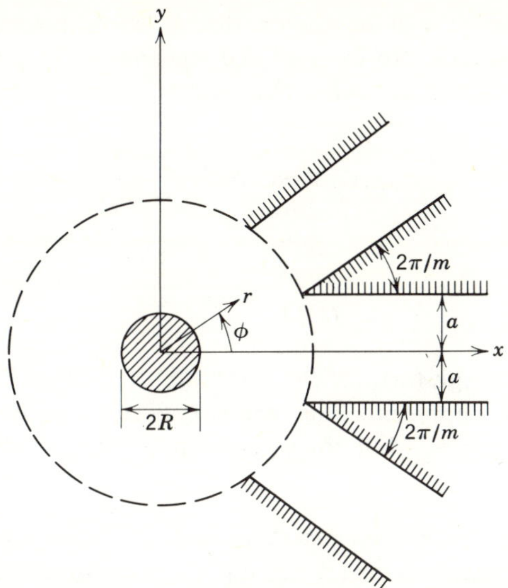

Figure 11.3.   *Configuration of ferrite post in the m-port waveguide junction.*

therefore reduces to that of finding the eigenvalues of the $\bar{S}$ matrix in terms of the e.m. problem. These may be obtained in a simple way by making use of the symmetry properties of the $\bar{S}$ matrix. This will now be illustrated in the case of the three port circulator.

The entries of the $\bar{S}$ matrix at the terminals of the rectangular waveguide are given in the usual way by

$$3\Gamma_{11} = \gamma_0 + \gamma_{+1} + \gamma_{-1} \tag{11.1}$$

$$3\Gamma_{12} = \gamma_0 + \gamma_{+1}e^{j2\pi/3} + \gamma_{-1}e^{-j2\pi/3} \tag{11.2}$$

$$3\Gamma_{13} = \gamma_0 + \gamma_{+1}e^{-j2\pi/3} + \gamma_{-1}e^{j2\pi/3} \tag{11.3}$$

If the eigenvalues can be calculated or measured, the amplitude and phase of the scattering elements can be determined. One way in which they may be obtained one at a time is by determining the reflection coefficient at any port with incident fields proportional to the three eigenvectors of the junction. This technique has been discussed in Chapter 2. It is a general statement which applies to the *m*-port junction also. It is therefore sufficient to satisfy the boundary condition of the electromagnetic problem at one port only.

In what follows it will be shown that if the evanescent modes in the rectangular waveguide are omitted the eigenvalues can be approximately evaluated by using the following equation.*

$$j\tan(\psi_n + \pi/2) = \frac{jC_n + D_n y_n}{A_n + jB_n y_n} \tag{11.4}$$

where ψ_n is defined by

$$\gamma_n = e^{-j2(\psi_n + \pi/2)} \tag{11.5}$$

Equation 11.4 has the form of the eigenadmittances in Chapter 16. In the above $A_n B_n C_n D_n$ represents the region between the ferrite post and the rectangular waveguide, while the eigenadmittance y_n is defined in Eq. 10.54 with $a = 1$.

11.2.1. *Fields Inside Waveguide Junction.* Circular cylindrical and rectangular co-ordinate systems (r, ϕ, z) and (x, y, z), are employed (see Figure 11.3), the z axis coincides with the junction axis of symmetry.

The longitudinal axis of a particular waveguide is taken as $\phi = 0$ and $y = 0$. The ferrite post radius is R and the waveguide width is $2a$.

Because fields throughout the junction are excited by TE_{10} modes at the terminal planes, and as there are no z-dependent boundary conditions, solutions are sought with electric fields purely in the z direction and magnetic fields purely in the x–y plane. The fields are independent of z and are taken to have time dependence $\exp(j\omega t)$. At the edge of the junction they, therefore, have the form of Eqs. 6.69 and 6.70.

$$E_z(k_0 R_0) = \sum_{n=-\infty}^{\infty} [A_n' E_{zfn} + jB_n' H_{\phi fn}] e^{jn\phi} \tag{11.6}$$

$$H_\phi(k_0 R_0) = \sum_{n=-\infty}^{\infty} [jC_n' E_{zfn} + D_n' H_{\phi fn}] e^{jn\phi} \tag{11.7}$$

where

$$R_0 = \frac{a}{\sin(\pi/m)} \tag{11.8}$$

and $A_n' B_n' C_n' D_n'$ are given in Chapter 5.

*The author would like to acknowledge many helpful discussions with Dr. J. B. Davies.

11.2.2.   *Fields Inside Rectangular Waveguide.*   One now considers fields throughout the *m*-port junction due to excitation of one particular eigenvector of the scattering matrix. By treating the eigensolutions individually it is sufficient to consider the fields in one waveguide together with the sector joining it to the junction axis, as shown in Figure 11.3.

A complete expansion for the fields in the waveguide is

$$E_z = \sum_{p=-1}^{\infty} D_p \sin\left[\frac{(1+y/a)p\pi}{2} \right] \exp(j\beta_p x) \tag{11.9}$$

$$H_x = j\eta_0 \left(\frac{\lambda_0}{\lambda_c} \right) \sum_{p=-1}^{p=\infty} p D_p \cos\left[\frac{(1+y/a)p\pi}{2} \right] \exp(j\beta_p x) \tag{11.10}$$

$$H_y = \eta_0 \left(\frac{\lambda_0}{\lambda_g} \right) \sum_{p=-1}^{\infty} D_p \sin\left[\frac{(1+y/a)p\pi}{2} \right] \exp(j\beta_p x) \tag{11.11}$$

where

$$\beta_{+1} = \left[\omega^2 \mu_0 \epsilon_0 - \left(\frac{\pi}{2a} \right)^2 \right]^{\frac{1}{2}} = \frac{2\pi}{\lambda_g} \tag{11.12}$$

$$\beta_{-1} = -\beta_{+1}$$

and

$$\beta_p = j\left[\left(\frac{p\pi}{2a} \right)^2 - \omega^2 \mu_0 \epsilon_0 \right]^{\frac{1}{2}} \quad \text{for } p \geqslant 2 \tag{11.13}$$

λ_g is the guide wavelength of the TE_{10} mode. D_{+1} and D_{-1} are field amplitudes of the inward and outward traveling waves, referred to $x=0$ or equivalently to the terminal plane.

If the above field correspond to excitation of the *n*th eigensolution, then the eigenvalue is

$$\gamma_n = \frac{D_{-1}}{D_{+1}} \tag{11.14}$$

It is assumed that only the dominant waveguide mode exists in the waveguide away from the discontinuity at the junction boundaries.

11.2.3.   *The Waveguide Boundary Condition.*   The next step is to match the tangential field components of one of the cavity modes to the dominant waveguide mode along the arc of radius R. Compared with the exact analysis the adopted approximation is equivalent to ignoring the evanescent modes of the rectangular waveguide arms. No approximation is made on the cylindrical modes, which are matched individually at the surface of the ferrite post and at the joint of the waveguide arm to the central junction.

Equating E_z from Eqs. 11.6 and 11.9, multiplying by $\exp(-jn\phi)$ and integrating from $-\pi/m$ to π/m gives

$$A_n' E_{zfn} + jB_n' H_{\phi fn}$$

$$= \frac{m}{2\pi} \int_{-\pi/m}^{\pi/m} \left[\exp(-jn\phi) \sum_{p=-1}^{p=+1} D_p \sin\left\{ \frac{p\pi}{2}\left|1 + \frac{\sin\phi}{\sin(\pi/m)}\right|\right\} \right.$$

$$\left. \times \exp\left(j\frac{\beta_p a\cos\phi}{\sin(\pi/m)}\right) \right] d\phi \tag{11.15}$$

Similarly matching H_ϕ to $(H_y\cos\phi - H_x\sin\phi)$ gives

$$jC_n' E_{zfn} + D_n' H_{\phi fn}$$

$$= \frac{m}{2\pi\omega\sqrt{(\mu_0\epsilon_0)}} \int_{-\pi/m}^{\pi/m} \left[\exp(-jn\phi) \right.$$

$$\times \sum_{p=-1}^{p=1} D_p \left\{ j\beta_p\cos\phi\sin\left[\frac{p\pi}{2}\left(1 + \frac{\sin\phi}{\sin(\pi/m)}\right)\right] \right.$$

$$\left. + \frac{p\pi}{2a}\sin\phi\cos\left[\frac{p\pi}{2}\left(1 + \frac{\sin\phi}{\sin(\pi/m)}\right)\right] \right\}$$

$$\left. \times \exp\left(j\frac{\beta_p a\cos\phi}{\sin(\pi/m)}\right) \right] d\phi \tag{11.16}$$

Combining the last two equations gives

$$D_{+1}\int_{-\pi/m}^{\pi/m} e^{jM\cos\phi}\left\{\left[j\frac{\lambda_0}{\lambda_g}\cos\phi\cos(N\sin\phi)-\frac{\lambda_0}{\lambda_c}\sin\phi\sin(N\sin\phi)\right.\right.$$

$$\left.\left.-y_n'\cos(N\sin\phi)\right]\cos n\phi\right\}d\phi$$

$$+D_{-1}\int_{-\pi/m}^{\pi/m} e^{-jM\cos\phi}\left\{\left[j\frac{\lambda_0}{\lambda_g}\cos\phi\cos(N\sin\phi)+\frac{\lambda_0}{\lambda_c}\sin\phi\sin(N\sin\phi)\right.\right.$$

$$\left.\left.+y_n'\cos(N\sin\phi)\right]\cos n\phi\right\}d\phi=0 \qquad (11.17)$$

where

$$y_n'=\frac{jC_n'+D_n'y_n}{A_n'+jB_n'y_n} \qquad (11.18)$$

$$M=\frac{2\pi R_0}{\lambda_g} \qquad (11.19)$$

$$N=\frac{2\pi R_0}{\lambda_c} \qquad (11.20)$$

The calculation of the eigenvalues requires only the ratio of the reflected and incident modes, $p=-1$ and $p=+1$. It is also observed that

$$D_{+1}^*=D_{-1} \qquad (11.21)$$

Defining ψ_n by eq. 11.5 the approximation takes the form

$$j\tan(\psi_n+\pi/2)=\frac{jC_n''+D_n''y_n'}{A_n''+jB_n''y_n'} \qquad (11.22)$$

where

$$A_n''=-\int_0^{\pi/m}\cos n\phi\left\{\sin\phi\sin(M\cos\phi)\sin(N\sin\phi)\frac{\lambda_0}{\lambda_c}\right.$$

$$\left.-\cos\phi\cos(M\cos\phi)\cos(N\sin\phi)\frac{\lambda_0}{\lambda_g}\right\}d\phi \qquad (11.23)$$

$$B_n'' = -\int_0^{\pi/m} \cos n\phi \sin(M\cos\phi)\cos(N\sin\phi)d\phi \qquad (11.24)$$

$$C_n'' = -\int_0^{\pi/m} \cos n\phi \left\{ \sin\phi\cos(M\cos\phi)\sin(N\sin\phi)\frac{\lambda_0}{\lambda_c} \right.$$

$$\left. + \cos\phi\sin(M\cos\phi)\cos(N\sin\phi)\frac{\lambda_0}{\lambda_g} \right\} d\phi \qquad (11.25)$$

$$D_n'' = \int_0^{\pi/m} \cos n\phi \cos(M\cos\phi)\cos(N\sin\phi)d\phi \qquad (11.26)$$

The transfer matrix $A_n'' B_n'' C_n'' D_n''$ represents the junction of the radial and rectangular waveguides.

Rewriting eqn. 11.22 in terms of y_n instead of y_n' gives eq. 11.4 where

$$\begin{bmatrix} A_n & jB_n \\ jC_n & D_n \end{bmatrix} = \begin{bmatrix} A_n'' & jB_n'' \\ jC_n'' & D_n'' \end{bmatrix} \times \begin{bmatrix} A_n' & jB_n' \\ jC_n' & D_n' \end{bmatrix} \qquad (11.27)$$

11.2.4. *3-Port Waveguide Circulator.* The approximate theory will now be applied to the 3-port circulator depicted in Figure 11.4.

Writing the eigenvalues in the form given by equation 11.4 gives

$$j\tan\psi_0 = \frac{jC_0 + D_0 y_0}{A_0 + jB_0 y_0} \qquad (11.28)$$

$$j\tan\left(\psi_{+1} + \frac{\pi}{2}\right) = \frac{jC_1 + D_1 y_{+1}}{A_1 + jB_1 y_{+1}} \qquad (11.29)$$

$$j\tan\left(\psi_{-1} + \frac{\pi}{2}\right) = \frac{jC_1 + D_1 y_{-1}}{A_1 + jB_1 y_{-1}} \qquad (11.30)$$

where the form of γ_0 has been modified to allow for the fact that its eigennetwork is an open-circuit line. The eigennetworks defined by the last three equations are depicted in equation 11.5.

For the empty waveguide one also has from Eqs. 11.29 and 11.30

$$\gamma_{+1} = \gamma_{-1} = \gamma_1$$

Hence the eigenvalues γ_{+1} and γ_{-1} are degenerate as they should be. Figures 11.6a and 11.6b illustrate the influence of the dielectric constant and magnetization upon the center frequency of an X-band circulator using such a ferrite post.

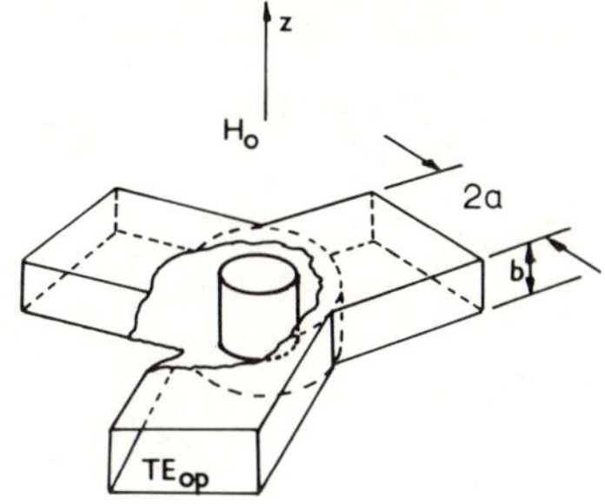

Figure 11.4. *Schematic of 3-port waveguide circulator.*

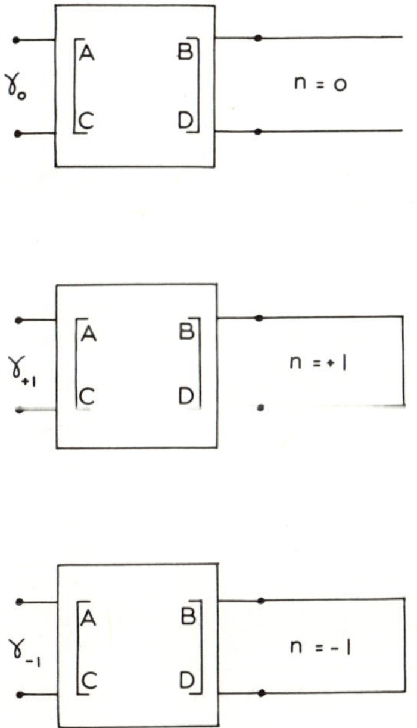

Figure 11.5. *Eigennetworks of simple post wave-guide circulator.*

11.2.5.   *Characteristic Plane of H-Plane Waveguide Circulator.*   Some simplification of the waveguide circulator boundary problem may be obtained by introducing the notion of the characteristic plane defined in Chapter 5. This relation is given graphically in Figure 11.7. The approximate result is obtained by setting Eq. 11.6 to zero with $kR = 1.84$ while the exact result is obtained with the help of Eq. 11.29 or 11.30 with $kR = 1.84$ and $K/\mu = 0$. The latter curve indicates the influence of the transition between the rectangular and radial waveguides.

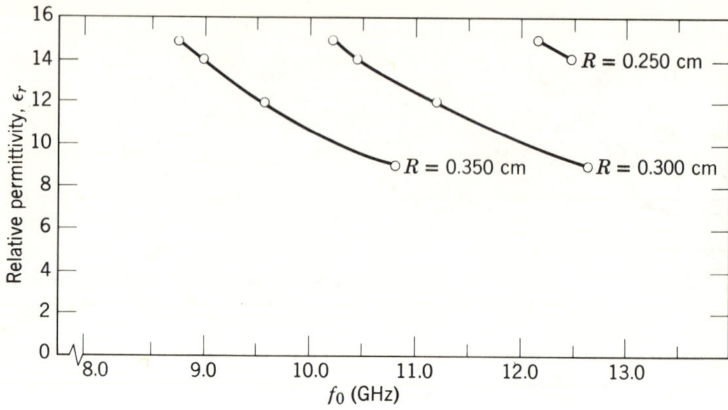

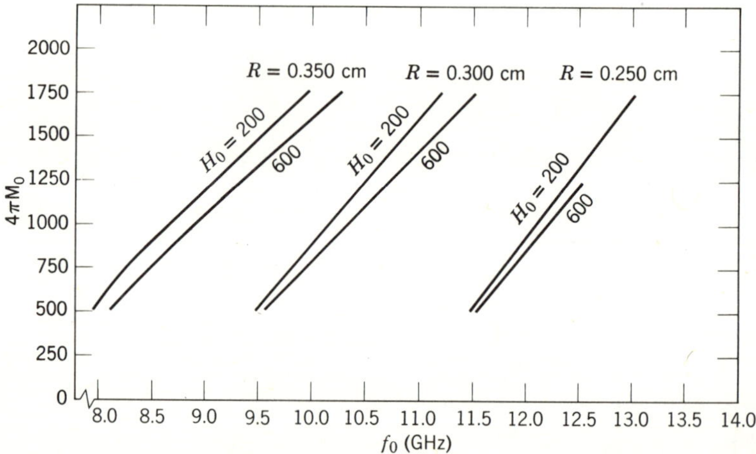

Figure 11.6. (*a*) *Center frequency versus relative dielectric constant* (*Ref.* 6). (*b*) *Center frequency versus saturation magnetization* (*Ref.* 6).

11.3. *H-PLANE JUNCTION CIRCULATOR USING*
PARTIAL HEIGHT FERRITE GEOMETRIES

The most useful of the experimental H-plane waveguide circulators with equal ripple Chebychev characteristics rely for their operations on partial height ferrite geometries between metal transformer plates. An equal ripple Chebyshev frequency response may be obtained with such configurations

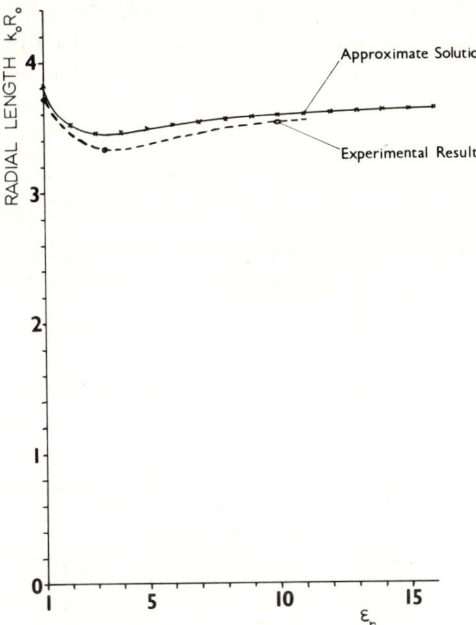

Figure 11.7.   *Characteristic plane of H-plane waveguide circulator.*

by establishing the proper phase angles and admittance levels of the three eigennetworks of the junction.

Circulators using partial height ferrite geometries are related to the original turnstile junction described in Section 11.1. Figures 11.1*a* through *c* depicts three commonly used structures with symmetrical quarterwave admittance transformers, and Figures 11.1*d* through *f* illustrates the same devices with asymmetrical ones, which is a more recent development. All these geometries use circular or triangular cross sections for both the ferrite and transformer plates or a mixture of both. The frequency of these circulators is determined by open dielectric resonances of the hybrid $HE_{11\delta}$ dielectric waveguide mode for which a mode chart is derived in the section. Circulators with $1/4, 1/2, 3/4$, and $5/4$ wavelength axial dimensions have now been described. Geometries with the odd number of quarterwave dimensions require one short-circuited and one open-circuited boundary condition at the end faces, while the ones with an even number of quarter wave dimensions require both end faces to be opencircuited. The higher order ones being particularly suitable at millimeter waves because appropriate ferrite materials are less readily available at these frequencies. Suitable eigennetworks for the junctions illustrated in Figure 11.1d and e are depicted in Figure 11.8 and for that in Figure 11.1f in Figure 11.9.

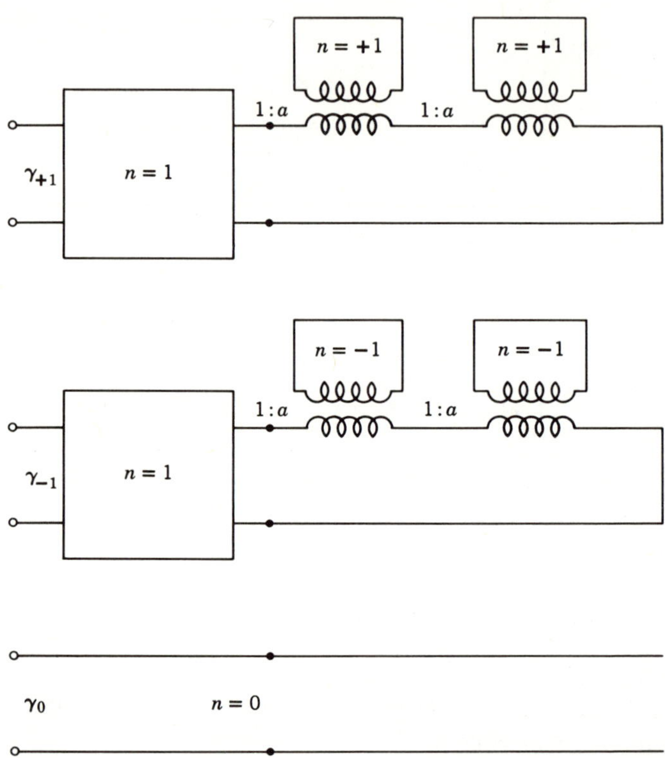

Figure 11.8.   *Eigennetworks of coupled disk H-plane waveguide circulator (Ref. 41).*

11.3.1.   *Images Planes of Partial Height Ferrite Waveguide Circulators.* Figures 11.1a through c give three commonly used partial height ferrite geometries whose path lengths may all be related to quarterwave-long short-circuited transmission lines by the method of images. This is done by introducing symmetry planes into the two configurations in Figures 11.1a and b, and relating them to that in Figure 11.1c. The signs of the field polarities of the two images can be assigned by noting that the tangential electric field is zero at the image plane (plane of symmetry). It is im-

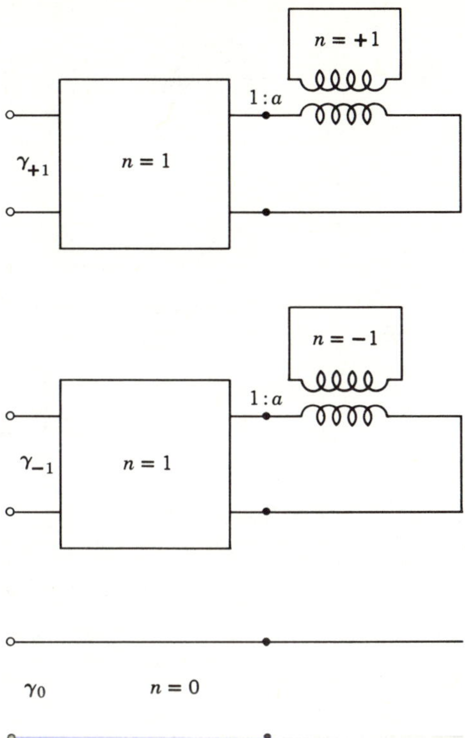

Figure 11.9. *Eigennetworks of single disk H-plane waveguide circulator (Ref. 41).*

mediately observed that it is possible to introduce a conducting plane at the plane of symmetry of the halfwave-long geometry shown in Figure 11.1*a*, since the tangential electric field is zero at that plane. The field polarities are in this case illustrated in Fig. 11.10*a*. The symmetry plane and the field polarities in the case of the geometry in Figure 11.1*b* is indicated in Figure 11.10*b*. The fields are oppositely orientated in the two image disks in Figure 11.10*a* and *b* as required from the Faraday rotation description, since otherwise propagation is in opposite direction along the two disks.

This result indicates that a single set of radial dimensions can be used to describe all three geometries in Figures 11.1*a* through *c* provided the spacing between the ferrite disk is dimensioned with respect to the waveguide wall in Figure 11.1*c* and with respect to the image planes in the cases of Figures 11.1*a* and *b*. The equivalence between these geometries

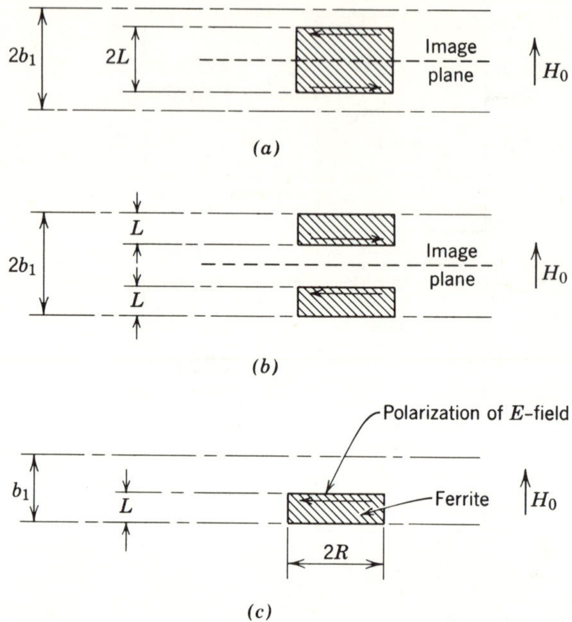

Figure 11.10.   *Image planes of partial height ferrite geometries (Ref. 41).*

also applies to the frequency shift of the open dielectric resonator due to the proximity of the waveguide wall or the image plane as the case may be.

11.3.2.   *Counter-rotating Modes.*   The mode chart for the open dielectric resonator is given in Figure 11.11 in the form $k_0 R$ versus R/L. It applies to the single cylinder of length $2L$ and diameter $2R$, which is open-circuited at both ends. It also applies to the disk L long, open-circuited at one end and short-circuited at the other. The mode chart for the open dielectric resonator is constructed by first forming the field equations of the HE_{11} unbounded dielectric waveguide. Inside the dielectric region the longitudinal components of electric and magnetic fields are given by

$$E_z = a_n^i J_n(kr) e^{jn\theta} \tag{11.31}$$

$$H_z = b_n^i J_n(kr) e^{jn\theta} \tag{11.32}$$

Outside the dielectric region the fields are

$$E_z = a_n^0 H_n^{(1)}(k_1 r) e^{jn\theta} \tag{11.33}$$

$$H_z = b_n^0 H_n^{(1)}(k_1 r) e^{jn\theta} \tag{11.34}$$

Where J_n and H_n are Bessel's functions of the first and third kinds

$$k^2 = \left(\frac{2\pi\sqrt{\epsilon_r}}{\lambda_0}\right)^2 - \left(\frac{2\pi}{\lambda_g}\right)^2 \tag{11.35}$$

$$k_1^2 = \left(\frac{2\pi}{\lambda_0}\right)^2 - \left(\frac{2\pi}{\lambda_g}\right)^2 \tag{11.36}$$

A factor of $e^{j(\omega t - \beta z)}$ is omitted in the field equations. The transverse components of the electric and magnetic fields for the inside and outside of the unbounded dielectric waveguide are given from Maxwell's equations in the usual way. Inside the dielectric region the result is

$$H_\theta = \frac{r}{(kr)^2}\left\{-j\omega\epsilon_r\epsilon_0(kr)a_n^i J_n'(kr) + n\beta b_n^i J_n(kr)\right\}e^{jn\theta} \tag{11.37}$$

$$E_\theta = \frac{r}{(kr)^2}\left\{n\beta a_n^i J_n(kr) + j\omega\mu_0(kr)b_n^i J_n'(kr)\right\}e^{jn\theta} \tag{11.38}$$

where

$$\beta = \frac{2\pi}{\lambda_g} \tag{11.39}$$

The boundary condition here is $s_1 = +1$, which coincides with

$$H_\theta = \frac{\delta}{\delta r}(rE_\theta) = 0 \tag{11.40}$$

This condition is satisfied provided

$$k_0 R = (\epsilon_r)^{-1/2} \cdot n \tag{11.41}$$

$$b_n^i = J_n'(kR) = 0 \tag{11.42}$$

$$a_n^i = J_n(kR) = 0 \tag{11.43}$$

The first solution gives the radius where $H_\theta = 0$ for the hybrid HE_{11} mode. The second and third equations represent TM and TE solutions. Since the radius associated with the hybrid solution is much smaller than that encountered in practice, it is disregarded. For $n = 1$ the TM mode is the dominant one, and its solution is found to be in good agreement with experiment. However, the TM solution adopted here does not exclude the possibility of operating partial height ferrite circulators with hybrid HE_{11}

mode for which $H_\phi \neq 0$. The result is

$$J_n'(kR) = 0 \qquad (11.44)$$

The solution is given by $(kR)\, n,j$ where n,j denotes jth root of the nth order equation. For $n=1$, the result is

$$(kR)_{1,1} = 1.84 = \left[\left(\frac{2\pi \sqrt{\epsilon_r}}{\lambda_0} \right)^2 - \left(\frac{2\pi}{\lambda_g} \right)^2 \right]^{1/2} R \qquad (11.45)$$

For a quarterwave dielectric resonator with an electric wall at one face and a magnetic wall at the other,

$$n\lambda_g = 4L \qquad (11.46)$$

Combining the last two equations gives

$$(k_0 R) = \frac{1}{\sqrt{\epsilon_r}} \left[\left(\frac{\pi R}{2L} \right)^2 + (1.84)^2 \right]^{1/2} \qquad (11.47)$$

where

$$k_0 = \frac{2\pi}{\lambda_0} \qquad (11.48)$$

Figure 11.11 gives the mode chart for such an open dielectric resonator. It shows that the ferrite dimensions are not unique in that there is a wide choice of possible solutions, which will establish the proper frequency with $H_\theta = 0$ at $r = R$. However, each shape will have its own Q-factor. Some experimental results obtained from the literature and in this work are superimposed on the mode chart for E-plane and both type of H-plane junctions suggest that they all use a TM_{11} open dielectric resonator mode. It is useful to note that for large R/L, the resonant frequency of the ferrite disk is primarily dependent upon L, which is determined from the gradient of the curve on the mode chart. However, the introduction of the image plane or waveguide wall in the vicinity of the open dielectric resonator perturbs its resonant frequency. A suitable correction factor for the ferrite length is

$$\Delta L = \frac{1}{\beta} \tan^{-1} \left[\frac{\alpha \epsilon_r}{\beta} \tanh(\alpha s) \right] - L \qquad (11.49)$$

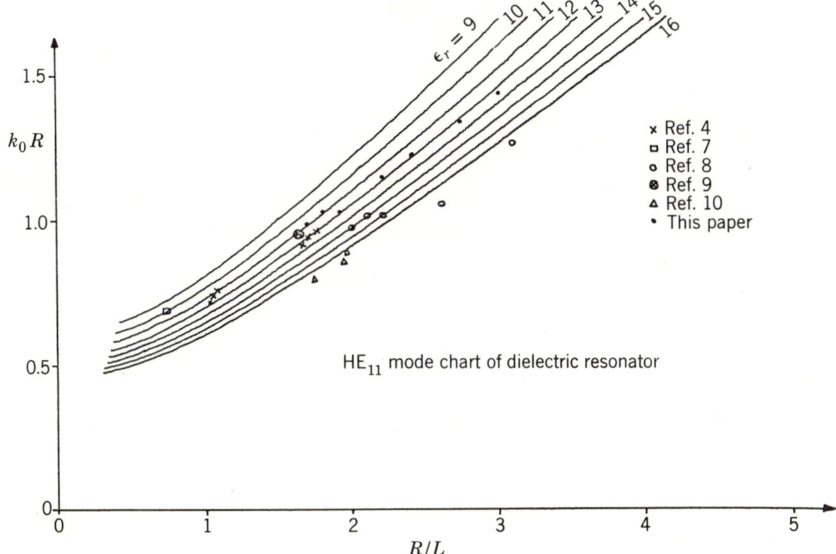

Figure 11.11.   *Mode chart of open dielectric resonator (Ref. 41).*

where $\alpha = \sqrt{k^2 - k_0^2}$. The spacing s is measured from the $0/C$ face of the ferrite to the image plane or waveguide wall depending on the geometry used.

Figure 11.12 depicts the phase angle of the in-phase eigennetwork as a function of L/b_1.

11.4.   *E-PLANE WAVEGUIDE CIRCULATOR*

Another junction circulator is the *E*-plane one in Figure 11.13. This geometry consists of two ferrite disks placed against the narrow walls at the plane of symmetry of a symmetrical 3-port *E*-plane junction. One advantage of the geometry over the more usual *H*-plane one is its ability to handle higher peak powers than is in the case for the latter version. The ferrite dimensions are here also determined by an open dielectric quarterwave-long resonance with open-circuited and short-circuited boundary conditions at the end faces of the resonator. Hence the mode chart for the partial height *H*-plane geometry applies to the *E*-plane circulator also, and may be used for design purposes. The relation between the frequency and the spacing between the two ferrite disks for parametric values of ferrite thickness is shown in Figure 11.14. It is seen from this illustration that two modes are obtained within the X-band frequency range. The frequency of

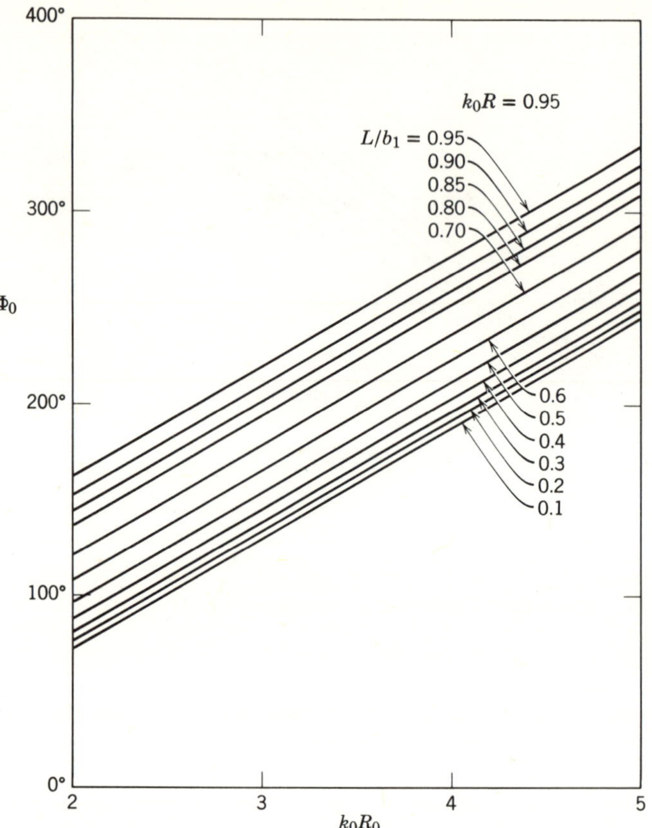

Figure 11.12.   *Mode chart of in-phase eigennetwork for partial height ferrite circulator (Ref.* 41).

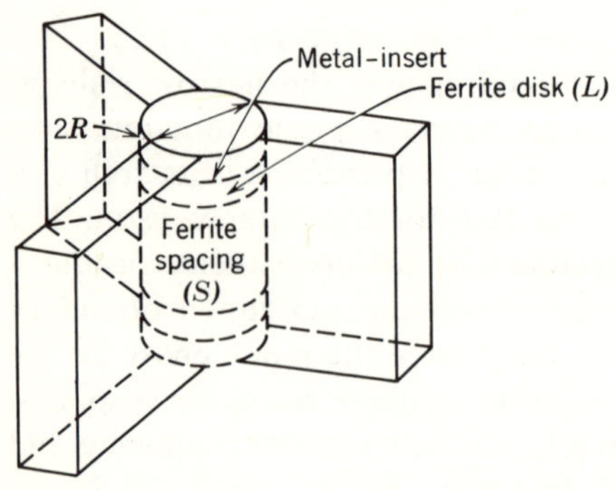

Figure 11.13.   *Cross section of E-plane waveguide circulator.*

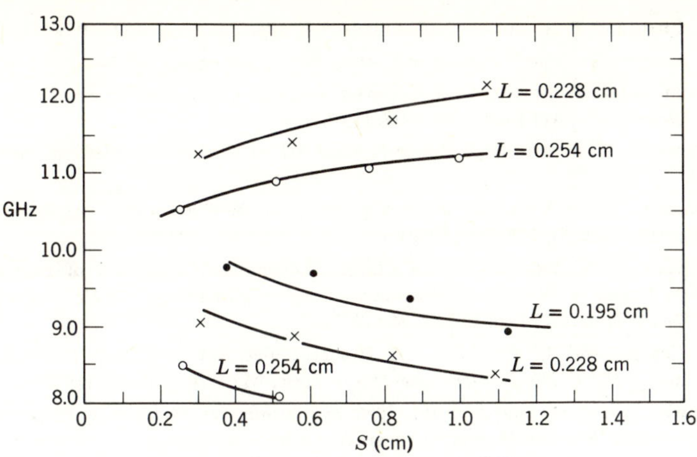

Figure 11.14.   *Mode chart for E-plane circulator (Ref.* 33).

one mode decreases as the spacing between the two ferrite disks increases. The frequency of the other increases with the spacing. Both modes become independent of ferrite spacing when it is large. It is also seen that the frequency of each mode is proportional to ferrite thickness.

REFERENCES

1. J. B. Davies, "An Analysis of the *m*-Port Symmetrical *H*-Plane Waveguide Junction with Central Ferrite Post," *IRE Trans. Microwave Theory Tech.*, **MTT-10**, 596–604 (1962).

2. J. B. Davies, "Theoretical Design of Wideband Waveguide Circulators," *Electron. Lett.*, **1**, 60–61 (1965).

3. L. E. Davis, "Theoretical Design of Static and Latching Ferrite 3-Port and 4-Port Symmetrical Waveguide Circulators," *Digest of 1966 Int. Microwave Symp.*, 281–285.

4. C. G. Parsonson, S. R. Longley, and J. B. Davies, "The Theoretical Design of Broadband 3-Port Waveguide Circulators (Corresp.), *IEEE Trans. Microwave Theory Tech.*, **MTT-16**, 256–258 (1968).

5. L. E. Davis, "Central-Pin Tolerances in Broadband 3-Port Waveguide Circulators," *Electron. Lett.*, **4** (15) (1968).

6. J. B. Castillo and L. E. Davis, "Computer-Aided Design of 3-Port Waveguide Junction Circulators," *IEEE Trans. Microwave Theory Tech.*, **MTT-18**, 25–34 (1970).

7. J. B. Castillo and L. E. Davis, "Identification of Spurious Modes in Circulators," *IEEE Trans. Microwave Theory Tech.*, **MTT-19**, 112–113 (1972).

8. B. Owen, "The Identification of Modal Resonances in Ferrite Loaded Waveguide Y-Junctions and their Adjustment for Circulation," *Bell Sys. Tech. J.*, **51** (3), 595–627 (1972).

9. H. N. Chait and T. R. Curry, "Y-Circulator," *J. Appl. Phys.*, **30**, 152 (1959).

10. H. J. Butterweck, "The Y-Circulator," *Arch. Elek. Ubertragung*, **17** (4), 163–176 (1963).

11. B. Owen and C. E. Barnes, "The Compact Turnstile Circulator," *IEEE Trans. Microwave Theory Tech.*, **MTT-18**, 1096–1100 (1970).

12. E. Wantuch, "Extremely High Power X Band 3 Port Circulator," *Electron. Lett.*, **1**, 45 (1965).

13. J. Helszajn, "Three Resonant Mode Adjustment of the Waveguide Circulator," *Radio Electron. Eng.*, **42** (5), 213–216 (1972).

14. R. Roveda, C. Camillo, and G. Cattarin, "Dissipative Parameters in Ferrites and Insertion Losses in Waveguide Y-Circulators Below Resonance," *IEEE Trans. Microwave Theory Tech.*, **MTT-20** (2), 89–96 (1972).

15. J. B. Castillo and L. E. Davis, "A Higher Order Approximation for Waveguide Circulators," *IEEE Trans. Microwave Theory Tech.*, **MTT-20**, 410–412 (1972).

16. H. Fowler, paper presented at the 1956 Symposium on Microwave Properties and Applications of Ferrites, Harvard University, Cambridge, Mass.

17. F. M. Aitken and R. McLean, "Some Properties of the Waveguide Y Circulator," *Proc. IEE*, **110** (2), 256–260 (1963).

18. J. H. Collins and A. R. Watters, "A Field Map for the Waveguide Y-Junction Circulator," *Electron. Eng. London*, **35**, 540–543 (1963).

19. E. Wantuch and D. Lepore, "An Extremely High Power X-Band Three Port Circulator," *J. Appl. Phys.*, **37**, 1079–1080 (1966).

20. K. Chandra, R. Parshad, and V. K. Argawal, "Broadbanding of Waveguide Junction Circulator," *J. Inst. Telecommun. Eng. New Delhi*, **13**, 145–148 (1967).

21. A. P. Darwent, "Symmetry in High Power Circulator for 35 GHz," *IEEE Trans. Microwave Theory Tech.*, **MTT-15**, 120 (1967).

22. A. K. Stolyarov and I. P. Tyukov, "On the Analysis of Waveguide Y-Circulators in Terms of their Extrinsic Characteristics," *Radio Eng. Electron. Phys.*, **12**, 346–348 (1967).

23. A. K. Stolyarov and I. P. Tyukov, "Electrical Parameters of Waveguide Y-Circulators," *Radio Eng. Electron. Phys. (USSR)*, **12**, 1954–1962 (1967).

24. I. P. Tyukov, "Theory of Symmetrical Waveguide Circulators," *Radio Eng. Electron. Phys. (USSR)*, **12**, 244–253 (1967).

25. I. P. Tyukov, "Analysis of Symmetrical Y-Junctions Filled with Ferrite Dielectrics," *Radio Eng. Electron. Phys. (USSR)*, **12**, 341–345 (1967).

26. S. Yoshida, "E-Type T Circulator," *Proc. IRE*, **47**, 208 (1959).

27. J. Q. Owen, "E-Plane T Circulator," Unpublished memorandum, Raytheon Co., Waltham, Mass. (1962).

28. L. E. Davis and S. R. Longley, "E-Plane Three-Port X-Band Waveguide Circulators," *IEEE Trans. Microwave Theory Tech.*, **MTT-11**, 443 (1963).

29. G. Buchta, "Miniaturized Broadband E-Tee Circulator at X-Band," *Proc. IEEE*, **54**, 1607 (1966).

30. J. W. McGrown and W. H. Wright, Jr., "A High Power, *Y*-Junction, *E*-Plane Circulator," *G-MTT Int. Microwave Symp. Digest*, 85 (1967).

31. C. E. Fay and R. L. Comstock, "Operation of the Ferrite Junction Circulator," *IEEE Trans. Microwave Theory Tech.*, **MTT-13**, 15 (1965).

32. M. Omori, "An Improved E-Plane Waveguide Circulator," *G-MTT Int. Microwave Symp. Digest*, 228 (1968).

33. J. Helszajn and M. McDermott, "Mode Chart for E-Plane Circulators," *IEEE Trans. Microwave Theory Tech.*, **MTT-20**, 187–188 (1972).

34. S. Yoshida, "X-Circulator," *Proc. Inst. Radio Eng.*, **47**, 1150 (1969).

35. D. H. Landry, "A Single Junction Four-Port Coaxial Circulator," 1963 Wescon, Part 5, Session 4.2.

36. L. E. Davis, M. D. Coleman, and J. J. Cotter, "Four-Port Crossed Waveguide Junction Circulators" (corresp.), *IEEE Trans. on Microwave Theory Tech.*, **MTT-12**, 43–47 (1964).

37. C. E. Fay and W. A. Dean, "The Four-Port Single Junction Circulator in Stripline," *Int. Symp. Digest PG-MTT IEEE* (1966).

38. S. R. Longley, "Experimental 4-Port E-Plane Junction Circulator" (corresp.), *IEEE Trans. Microwave Theory Tech.*, **MTT-15**, 378–380 (1967).

39. J. Helszajn and C. R. Buffler, "Adjustment of the 4-Port Single-Junction Circulator," *Radio Electron. Eng.*, **35**, 357–360 (1968).

40. A. G. Bogdanov, "Design of Waveguide X-Circulators," *Radio Eng. Electron. Phys.*, **14** (1969).

41. J. Helszajn and F. C. F. Tan, "Radial H-Plane Waveguide Junction Circulators," *IEEE Trans. Microwave Theory Tech.*, March 1975.

CHAPTER TWELVE

Circuit Theory of Junction Circulator

It is well-known experimentally that the reactance function of a junction circulator approximates that of a shunt network at a suitable pair of reference terminals. It is usually located at the edge of the junction. In the vicinity of the operating frequency, the junction can then always be characterized by a lumped element equivalent circuit from this reactance function. Such a 1-port approximation to the junction circulator has been derived in Chapter 5. The notion of the loaded Q factor can also be introduced in connection with this equivalent circuit. The bandwidth is defined once the loaded Q factor is known. The purpose of this chapter is to derive general circulator equations in terms of these circuit elements, which lead to a universal gyrator equation for circulators which rely on $\pm 30°$ splitting between the counter-rotating modes. This equation relates the susceptance slope parameter and the difference between the split frequencies to the gyrator admittance of the circulator.

12.1. *NETWORK EQUIVALENT CIRCUIT FROM REACTANCE FUNCTION*

A useful formula for the loaded Q_L of a resonant circuit in terms of its admittance function is

$$Q_L = \frac{\omega_0}{2G} \frac{\partial B}{\partial \omega}\bigg|_{\omega_0} \tag{12.1}$$

where B is the shunt susceptance and G is the shunt conductance. This representation is particularly useful when working with a Smith chart. It is worthwhile noticing that Q_L is determined by the susceptance slope parameter and the shunt conductance of the circuit.

The equivalent shunt elements are related to the susceptance slope parameter by

$$B' = \frac{\omega_0}{2} \left.\frac{\partial B}{\partial \omega}\right|_{\omega_0} = \omega_0 C = \frac{1}{\omega_0 L} = \sqrt{\frac{C}{L}} \tag{12.2}$$

It has been noted experimentally in a number of devices that for a given geometry the susceptance slope remains essentially constant over fairly large variations in shunt conductance. The conductance can be adjusted by varying the ferrite parameters or the magnetic field. The admittance plot of a ring circulator is indicated in Figure 13.2. Its equivalent circuit is depicted in Figure 12.1.

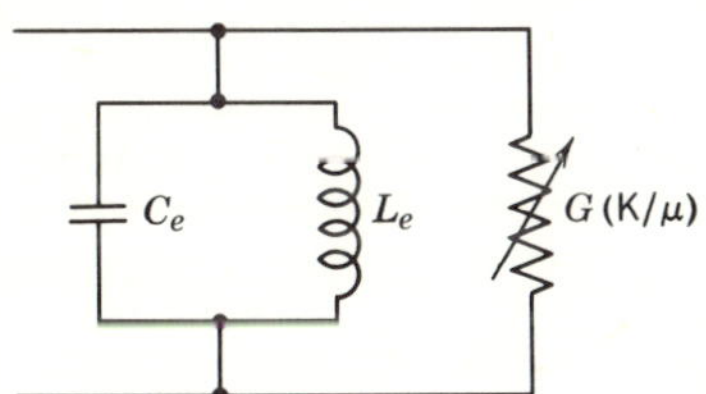

Figure 12.1.   *Equivalent circuit at reference plane of junction.*

12.2.   NORMAL MODE DESCRIPTION OF SHUNT RESONATOR

In this section, the normal mode description of a shunt resonator will be developed. According to Louisell, this can be done by taking a linear combination of the voltage current relations of the LC circuit that will decouple them. From Figure 12.2, one has

$$\frac{di}{dt} = \left(\frac{-1}{L}\right)V \tag{12.3}$$

$$\frac{dV}{dt} = \left(\frac{1}{C}\right)i \tag{12.4}$$

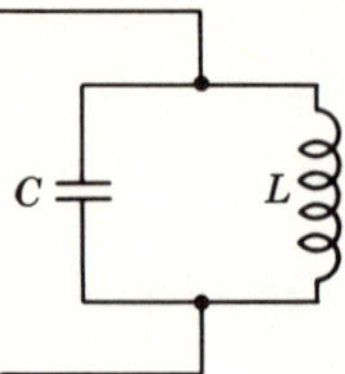

Figure 12.2. *Shunt resonator.*

If Eq. 12.4 is multiplied by an arbitrary constant Y and added to Eq. 12.3, the result is

$$\frac{d}{dt}(i+YV)=\frac{Y}{C}\left(i-\frac{C}{LY}V\right) \tag{12.5}$$

This equation is satisfied by making $Y=\pm j\sqrt{C/L}$. Equation 12.5 now becomes

$$\left(\frac{d}{dt}-j\omega\right)\alpha=0 \tag{12.6}$$

$$\left(\frac{d}{dt}+j\omega\right)\alpha^*=0 \tag{12.7}$$

where

$$\alpha=\frac{\sqrt{L}}{2}(i+j\omega CV) \tag{12.8}$$

and

$$\alpha^*=\frac{\sqrt{L}}{2}(i-j\omega CV) \tag{12.9}$$

The solution of the normal mode equations are

$$\alpha(t)=\alpha(0)e^{j\omega t} \tag{12.10}$$

$$\alpha^*(t)=\alpha^*(0)e^{-j\omega t} \tag{12.11}$$

The sum of the normal modes gives the total energy in the system

$$E=\left[\frac{CV^2(t)+Li^2(t)}{2}\right]=|\alpha(t)|^2+|\alpha^*(t)|^2 \tag{12.12}$$

From Eqs. 12.10 and 12.11 it is noted that α and α^* are a pair of normal degenerate modes rotating in opposite directions. It is also noted that each mode is circularly polarized and has half the total energy of the system. This result is satisfying because it phenomenologically describes the uncoupled disks, of the classical junction. In terms of the normal mode representation, the equivalent circuit is now given in Figure 12.3.

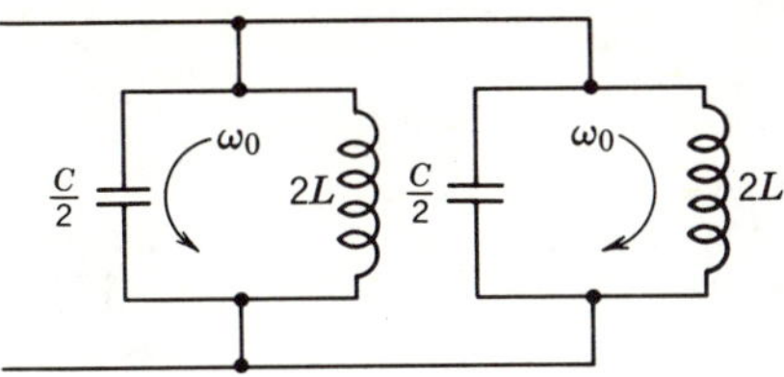

Figure 12.3.   *Normal mode of form shunt resonator.*

When the junction is magnetized, the degeneracy between the modes is removed and their resonant frequencies split. This situation can be represented by replacing the shunt inductances in Figure 12.3 by new inductances $L_{\pm}$. It is also possible to write $L_{\pm} = L(\mu \mp K)$ because the normal modes are circularly polarized. It will be found that this representation gives answers consistent with the known results.

The equivalent circuit of the circulator is now depicted in Figure 12.4 where the shunt conductance is a function of K/μ, the splitting factor.

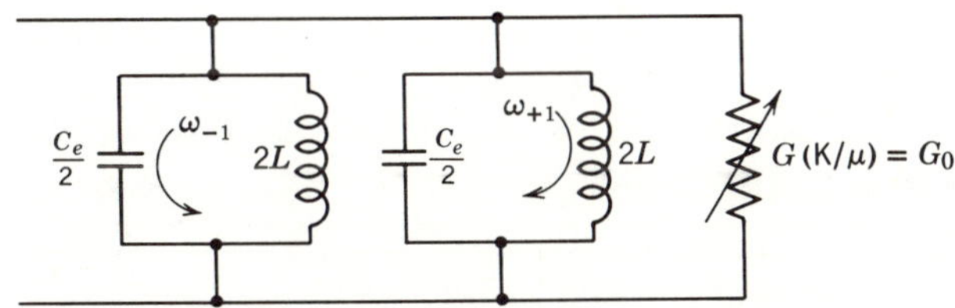

Figure 12.4.   *Normal mode form of junction circulator at reference plane of junction.*

12.3.   *CIRCULATION ADJUSTMENT*

The total admittance of the split modes at the plane of the ferrite disks can now be obtained from Figure 12.4:

$$Y_{in} = Y_+ + Y_- = G\left(\frac{K}{\mu}\right)$$

$$+ j\left[\left(\frac{\omega C_e}{2} - \frac{1}{2\omega L_+}\right) + \left(\frac{\omega C_e}{2} - \frac{1}{2\omega L_-}\right)\right] \qquad (12.13)$$

The two circulation conditions will now be obtained by matching the input admittance Y_{in}.

One of the two circulation conditions is satisfied by adjusting the operating frequency to lie between the two split resonant frequencies ω_{+1} and ω_{-1}. This condition is met by setting the imaginary parts of Y_{in} to zero:

$$\left(\frac{\omega C_e}{2} - \frac{1}{2\omega L_+}\right) + \left(\frac{\omega C_e}{2} - \frac{1}{2\omega L_-}\right) = 0 \tag{12.14}$$

The circulation frequency is then given by

$$\omega_0^2 L_e C_e = 1 \tag{12.15}$$

where

$$L_e = \frac{2L_+ L_-}{L_+ + L_-} \tag{12.16}$$

Writing $L_\pm = L(\mu \mp K)$ gives

$$L_e = \left(\frac{\mu^2 - K^2}{\mu}\right) L = \mu_e L \tag{12.17}$$

It is also possible to write

$$C_e = \epsilon_r C \tag{12.18}$$

where ϵ_r is the relative dielectric constant of the ferrite material.

The variation of μ_e as a function of p and σ is shown in Figure 19.2. The junction can in general be biased either above or below ferrimagnetic resonance. It is further noted that

$$\omega_0 = \frac{\omega_{-1} + \omega_{+1}}{2} \tag{12.19}$$

Here

$$\omega_{\pm 1}^2 L_\pm C_e = 1 \tag{12.20}$$

The second circulation condition requires that the standing wave pattern be rotated until port 3 is decoupled. From symmetry, this condition is met by making the phase angles between the split admittances $\pm 30°$. The result is

$$\frac{\omega_0 C_e}{2} - \frac{1}{2\omega_0 L_-} = \frac{Y_0}{2\sqrt{3}} \tag{12.21}$$

$$\frac{\omega_0 C_e}{2} - \frac{1}{2\omega_0 L_+} = -\frac{Y_0}{2\sqrt{3}} \tag{12.22}$$

Subtracting Eq. 12.22 from Eq. 12.21 and rearranging gives

$$Y_0 = \sqrt{3}\,(\omega_0 C_e)\left(\frac{\omega_{+1} - \omega_{-1}}{\omega_0}\right) \tag{12.23}$$

The quantity $\omega_0 C_e$ is the susceptance slope parameter B' of the shunt network. Writing Eq. 12.23 in terms of the susceptance slope parameter gives the following universal gyrator equation for circulators that rely on $30°$ splitting between the degenerate eigennetworks:

$$Y_c = \sqrt{3}\,B'\left(\frac{\omega_{+1} - \omega_{-1}}{\omega_0}\right) \tag{12.24}$$

Equation 12.24 also leads to the following standard form when $G \neq Y_0$

$$\frac{1}{\sqrt{3}\,Q_L} = \left(\frac{\omega_{+1} - \omega_{-1}}{\omega_0}\right) \tag{12.25}$$

where Q_L is the loaded Q factor of the junction defined in equation 12.1. This last equation is approximately given with the help of Eq. 12.20 by

$$Q_L = \frac{1}{\sqrt{3}}\,\frac{\mu}{K} \tag{12.26}$$

The loaded Q-factor is equivalent to the normalized, susceptance slope parameter b' when the gyrator conductance G is equal to the characteristic admittance Y_0. Under this condition either notation is sometimes used.

12.4.   BANDWIDTH OF DIRECTLY COUPLED CIRCULATOR

The maximum VSWR r and the normalized bandwidth $2\delta_0$ of the 1-port network in Figure 12.1 is related to the susceptance slope parameter. The relation between the VSWR and reflection coefficient of the network is

$$\gamma = \frac{r-1}{r+1} \tag{12.27}$$

where

$$\gamma = \left|\frac{Y_0 - Y_{in}}{Y_0 + Y_{in}}\right| \tag{12.28}$$

Combining the above equations with Eq. 12.13 with $G(K/\mu) = Y_0$ gives the following relation

$$b' = \frac{r-1}{2\delta_0\sqrt{r}} \tag{12.29}$$

where

$$b' = \frac{B'}{Y_0} = \frac{\omega_0 C_e}{Y_0} \tag{12.30}$$

The first of these equations determines the susceptance slope parameter in terms of the junction specification while the second sets the physical variables of the junction.

The magnetic splitting necessary to satisfy the circulation condition is now obtained by combining Eq. 12.29 with Eq. 12.26 with Q_L replaced by b' which applies when $G = Y_0$.

$$\frac{1}{\sqrt{3}}\frac{\mu}{K} = \frac{r-1}{2\delta_0\sqrt{r}} \tag{12.31}$$

This equation gives the magnetic splitting required to satisfy the admittance equation once the susceptance slope parameter or the loaded Q factor of the device is established by the specification. It is also observed in passing that Eq. 12.31 allows K/μ to be obtained from the measurement of Q_L.

12.5.   *INPUT ADMITTANCE OF JUNCTION CIRCULATORS IN TERMS OF SUSCEPTANCE SLOPE PARAMETER*

A universal definition for the input admittance of junction circulators will now be given that applies to circulators which rely on 30° splitting between the degenerate eigennetworks. It is obtained by observing that the total normalized complex admittance is consistent with the definition of the loaded Q factor in Eq. 12.25 provided it has the following general form:

$$y_{in} = \sqrt{3}\,b'\left(\frac{\omega_{+1} - \omega_{-1}}{\omega_0}\right) + j2b'\left(\frac{\omega - \omega_0}{\omega_0}\right) \tag{12.32}$$

which may be compared with Eq. 12.13. A distributed equivalent circuit which satisfies this equation is indicated in Figure 12.5. A similar result is obtained by starting with the admittance eigenvalues of the circulator.

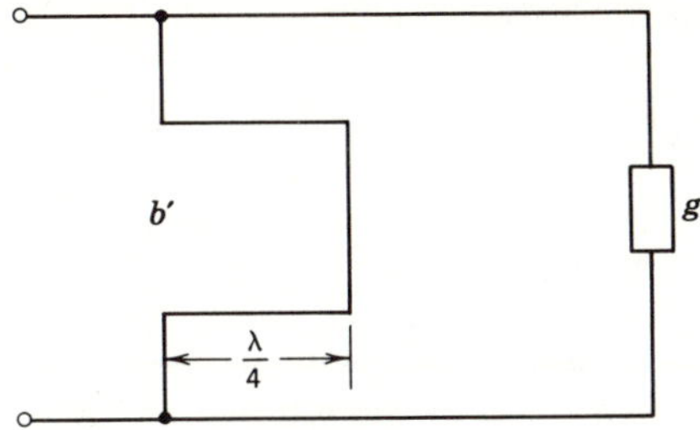

Figure 12.5.   Equivalent circuit of junction circulator in terms of g and b'.

For most junctions the difference between the frequencies ω_{+1} and ω_{-1} is related to the magnetic splitting K/μ by

$$\frac{\omega_{+1} - \omega_{-1}}{\omega_0} \approx \frac{K}{\mu} \tag{12.33}$$

Hence,

$$y_{in} \approx \sqrt{3}\, b' \frac{K}{\mu} + j2b'\left(\frac{\omega - \omega_0}{\omega_0}\right) \tag{12.34}$$

which is the form of equations 9.23 and 10.12.

The problem of the junction circulator, therefore, can be reduced to that of calculating the susceptance slope parameter and operating frequency of the junction. This will now be demonstrated for various devices. The calculation can be further simplified by relating the susceptance slope parameter of the overall network to that of the constituent resonator.

12.6.   RELATION BETWEEN SUSCEPTANCE SLOPE PARAMETER OF CONSTITUENT NETWORK AND CIRCULATOR

The susceptance slope parameter of circulators can be found by assuming that the unmagnetized junction consists of the connection of three constituent resonators. Each resonator being mutually coupled to the others. This is shown in Figure 12.6. This equivalent network stems directly from the impedance matrix of the junction. In the case of the stripline circulator, the constituent resonator consists of a halfwave-long open-circuited structure fed at one end by a TEM line. This is illustrated in Figure 12.7. Figures 12.8 and 12.9 indicate the constituent resonators for the *H*- and *E*-plane waveguide circulators. Because the three resonators are not linearly independent, a relation exists between them and the eigenresonator that defines the input resonator of the circulator. This relation is

$$b' = \frac{2}{3} b_0'$$

(12.35)

where b_0' is the normalized susceptance slope parameter of the constituent resonator.

This last relation applies to that part of the circuit that is contained within the junction region. For instance, in the case of the lumped element circulator, there is no mutual coupling between the lumped capacitances.

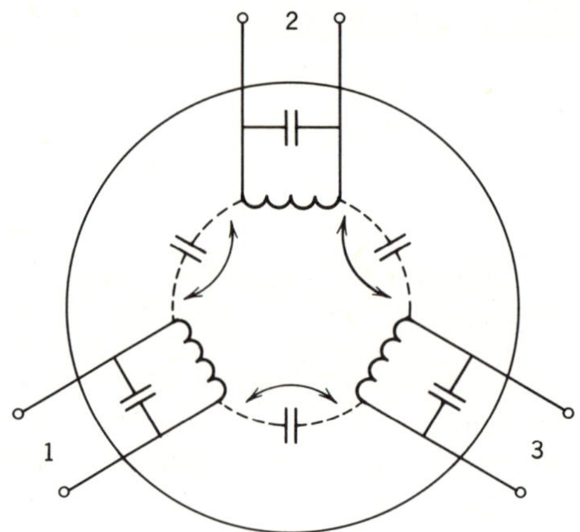

Figure 12.6. *Circulator model using three constituent resonators mutually coupled one to another (Ref. 4).*

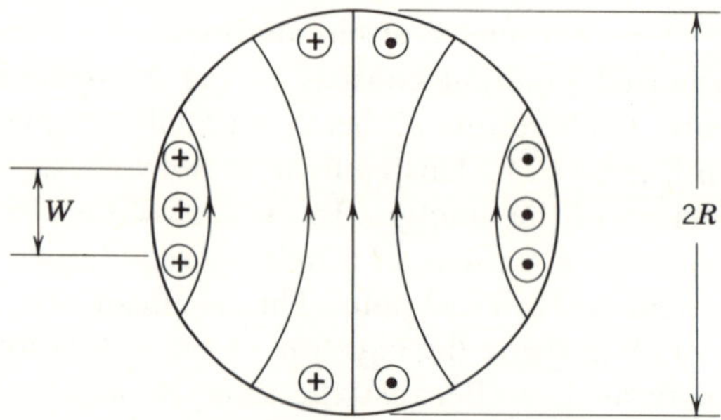

Figure 12.7. *Constituent resonator for stripline circulator (Ref. 4).*

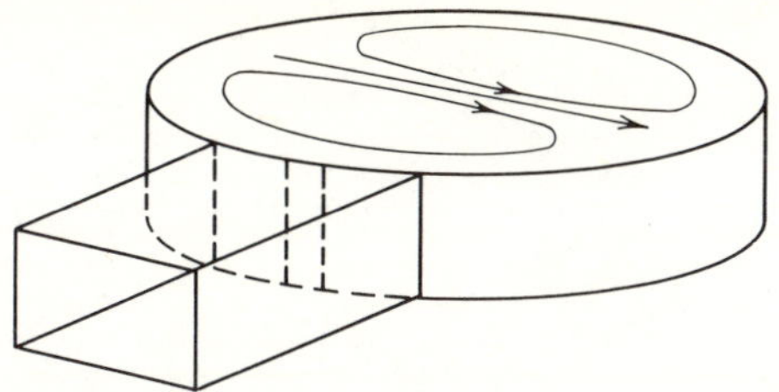

Figure 12.8.   Constituent resonator for H-plane waveguide circulator (Ref. 4).

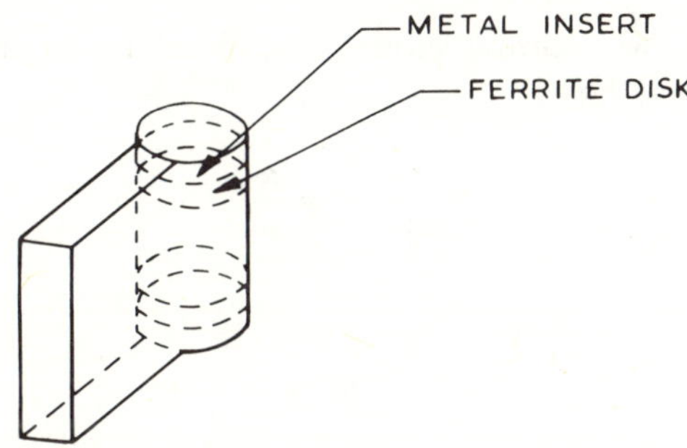

Figure 12.9.   Constituent resonator for E-plane waveguide circulator.

12.7.   *SUSCEPTANCE SLOPE PARAMETER OF LUMPED ELEMENT CIRCULATOR*

For the lumped element circulator the constituent resonator consists of a lumped element resonator. Hence, only the inductance is mutually coupled to the other two resonators, and so Eq. 12.35 applies to that part of the circuit only. The constituent resonator is shown in Figure 12.10.

In terms of the inductance the susceptance slope parameter of the constituent resonator is

$$b'_0 = \frac{R_0}{\omega L_0 \mu_e} \tag{12.36}$$

The susceptance slope parameter of the overall network is, therefore,

$$b' = \frac{2}{3} b'_0 = \frac{R_0}{\omega L} \tag{12.37}$$

where

$$L = \frac{3}{2} L_0 \mu_e \tag{12.38}$$

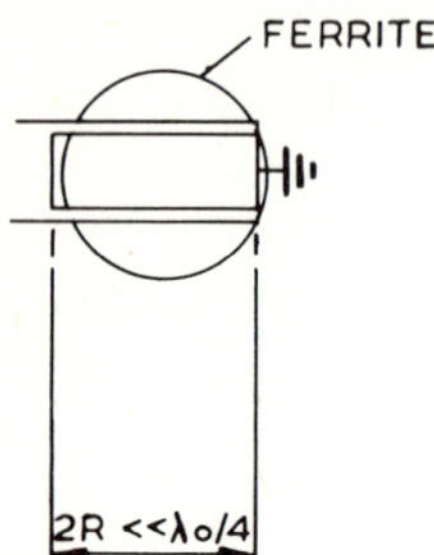

Figure 12.10.   *Constituent resonator for lumped element circulator.*

Assuming that the two normal modes are circularly polarized over the whole ferrite disk, one has

$$\frac{\omega_{+1}-\omega_{-1}}{\omega_0}=\frac{K}{\mu}$$
(12.39)

hence,

$$g=\frac{\sqrt{3}\,R_0}{\omega L}\frac{K}{\mu}$$
(12.40)

Equations 12.37 and 12.40 coincide with Eqs. 9.25 and 9.14 in Chapter 9.

12.8.   *SUSCEPTANCE SLOPE PARAMETER OF STRIPLINE CIRCULATOR*

Inside the ferrite disk the field components are given from Maxwell's equations by

$$H_z=E_\phi=E_r=0$$
(12.41)

$$E_z=A_nJ_n(k_er)\cos n\phi$$
(12.42)

$$H_r=\frac{j}{\omega\mu_0\mu_e}\left[\frac{1}{r}\frac{\partial E_z}{\partial\phi}\right]$$
(12.43)

$$H_\phi=\frac{-j}{\omega\mu_0\mu_e}\left[\frac{\partial E_z}{\partial r}\right]$$
(12.44)

with

$$k_e=\omega\sqrt{\mu_0\mu_e\epsilon_0\epsilon_r}$$
(12.45)

The susceptance slope parameter of the resonator shown in Figure 12.7 is now found by forming the input admittance of the network subject to the boundary conditions. For the electrical field it is assumed as an approximation that E_z has a cosinusoidal distribution with one period around the periphery of the disk, that is, only the $n=1$ mode in Eq. 12.42 is included for the electric field:

$$E_z = A_1 J_1(k_e R) \cos\phi \tag{12.46}$$

For the magnetic field it is assumed that the tangential component of the magnetic field at $r=R$ is a constant over the width of the stripline and zero elsewhere. The tangential component of magnetic field can therefore be expanded in a Fourier series:

$$H(R,\phi) = \frac{H_1\psi}{\pi} + 2H_1 \sum_{n=1}^{\infty} \frac{\sin n\psi}{n\pi}\cos\phi \tag{12.47}$$

where

$$\sin\psi = \frac{W}{2R}$$

Another solution for H_ϕ is given by Eq. 12.44 with E_z given by Eq. 12.46

$$H_\phi = \sum_{n=1}^{\infty} -j\zeta_e (A_1 \cos\phi) J_1'(k_e R) \tag{12.48}$$

where

$$\zeta_e = \sqrt{\frac{\epsilon_0 \epsilon_r}{\mu_0 \mu_e}} \tag{12.49}$$

Comparing Eq. 12.47 with $n=1$ to eq. 12.48, one has, for the coefficient A_1,

$$A_1 = \frac{j2H_1 \sin\psi}{\pi\zeta_e J_1'(k_e R)} \tag{12.50}$$

Substituting this last equation into Eq. 12.42 gives

$$E_z = \frac{j2H_1 \sin\psi}{\pi\zeta_e}\left[\frac{J_1(k_e R)}{J_1'(k_e R)}\right]\cos\phi \tag{12.51}$$

The input wave admittance at $\phi=0$ is, therefore, given by

$$\zeta_{in}=jB_0=\frac{-j\pi\zeta_e}{2\sin\psi}\left[\frac{J_1'(k_eR)}{J_1(k_eR)}\right] \tag{12.52}$$

The susceptance slope parameter of the network is now obtained from

$$B_0'=\frac{\omega_0}{2}\left.\frac{\partial B_0}{\partial\omega}\right|_{\omega_0} \tag{12.53}$$

The result is

$$B_0'=\frac{\pi\zeta_e}{2\sin\psi}\left[\frac{(k_eR)^2-1}{2k_eR}\right] \tag{12.54}$$

Making use of the relation between the susceptance slope parameter of the constituent resonator and that of the gyrator resonator, one has

$$B'=\frac{\pi\zeta_e}{3\sin\psi}\left[\frac{(k_eR)^2-1}{2k_eR}\right] \tag{12.55}$$

This is the result given by Eq. 10.59.

   The normalized susceptance slope parameter b' is now obtained by normalizing Eq. 12.55 to ζ_0.

12.9. *QUARTERWAVE MODE CIRCULATOR*

The constituent resonator for the quarterwave mode circulator discussed in Section 9.9 is shown in Figure 12.11. The input admittance for this transmission line is

$$Y_{in}=jY_{0e}\sqrt{\frac{\epsilon_r}{\mu_e}}\;\cot\beta_e1 \tag{12.56}$$

where Y_{oe} is the free space even mode impedance of two coupled striplines,

β_e is the phase velocity

$$\beta_e = \frac{2\pi \sqrt{\epsilon_r \mu_e}}{\lambda_0} \tag{12.57}$$

and

$$1 = 2R \tag{12.58}$$

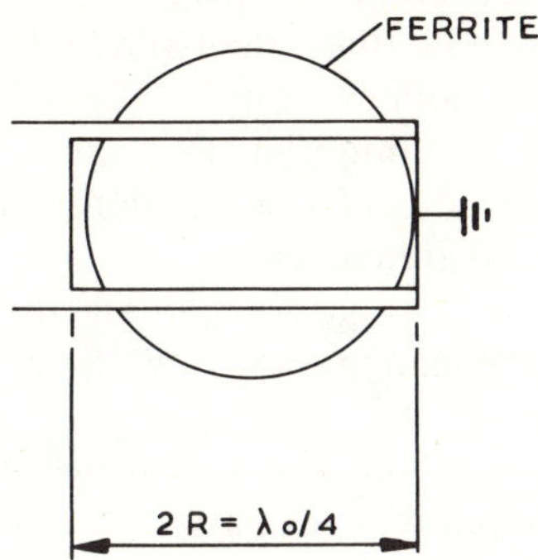

Figure 12.11.   *Constituent resonator for $\lambda/4$ circulator.*

At the operating frequency the input admittance of the constituent resonator is zero. Hence,

$$2R = \frac{\lambda_0}{4\sqrt{\epsilon_r \mu_e}} \tag{12.59}$$

In the vicinity of the operating frequency the susceptance slope parameter of the constituent resonator is

$$b_0' = \frac{\pi}{4}\left(\frac{Y_{0e}}{Y_0}\right)\sqrt{\frac{\epsilon_r}{\mu_e}} \tag{12.60}$$

and

$$b' = \frac{2b_0'}{3} = \frac{\pi}{6}\left(\frac{Y_{0e}}{Y_0}\right)\sqrt{\frac{\epsilon_r}{\mu_e}} \tag{12.61}$$

The split frequencies are related to the magnetic splitting in the usual way by Eq. 12.39.

12.10. *RANGE OF VALIDITY OF CIRCULATOR EQUIVALENT CIRCUIT*

It has been observed in a number of places in this text that one approximation to the equivalent circuit of junction circulators at the input terminals is a 1-port network consisting of a shunt lumped element resonator in parallel with the gyrator conductance of the device.

In order to combine this equivalent circuit with external matching networks, as was done for instance, in the case of the lumped-element circulator described in Chapter 9, it is necessary to establish the bandwidth over which this network applies. This will now be done in terms of the two split frequencies of the magnetized junction by assuming that the equivalent circuit applies so long as the overall bandwidth of the circulator is equal to or less than the bandwidth between these frequencies.

Provided that this constraint is satisfied, it is possible to increase the overall bandwidth of junction circulators in a meaningful way with the help of lumped or distributed matching networks.

In order to determine the bandwidth over which the assumed form of the circulator equivalent circuit applies, it is, therefore, necessary to establish the maximum possible splitting between the two degenerate junction frequencies. This may in general be done by using perturbation theory but for a number of simple junctions it is related to the magnetic variables by Eqs. 9.27 or 12.39.

The ratio K/μ usually lies between 0 and 0.75, and it is this factor that determines the difference between the two split frequencies and the allowable range of Q_L. Using this criterion the difference between the normalized split frequencies lies between

$$0 < 2\delta_{+1} < 0.75 \tag{12.62}$$

The form of the equivalent circuit used here for the circulator is assumed to be reliable between the two split frequencies of the magnetized junction. The assumption about this circuit is satisfied provided the normalized bandwidth $2\delta_0$ is approximately equal to or less than $2\delta_{+1}$. Reasonable agreement between theory and experiment is, therefore, expected provided

$$2\delta_0 \leqslant 2\delta_{+1} \tag{12.63}$$

Combining equations shows that the maximum bandwidth of quarterwave-coupled circulators is

$$2\delta_0 = 0.75 \tag{12.64}$$

The way in which the difference between the two split frequencies may be obtained experimentally is described in Chapter 13. Figure 12.12 indicates

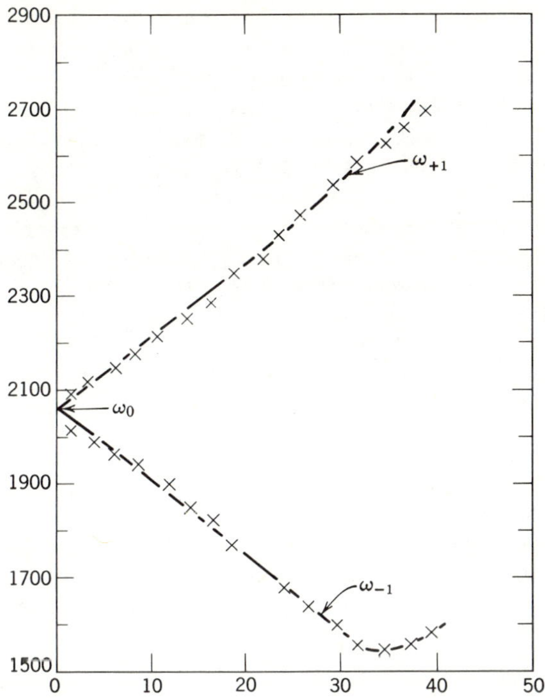

Figure 12.12.   *Split frequencies of 2 GHz stripline circulator (Ref. 14-7).*

one example obtained on a stripline circulator at 2GHz. Here $2\delta_{+1}=0.54$. Making use of the relation between the loaded Q factor and $2\delta_{+1}$ gives

$$0.77 > Q_L > \infty \tag{12.65}$$

The above results show that the minimum value of $Q_L = b'/g$ is a constant determined by the magnetic variables of the ferrite material.

In quarterwave-coupled circulators there is also a constraint on the maximum value that the gyrator conductance may assume because it is difficult to realize admittance levels for the transformers much in excess of about $y_t = 5.5$. Mathematical solutions for quarterwave-coupled circulators to be discussed in Chapter 14, for which either $Q_L < 0.77$, $2\delta_0 > 2\delta_{+1}$, or $y_t > 5.5$, do not, therefore, lead to physically realizeable devices.

REFERENCES

1. B. K. Anderson, "An Analysis of Broadband Circulators with External Tuning Elements" *IEEE Trans. Microwave Theory Tech.,***MTT-15**, 42–47 (1967).

2. J. Helszajn, "The Susceptance Slope Parameter of Junction Circulators," *Proc. IEE* **120**, 1257–1261 (1972).

3. J. T. Allanson, R. Cooper, and Z. G. Cowling, *J. IEE*, **93**, Part III (23),177–187 (1946).

4. H. Bosma, "A General Model for Junction Circulators: Choice of Magnetization and Bias Field," *IEEE Trans. Magn.*, **Mag 4** (3), 587–596 (1968).

5. H. J. Butterweck, "Der Y-Zirkulator," *Arch. Elekt. Ubertragung*, **17**, 163–176 (1963).

6. C. E. Fay and R. L. Comstock, "Operation of the Ferrite Junction Circulator," *Trans. IEEE Microwave Theory Tech.*, **MTT-13**, 15–27 (1965).

7. Y. Konishi, "Lumped Element Y-Circulators," *IEEE Trans. Microwave Theory Tech.*, **MTT-13**, 852–864 (1965).

8. J. O. Bergman and C. Christensen, "Equivalent Circuit for a Lumped Element Y-Circulator," *IEEE Trans. Microwave Theory Tech.*, **MTT-16**, 308–310 (1968).

9. J. Helszajn and M. McDermott, "Junction Inductance of a Lumped Constant Circulator," *IEEE Trans. Microwave Theory Tech.*, **MTT-17**, 50–52 (1970).

10. H. Bosma, "On Stripline Y-Circulators at UHF," *IEEE Trans. Microwave Theory Tech.* **MTT-12**, 61–72 (1964).

11. W. Hilberg, "From Approximations to Exact Relations for Characteristic Impedances," *IEEE Trans. Microwave Theory Tech.*, **MTT-17** (5), 259–265 (1969).

12. J. Helszajn, "A Simplified Theory of the Three Port Junction Circulator," *Radio Electron. Eng.*, **33**, 283–288 (1967).

13. J. Helszajn, "Synthesis of Quarter Wave Coupled Circulators with Chebyshev Characteristics," *IEEE Trans. Microwave Theory Tech.*, **MTT-20**, 764–769 (1972).

14. W. H. Louisell, *Coupled Mode and Parametric Electronics*, 4–7, John Wiley, New York, 1960.

CHAPTER THIRTEEN

Microwave Measurement Techniques

In Chapter 12 the gyrator admittance was related to the product of the susceptance slope parameter, and the difference between the two split frequencies of the magnetized junction. This chapter describes various techniques whereby these variables can be obtained one at a time.

The chapter starts by relating the gyrator at the centre frequency of the junction to the amplitude of the VSWR at the input port with the other two ports terminated in matched loads. This is done by relating the impedance and scattering matrices. The mode chart of circulators is next obtained by observing that the operating frequency of the device coincides with that at which the VSWR passes through two-to-one. It is next shown, with reference to the eigenvalue diagram of the scattering matrix, that for a lossless junction the two split frequencies always coincide with those at which the VSWR passes through two-to-one. The modes need not be distinguishable to obtain this result. The susceptance slope parameter is next derived from the frequency response of the demagnetized and matched junctions. This means that the demagnetized junction contains all the bandwidth information of the circulator. Finally, the susceptance slope parameter, the shunt conductance, and the frequency splitting between the resonant modes are all related to the universal admittance equation. This allows any one of these three to be obtained by measuring the other two.

One measurement which allows the resonator model of the junction circulator to be verified as a function of frequency and magnetic field is also included in this chapter. It consists of measuring the input gyrator admittance of the device by first decoupling one port. It is, of course, possible to directly measure the reflection eigenvalues of the junction by applying the appropriate excitations at the terminals of the device.

229

13.1. *INTERDEPENDENCE OF ISOLATION AND VSWR*

For a lossless, but not perfectly matched symmetrical 3-port junction with S_{12} close to unity and S_{11} and S_{13} small, Eqs. 2.50 and 2.51 apply

$$|S_{11}| \approx |S_{13}| \tag{13.1}$$

and

$$|S_{12}| \approx 1 - 2|S_{11}|^2 \tag{13.2}$$

Thus, minimum insertion loss corresponds to both maximum isolation and minimum VSWR looking into any of the three ports.

Equation 13.1 also states that the isolation between ports 1 and 3 is determined by the VSWR at port 2. The interdependence of isolation and VSWR is depicted in the introduction.

13.2 *COMPLEX GYRATOR ADMITTANCE*

As already seen, one model for a 3-port asymmetric circulator consists of an ideal 3-port circulator with 2-port networks connected to each port:

$$\bar{S}^{(i)} = \begin{bmatrix} S_{11}^{(i)} & S_{12}^{(i)} \\ S_{21}^{(i)} & S_{22}^{(i)} \end{bmatrix} \tag{13.3}$$

where $i = 1$, 2, or 3 denotes the port to which the network is connected.

For the overall network the reflection coefficient S_{11} is given in terms of the scattering coefficients of the 2-ports by

$$S_{11} = S_{11}^{(1)} + \frac{S_{12}^{(1)} S_{21}^{(1)} S_{22}^{(2)} S_{22}^{(3)} t_1 t_2 t_3}{1 - S_{22}^{(1)} S_{22}^{(2)} S_{22}^{(3)} t_1 t_2 t_3} \tag{13.4}$$

where t_1, t_2, and t_3 represent the phase shifts through the ideal circulator. Equation 13.4 applies to a circulator, which circulates in the direction 1-2-3-1. It indicates the way the scattering coefficient S_{11} of the 3-port network obtained by placing matched terminations at ports 2 and 3 is related to those of the 2-ports. This equation shows that $S_{11}^{(1)}$ which defines the gyrator admittance can be measured by setting either $S_{22}^{(2)}$ or $S_{22}^{(3)}$ to zero. Equation 13.4 reduces to

$$S_{11} = S_{11}^{(1)} \tag{13.5}$$

Since the 2-port networks are connected by an ideal circulator which circulates in the direction 1-2-3-1 it is sufficient to set $S_{22}^{(2)}$ to zero to obtain this result. This condition coincides with minimum transmission to port 3.

The experimental arrangement used is illustrated schematically in Fig. 13.1. It consists of a triple stub tuner at port 2 that is used to tune out the mismatch at port 2 and a crystal detector at port 3 to establish when transmission to port 3 becomes zero. Figure 13.2 gives the response of a ring circulator as a function of the insert diameter for the junction in Figure 13.3. The susceptance slope parameter can be obtained from this chart from the admittance function with the help of Eq. 12.2. The experimental arrangement used in this section is also that employed to derive the complex gyrator admittance in Chapter 5.

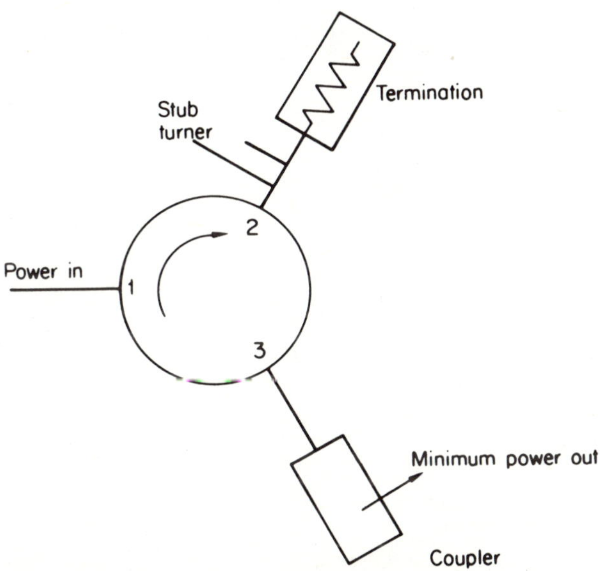

Figure 13.1.   *Use of triple-stub tuner in admittance measurement.*

13.3.   *MEASUREMENT OF GYRATOR CONDUCTANCE*

A simple way of measuring the gyrator conductance of the junction that does not involve decoupling port 3 is also possible. The approach to be described now relies on the relation between the scattering coefficient S_{11} and the gyrator admittance Y_{12}. The junction is tuned to its center frequency and ports 2 and 3 are terminated in matched loads.

The scattering coefficient S_{11} is given in terms of its eigenvalues by

$$S_{11} = \frac{s_0 + s_{+1} + s_{-1}}{3} \tag{13.6}$$

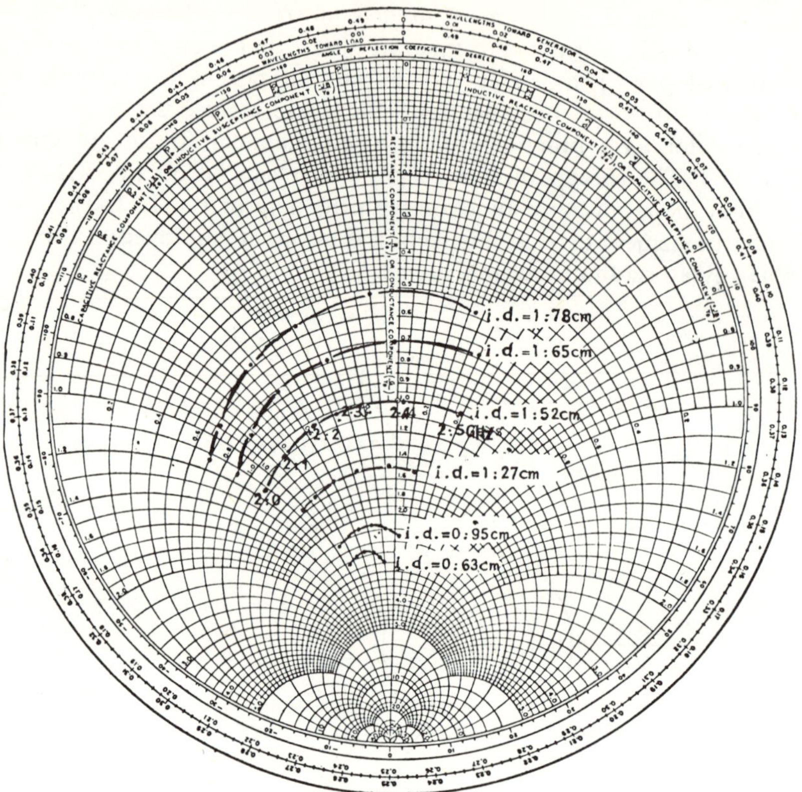

Figure 13.2. *Admittance of ring circulator (Ref. 5).*

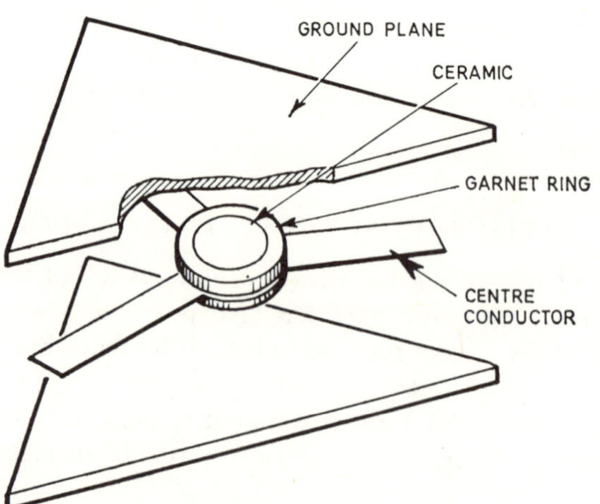

Figure 13.3. *Schematic of ring circulator.*

where for $S_{12} = -1$ Eqs. 5.1 through 5.3 apply with $\theta_0 = \theta_1 = \pi/2$

$$s_0 = -1 \tag{13.7}$$

$$s_{+1} = e^{-j2\theta_{+1}} \tag{13.8}$$

$$s_{-1} = e^{-j2\theta_{-1}} \tag{13.9}$$

The angles $\theta_{\pm 1}$ are defined in Figure 13.4. They normally lie between $0°$ and $\pm 30°$. It is observed from this illustration that, in the complex plane, the resultant always lies along the real axis provided the splitting is symmetric ($\theta_{+1} = \theta_{-1}$). This means that the reference plane remains unchanged as $\theta_{\pm 1}$ is varied and always coincides with that of an ideal

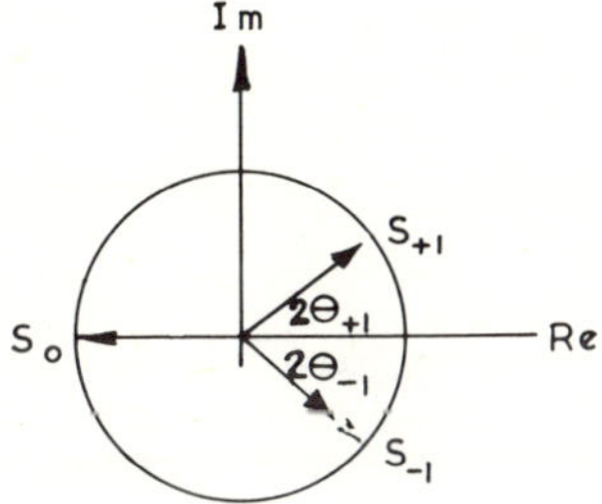

Figure 13.4.   Angle $\theta_{\pm 1}$ between s_0 and s_{+1}, s_{-1}.

circulator. This observation is in agreement with experiment. Making use of the relation between s and y in equation 13.6 gives

$$S_{11} = \frac{1 - 3\tan^2\theta_{+1}}{3 + 3\tan^2\theta_{+1}} \tag{13.10}$$

For the normalized gyrator resistance Z_{12} at the center frequency, one has

$$z = \frac{Z_{12}}{Z_0} = \frac{z_0 + z_{+1}e^{j2\pi/3} + z_{-1}e^{-j2\pi/3}}{3} \tag{13.11}$$

where

$$z_0 = \frac{1 + s_0}{1 - s_0} = 0 \tag{13.12}$$

$$z_{+1} = \frac{1 + s_{+1}}{1 - s_{+1}} = j\cot\theta_{+1} \tag{13.13}$$

$$z_{-1} = \frac{1 + s_{-1}}{1 - s_{-1}} = -j \cot\theta_{+1} \qquad (13.14)$$

Using the above relations, z becomes

$$z = \frac{\cot\theta_{+1}}{\sqrt{3}} \qquad (13.15)$$

and

$$g = \sqrt{3}\,\tan\theta_{+1} \qquad (13.16)$$

It is now possible to eliminate θ_{+1} between Eqs. 13.10 and 13.16. This gives a relation between r and g.

The result in terms of the VSWR r is

$$r = \frac{|3 + g^2| + |1 - g^2|}{|3 + g^2| - |1 - g^2|} \qquad (13.17)$$

For g larger than unity, this function is

$$g = \sqrt{2r - 1} \qquad (13.18)$$

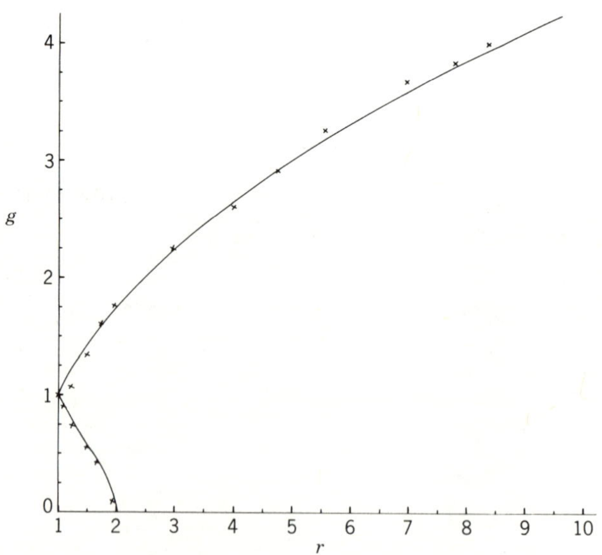

Figure 13.5.   *Relation between r and g for magnetized junction (Ref. 9).*

For g smaller than unity it is

$$g = \sqrt{\frac{2-r}{r}} \tag{13.19}$$

The relation between g and r is shown in Figure 13.5. The experimental data superimposed on this illustration was obtained on a 2 GHz circulator by separately making the experiments described in the last section and in this one. Whether g is larger or smaller than unity is determined by whether there is a voltage maxima or minima at the load. For a demagnetized junction the gyrator conductance is zero.

13.4   MODE CHARTS

It has been noted in Section 8.6 that the minimization of the reflection coefficient S_{11} can be taken as one definition of an electromagnetic resonance within the junction. For a 3-port junction with $s_1 = s_{-1} = s_{+1}$ the reflection coefficient is

$$S_{11} = \frac{s_0 + 2s_1}{3} \tag{13.20}$$

This relation can be effectively applied in practice to study the mode spectrum of junctions. Using this principle it is only necessary to measure the frequency at which the reflection coefficient of the junction passes through a minimum. Figure 11.14 depicts a mode chart obtained this way in the case of the E-plane junction in Figure 11.13.

13.5   MEASUREMENT OF SPLIT RESONANT FREQUENCIES OF JUNCTION CIRCULATORS

An essential quantity that determines the behavior of junction circulators is the split frequencies of the device. One simple way in which they can be measured is by determining the two frequencies at which the VSWR passes through two-to-one. These frequencies then coincide with the split frequencies of the device. It has already been shown that the two degenerate frequencies of the reciprocal junction coincide with the same value of VSWR. This can be demonstrated with reference to the eigenvalue diagrams in Figure 13.6, which give the eigenvalue arrangements of an arbitrarily magnetized junction at ω_0, ω_{+1}, and ω_{-1}. It is immediately apparent from these diagrams that the magnitude of the reflection coefficient S_{11} has the value of $|S_{11}| = 1/3$ at both ω_{+1} and ω_{-1}. At ω_{+1}

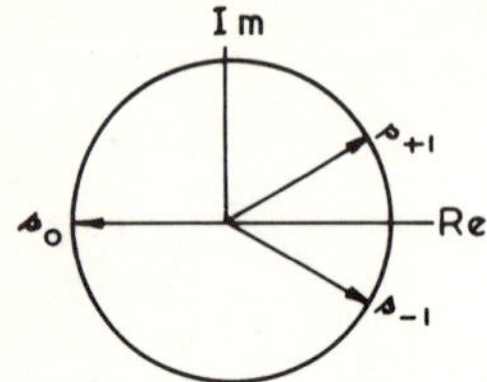

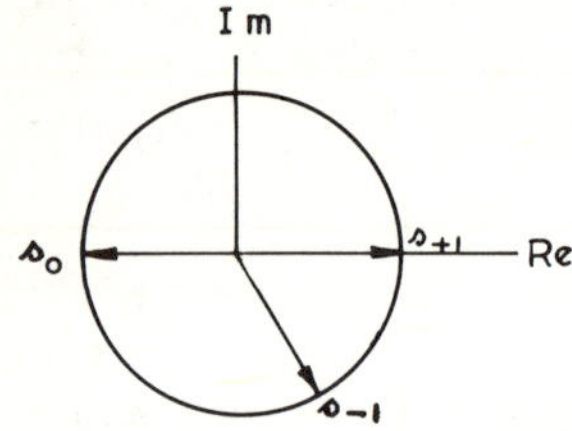

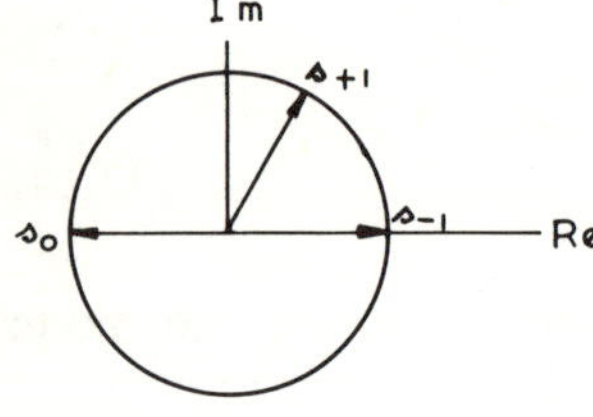

Figure 13.6. *Eigenvalue diagrams of a partially magnetized junction at* ω_0, ω_{+1}, *and* ω_{-1}.

Eq. 13.6 gives

$$|S_{11}| = \left|\frac{s_{-1}}{3}\right| = \frac{1}{3} \tag{13.21}$$

Similarly at ω_{-1} one has

$$|S_{11}| = \left|\frac{s_{+1}}{3}\right| = \frac{1}{3} \tag{13.22}$$

It is also observed that the reflection coefficient at ω_{+1} is s_{-1} and the one at ω_{-1} is s_{+1}. It is, therefore, possible to measure these quantities by the same method. Figure 13.7 illustrates split frequencies for one composite dielectric ferrite stripline circulator.

If there are losses within the junction the return loss at the center frequency of the demagnetized junction will depart from its ideal value of

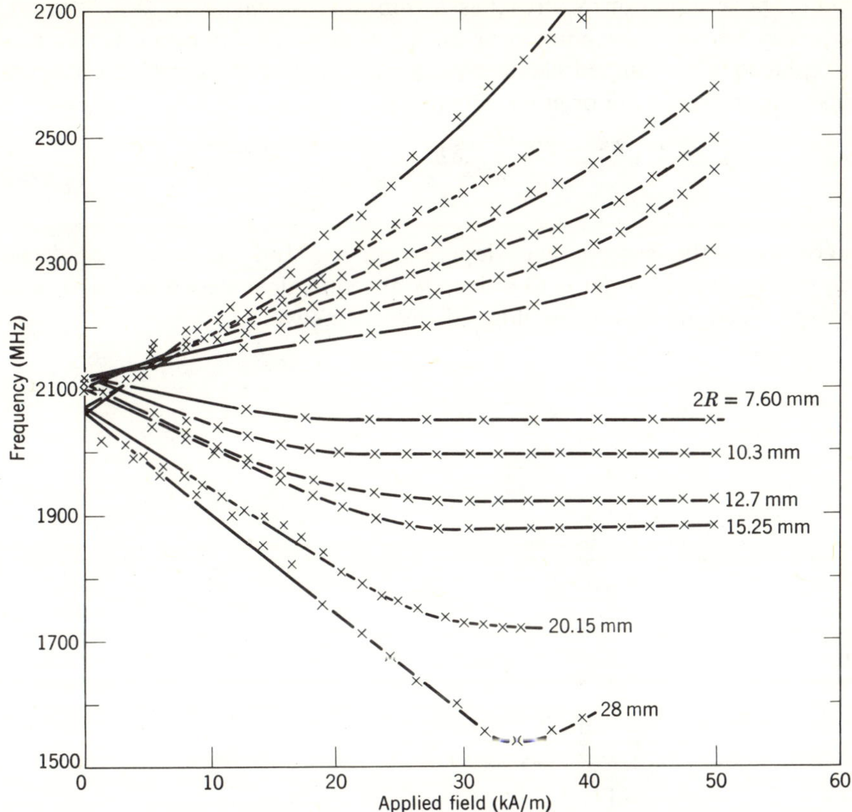

Figure 13.7.   *Split frequencies for ring stripline junction circulator* (*Ref.* 9).

9.5 dB. It can then be shown that, provided the magnetic splitting is small, the two split frequencies now coincide with the actual value of return loss obtained from the demagnetized junction instead of the ideal one.

13.6.   *SUSCEPTANCE SLOPE PARAMETER OF UNMAGNETIZED 3-PORT JUNCTION*

To obtain the susceptance slope parameter of the unmagnetized junction one starts with the reflection coefficient S_{11} given by Eq. 5.49

$$S_{11} = \frac{1 - 3y_1}{3 + 3y_1} \tag{13.23}$$

In the above equation S_{11} reaches a minimum equal to $S_{11}=1/3$ at the frequency where y_1 becomes zero. In the vicinity of this frequency y_1 can be replaced by a lumped element network that has the same susceptance slope parameter as the original network.

$$y_1 = \frac{j2\delta B'}{Y_0} = j2\delta b' \tag{13.24}$$

where B' is the susceptance slope parameter and 2δ is a normalized frequency variable. Figure 13.8 depicts an equivalent network of a 3-port unmagnetized junction using lumped element networks.

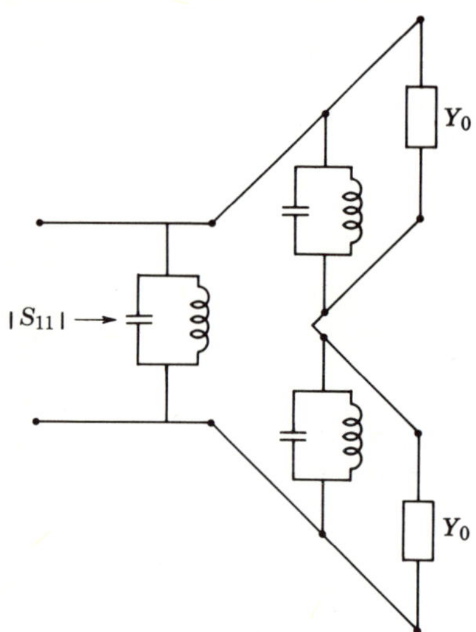

Figure 13.8.   *Equivalent circuit of unmagnetized 3-port junction using lumped element networks.*

Substituting Eq. 13.24 into Eq. 13.23 gives

$$S_{11} = \frac{1 - 3(j2\delta b')}{3 + 3(j2\delta b')} \tag{13.25}$$

Writing S_{11} in terms of the VSWR r gives

$$|S_{11}|^2 = \frac{1 + 9(2\delta b')^2}{9 + 9(2\delta b')^2} = \frac{(r-1)^2}{(r+1)^2} \tag{13.26}$$

Rewriting the last equation gives

$$b' = \frac{\dfrac{2}{3}\left[\dfrac{r^2 - 2.5r + 1}{2r}\right]^{1/2}}{2\delta} \tag{13.27}$$

Figure 13.9 gives the frequency response of the partially loaded unmagnetized stripline junction shown in Figure 13.10 as a function of the ferrite filling factor. Equation 13.27 can now be used to obtain the susceptance slope parameter of these different geometries. The results are $b'_1 = 5.65$, $b'_2 = 3.93$, $b'_3 = 3.33$, and $b'_4 = 1.84$.

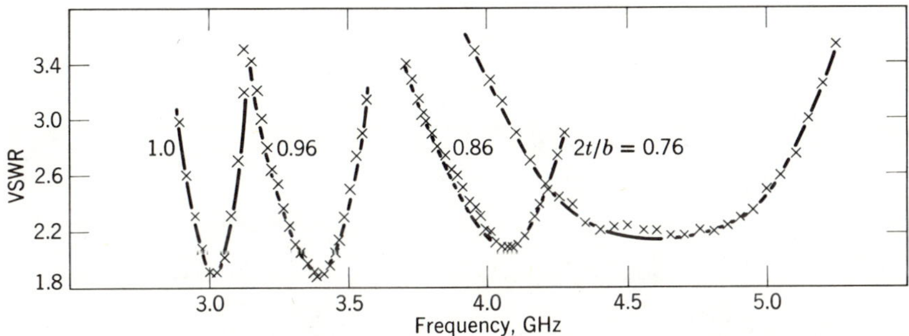

Figure 13.9.   *Frequency response of partially loaded unmagnetized stripline junction (Ref. 3).*

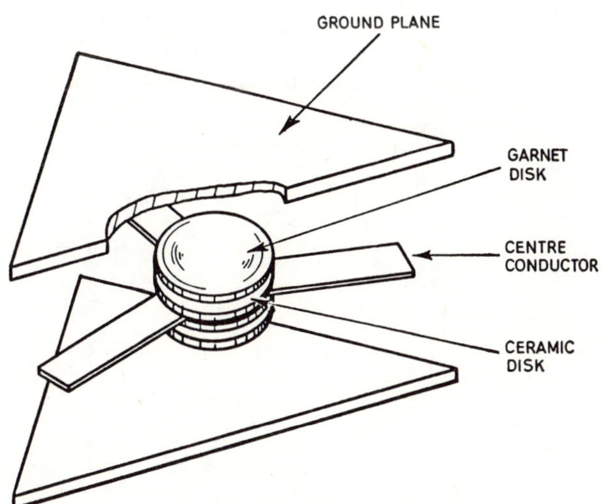

Figure 13.10.   *Partially loaded stripline circulator.*

13.7.　*SUSCEPTANCE SLOPE PARAMETER OF IDEAL CIRCULATOR*

In order to obtain the susceptance slope parameter of an ideal circulator one starts with the reflection coefficient for the 1-port approximation given in Chapter 5

$$S_{11} \approx \frac{y_1}{2} \tag{13.28}$$

The lumped element equivalent circuit is given in Figure 13.11. Writing S_{11} in terms of the VSWR now gives

$$S_{11} = \frac{2\delta b'}{2} = \frac{r-1}{r+1} \tag{13.29}$$

Hence

$$b' = \frac{2(r-1)}{2\delta(r+1)} \tag{13.30}$$

This last equation differs slightly from that derived in Chapter 12 because Eq. 13.28 is only an approximation to the reflection coefficient of a true lumped element resonator.

Figure 13.12 shows the frequency response of the magnetized partially loaded junction illustrated in Figure 13.10. The results obtained here are $b_1' = 6.6$, $b_2' = 3.89$, $b_3' = 3.51$, and $b_4' = 1.62$. It is seen that the two sets of results obtained from Figures 13.9 and 13.12 are in good agreement.

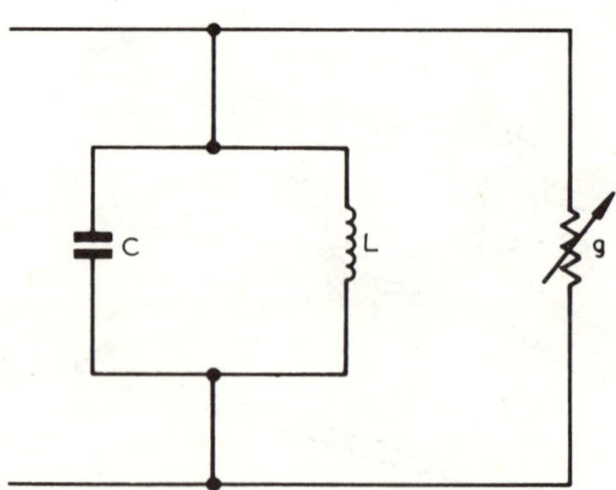

Figure 13.11.　*Lumped element equivalent network of matched circulator.*

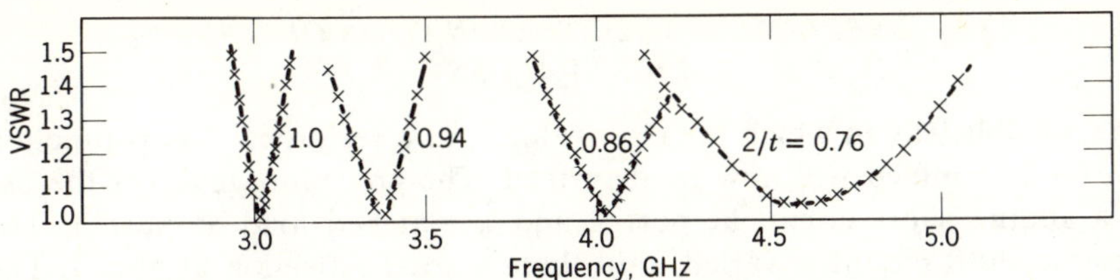

Figure 13.12.   *Frequency response of partially loaded stripline circulator (Ref. 3).*

13.8.   *SUSCEPTANCE SLOPE PARAMETER OF MAGNETIZED JUNCTION*

One way in which the susceptance slope parameter of a partially magnetized junction may be obtained is by using the universal gyrator equation

$$g = \sqrt{3}\, b'\left(\frac{\omega_{+1} - \omega_{-1}}{\omega_0}\right) \tag{13.31}$$

The susceptance slope parameter is immediately obtained from this last equation by independently measuring the gyrator admittance and the two split frequencies. Methods of measuring these two quantities have already been described. Figure 13.13 shows the susceptance slope parameter as a function of magnetic field for a ring junction circulator.

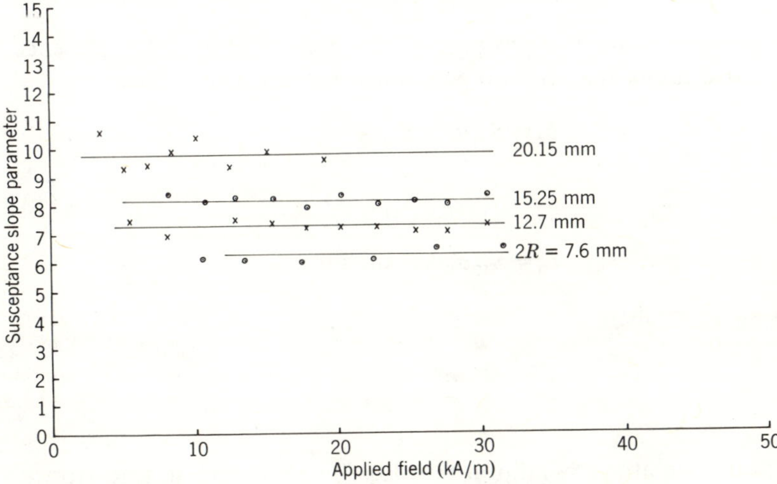

Figure 13.13.   *Susceptance slope parameter of ring stripline circulator (Ref. 9).*

It is also observed from this last equation that g is an increasing function of the magnetic field as long as the splitting of the resonant modes is widening.

13.9.   *EXPERIMENTAL DETERMINATION OF s_0 AND s_1 EIGENVALUES*

One possible way in which the eigenvalue s_1 of a reciprocal 3-port junction can be determined will now be described. The procedure involves the use of a sliding short circuit at port 2 and a matched load at port 3. The variable short circuit is varied until there is total reflection at port 1. The reflection coefficient at port 1 is then the eigenvalue s_1.

The derivation starts once more with the relation between the incoming and outgoing waves at the three ports of the junction.

$$\begin{bmatrix} S_{11} & S_{12} & S_{12} \\ S_{12} & S_{11} & S_{12} \\ S_{12} & S_{12} & S_{11} \end{bmatrix} \cdot \begin{bmatrix} a_1 \\ a_2 \\ a_3 \end{bmatrix} = \begin{bmatrix} b_1 \\ b_2 \\ b_3 \end{bmatrix} \tag{13.32}$$

In the arrangement considered here one places a variable short circuit at port 2, and a matched termination at port 3; hence,

$$b_2/a_2 = e^{-j2(\phi + \pi/2)} \tag{13.33}$$

and

$$a_3 = b_3 = 0 \tag{13.34}$$

where ϕ depends on the position of the short circuit with reference to b_2/a_2. Substituting the last two equations into Eq. 13.32 gives

$$b_1 = S_{11}a_1 + S_{12}a_2 \tag{13.35}$$

$$b_2 = S_{12}a_1 + S_{11}a_2 = a_2 e^{-j2(\phi + \pi/2)} \tag{13.36}$$

$$b_3 = S_{12}a_1 + S_{12}a_2 = 0 \tag{13.37}$$

Rewriting 13.37 gives

$$\frac{a_1}{a_2} = -1 \tag{13.38}$$

Substituting the above condition into Eqs. 13.35 and 13.36 leads to

$$\frac{b_1}{a_1} = S_{11} - S_{12} = s_1 \tag{13.39}$$

where

$$s_1 = e^{-j2(\phi + \pi/2)} \tag{13.40}$$

This measurement, therefore, defines both the characteristic plane and provides a means of measuring s_1 with reference to the short circuit position. The characteristic plane is defined by the short-circuit positions given by $\phi = 0, \pi, 2\pi$, etc.

The eigenvalues s_0 may now be obtained by measuring S_{11} at port 1 with ports 2 and 3 terminated in matched loads. s_0 is then determined from the following relation:

$$S_{11} = \frac{s_0 + 2s_1}{3} \tag{13.41}$$

REFERENCES

1. C. E. Barnes, "Integrated Circulator Design for Parametric Amplifier Application," *IEEE Int. Symp. Tech. Program Digest*, (1964).

2. F. M. Aitken and R. McLean, "Some Properties of the Waveguide *Y* Circulator," *Proc. IEE*, **110**, No. 2 256–260 (1963).

3. J. Helszajn, "Frequency and Bandwidth of the *H*-Plane Circulator," *Proc. IEE*, **118** (1970).

4. J. Helszajn, "Susceptance Slope Parameter of the Junction Circulator," *Proc. IEE*, **120** (1972).

5. J. Helszajn, "A Ferrite Ring Stripline Junction Circulator," *Radio Electron. Eng.*, **32** 55–60 (1966).

6. S. Hagelin, "A Flow Graph Analysis of 3 and 4-Port Junction Circulators," *IEEE Trans. Microwave Theory Tech.*, **MTT-14** 243–249 (1966).

7. J. Simon, "Broadband Strip-Transmission Line *Y* Junction Circulators," *Trans. Microwave Theory Tech.*, **MTT-13** 335–345 (1965).

8. J. Helszajn and M. McDermott, "Mode Chart for *E*-plane Circulations," *IEEE Trans. Microwave Theory Tech.*, **MTT-20**, 187–188 (1972).

9. J. Helszajn, "Microwave Measurement Techniques for Junction Circulators," *IEEE Trans. Microwave Theory Tech.*, **MTT-21**, 347–351 (1973).

10. S. J. Salay and H. J. Peppiatt, "Input Impedance Behavior of Stripline Circulators" (corresp.), *IEEE Trans. Microwave Theory Tech.*, **MTT-19**, 109–110 (1971).

11. S. J. Salay and H. J. Peppiatt, "An Accurate Junction Circulator Design Procedure" (corresp.), *IEEE Trans. Microwave Theory Tech.*, **MTT-20**, 192–193 (1972).

12. B. Owen and C. E. Barnes, "Identification of Field Modes in Ferrite-loaded Waveguide *Y*-Junctions and their Application to the Synthesis of Millimeter wave Circulators," European Microwave Conference, 1972.

CHAPTER FOURTEEN

Junction Circulators with Chebyshev Characteristics

An important property of the 3-port junction circulator is that an ideal circulator is obtained when it is matched. One well-known method of broadbanding this device is, therefore, to use external matching networks. One arrangement that is often used consists of a cascade of quarterwave transformers.

The purpose of this chapter is to give an exact synthesis procedure for the case of one- and two-step transformers that will give an equal ripple Chebyshev response for the reflection coefficient of the overall circulator network. It is assumed, in the usual way, that the equivalent network at the reference terminals of the junction consists of a shunt lumped element resonator in parallel with the gyrator conductance of the circulator. The synthesis procedure then starts by replacing the lumped element resonator by a distributed one consisting of a quarterwave short-circuited transmission line which has the same susceptance slope parameter as the original circuit. The two circuits are equivalent provided their susceptance slope parameters are the same. The equivalent circuit of the device can therefore be represented by a quarterwave short-circuited transmission line in parallel with the gyrator conductance of the circulator. The admittance of the distributed network is uniquely related to the susceptance slope parameter, once the nature of the network is stated. The equivalent circuit of the complete network now involves commensurate quarterwave transmission lines only. It can be made to have a Chebyshev response by forming the reflection coefficient of the circuit and equating its zeros and maxima to give a characteristic that is a Chebyshev polynomial of the first kind. The results have been obtained with the help of a computer program and are presented in tabular form for $n=2$ and $n=3$ in terms of the necessary bandwidth and VSWR of the overall network. These tables give the susceptance slope parameter, the gyrator conductance, and the loaded Q factor of the circulator network. They also give the admittance values of the matching transformers.

The bandwidth over which the assumed form for the circulator equivalent circuit applies has been discussed in Chapter 12. The assumption about the equivalent circuit is satisfied provided the overall bandwidth of the circulator is equal to or less than the bandwidth between the two split frequencies. This puts one constraint on the realizable frequency response of the overall junction.

Physically realizable values that the susceptance slope parameter and gyrator conductance of practical circulators can take lead to a second constraint. The susceptance slope parameter is primarily determined by the geometry of the device and the gyrator conductance by the magnetization and the direct magnetic field. The ratio of these two quantities defines the loaded Q factor of the junction. It is related to the split frequencies of the magnetized device and to the magnetic variables of the junction. The latter factor places a lower limit on the value of the loaded Q-factor and imposes a further constraint on the frequency response of the overall network. The permissible values of these variables are thus limited to a much smaller range than depicted in the tables given here.

14.1.   *THE CHEBYSHEV POLYNOMIALS*

The design approach used here consists of adjusting the overall reflection coefficient of the network to have an equal ripple Chebyshev characteristic. A Chebyshev response is obtained by putting

$$|\Gamma| = \gamma T_n(x) \tag{14.1}$$

where $|\Gamma|$ is the overall reflection coefficient of the network, γ is the maximum reflection coefficient in the band, and x is a frequency variable that satisfies a Chebyshev function of the first kind of order n. This Chebyshev function is defined as

$$T_n(x) = \begin{array}{l} \cos(n \arccos x) \cdots x \leqslant 1 \\[6pt] \cosh(n \operatorname{arc cosh} x) \cdots \geqslant 1 \end{array} \tag{14.2}$$

In the interval $|x| < 1$ this function oscillates between ± 1 and has $(n-1)$ equal maxima or minima in the interval. The function also assumes the values ± 1 at $x = \pm 1$. Outside this region it exceeds unity in value and approaches $\pm\infty$. Diagrams of five Chebyshev functions in the interval $|x|<1$ are shown in Figure 14.1. The Chebyshev polynomials may be expanded as a power series in x by use of the recurrence formula:

$$T_{n+1}(x)+T_{n-1}(x)=2xT_n(x) \tag{14.3}$$

and the first five Chebyshev functions expressed as polynomials in x are

$$T_0(x)=1$$

$$T_1(x)=x$$

$$T_2(x)=2x^2-1 \tag{14.4}$$

$$T_3(x)=4x^3-3x$$

$$T_4(x)=8x^4-8x^2+1$$

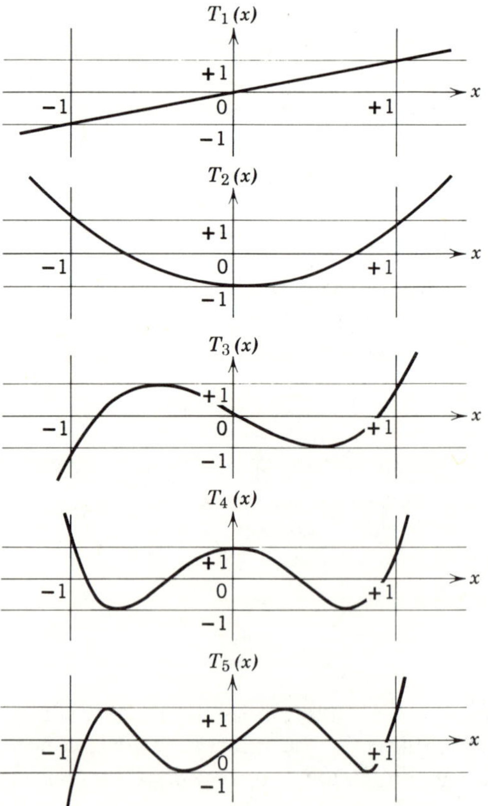

Figure 14.1.   *Chebyshev functions of the first kind.*

A suitable frequency variable is

$$x = \frac{\cos\theta}{\cos\theta_0} \tag{14.5}$$

where

$$\theta_0 = \frac{\pi}{2}(1 + \delta_0)$$

$$\theta = \frac{\pi}{2}(1 + \delta)$$

$$\delta_0 = \frac{\omega_1 - \omega_0}{\omega_0}$$

$$\delta = \frac{\omega - \omega_0}{\omega_0}$$

Here δ_0 is a bandwidth parameter, δ is a bandwidth variable, ω_0 is the center frequency, $\omega_{1,2}$ are the bandedges, and ω is the frequency. The notation used is given in Figure 14.2, which applies for $n=2$.

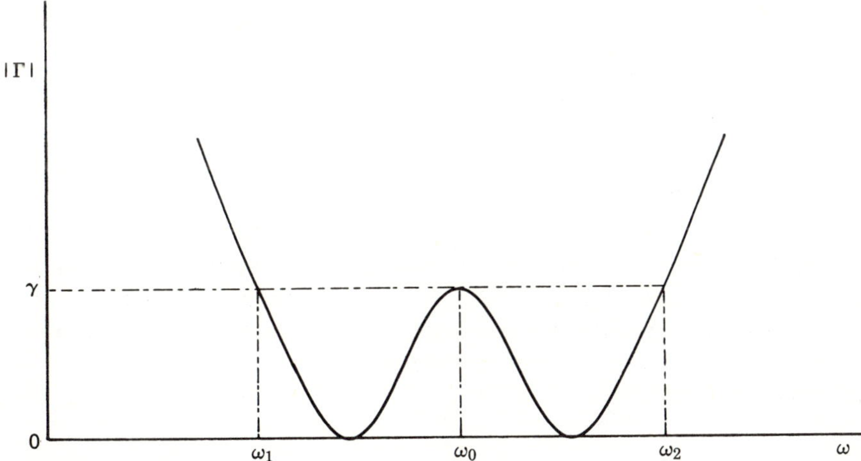

Figure 14.2.   *Chebyshev response of $n=2$ quarterwave coupled circulator.*

14.2.   *QUARTERWAVE COUPLED CIRCULATOR WITH CHEBYSHEV FREQUENCY RESPONSE*

One well known method of overcoming the physical limitations of the ferrite parameters is to use external matching transformers to improve the bandwidth. The purpose of this chapter is to obtain the values of b' and g and the external parameters for which the overall response has an equal ripple Chebyshev characteristic.

The synthesis procedure starts by replacing the lumped element resonator in Figure 14.3 by a distributed one consisting of a quarterwave short-circuited transmission line that has the same susceptance slope parameter as the original circuit. The equivalent circuit of the circulator now has the form depicted in Figure 14.4. The complete network now involves commensurate quarterwave transmission lines only, for which an exact synthesis procedure can be developed.

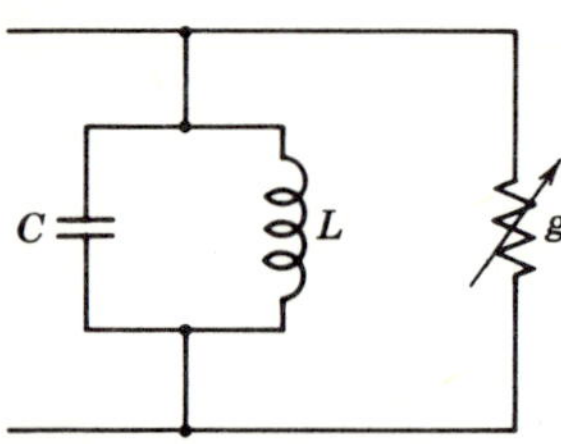

Figure 14.3. *Lumped element equivalent resonator circuit of circulator.*

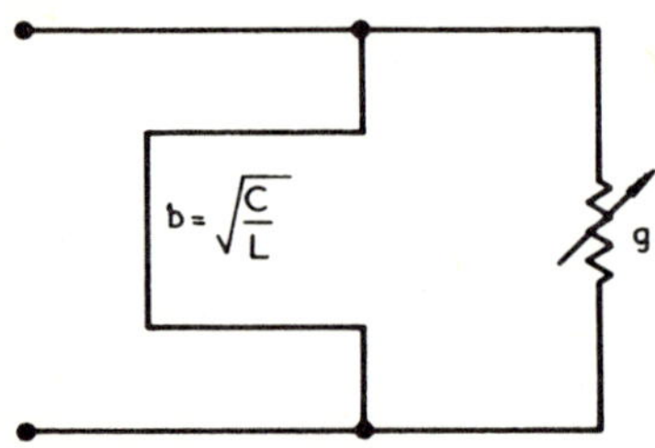

Figure 14.4. *Distributed element equivalent resonator circuit of circulator.*

The equivalent network for matching each of the three ports of the circulator is shown in Figure 14.5 in terms of the $ABCD$ matrix of the transformer. By straightforward calculations the admittance of the network at the input terminals is given in terms of the overall normalized $ABCD$ matrix of the circuit by

$$y_{in} = \frac{Y_{in}}{Y_0} = \frac{jC + Dy_L}{A + jBy_L} \tag{14.6}$$

The input admittance of the circulator y_L is

$$y_L = \frac{Y_L}{Y_0} = g - jy\cot\theta \tag{14.7}$$

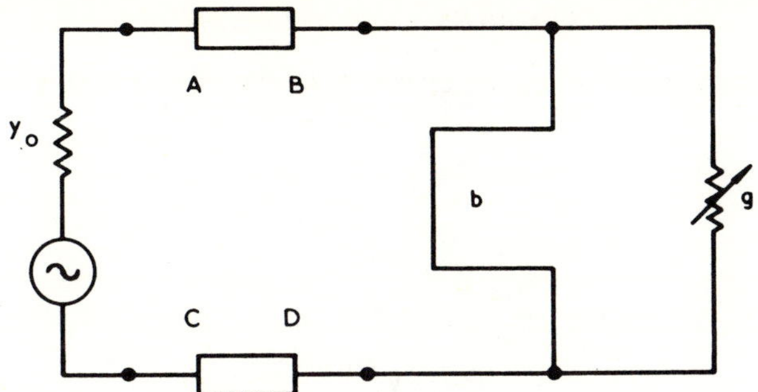

Figure 14.5. *Equivalent network in terms of overall ABCD matrix.*

The normalized susceptance slope parameter b' is related to the normalized characteristic admittance y of the network by

$$b' = \frac{\pi}{4}y \tag{14.8}$$

In terms of the original variables the magnitude of the reflection coefficient is

$$|\Gamma| = (\Gamma\Gamma^*)^{1/2} = \frac{\left[(Dg - A - By\cot\theta)^2 + (C - Bg - Dy\cot\theta)^2\right]^{1/2}}{\left[(Dg + A + By\cot\theta)^2 + (C + Bg - Dy\cot\theta)^2\right]^{1/2}} \tag{14.9}$$

The design proceeds by having the frequency at which the reflection coefficient passes through its zeros and maxima coincide with those of the Chebyshev polynomial.

When the reflection coefficient passes through zero,

$$g = \frac{AD + BC}{B^2 + D^2} \tag{14.10}$$

$$y\cot\theta = \frac{CD - AB}{B^2 + D^2} \tag{14.11}$$

The reflection coefficient passes through a maxima when

$$|\Gamma| = \gamma \tag{14.12}$$

In terms of the $VSWR$, γ is given by

$$\gamma = \frac{(r - 1)}{(r + 1)} \tag{14.13}$$

14.3. *SYNTHESIS OF $n=2$ NETWORK*

For a single step transformer the elements of the normalized $ABCD$ matrix are

$$A = \cos\theta$$

$$B = \frac{\sin\theta}{y_{01}} \tag{14.14}$$

$$C = y_{01}\sin\theta$$

$$D = \cos\theta$$

For $n=2$ the Chebyshev polynomial passes through zero when

$$2x^2 - 1 = 0 \tag{14.15}$$

This occurs when the frequency variable is

$$x = \pm\frac{1}{\sqrt{2}} \tag{14.16}$$

The Chebyshev polynomial passes through a maxima when

$$2x^2 - 1 = \pm 1 \tag{14.17}$$

This occurs when the frequency variable is

$$x = 0 \tag{14.18}$$

and

$$x = \pm 1 \tag{14.19}$$

The transformer admittance y_{01}, the normalized susceptance slope parameter b' and the normalized shunt conductance g are now obtained by adjusting the zeros and maxima of the reflection coefficient to coincide with those of the Chebyshev polynomial. This yields three equations from which y_{01}, b', and g are obtained.

At $x=0$, Eq. 14.12 applies so that

$$y_{01}^2 = rg \tag{14.20}$$

when Eq. 14.12 is evaluated with

$$\cos\theta = 0 \tag{14.21}$$

At $x = \pm\dfrac{1}{\sqrt{2}}$ the reflection coefficient $|\Gamma| = 0$. Here, the two independent equations given by 14.10 and 14.11 apply. The result for y and g in terms of the original variables is

$$g = \frac{r - \sin^2\theta}{r\cos^2\theta} \tag{14.22}$$

and

$$y = \sqrt{\frac{g}{r}} \; (rg - 1)\sin^2\theta \tag{14.23}$$

which must be evaluated with

$$\cos\theta = \frac{1}{\sqrt{2}}\cos\theta_0 \tag{14.24}$$

g and b' are given in tabular form as a function of r and $2\delta_0$ with the help of a computer program in Table 14.1. Figure 14.6 depicts the relation

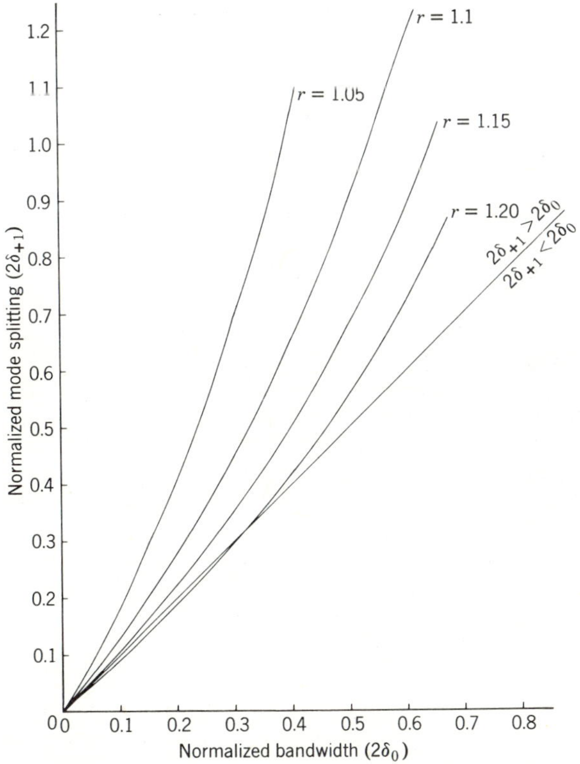

Figure 14.6.   *Relation between* $2\delta_0$ *and* $2\delta_{+1}$ *for* $n=2$ *and parametric values of* r (*Ref.* 7).

Table 14.1.   *Network Parameters for n* $= 2$.

$2\delta_0$	b'	g	Q	y_{01}
		$R = 1.05$		
0.100	50.304	16.424	3.063	4.153
0.150	15.433	7.846	1.967	2.870
0.200	6.809	4.844	1.406	2.255
0.250	3.672	3.455	1.063	1.905
0.300	2.248	2.700	0.833	1.684
0.350	1.501	2.245	0.669	1.535
0.400	1.067	1.950	0.547	1.431
0.450	0.795	1.747	0.455	1.355
0.500	0.614	1.603	0.383	1.297
0.550	0.488	1.496	0.326	1.253
0.600	0.397	1.414	0.280	1.219
0.660	0.317	1.340	0.237	1.186

$2\delta_0$	b'	g	Q	y_{01}
		$R = 1.10$		
0.100	133.831	30.445	4.396	5.787
0.150	40.384	14.070	2.870	3.934
0.200	17.457	8.339	2.093	3.029
0.250	9.205	5.686	1.619	2.501
0.300	5.508	4.245	1.297	2.161
0.350	3.598	3.377	1.065	1.927
0.400	2.505	2.813	0.890	1.759
0.450	1.831	2.427	0.754	1.634
0.500	1.390	2.151	0.645	1.538
0.550	1.088	1.946	0.559	1.463
0.600	0.872	1.791	0.487	1.404
0.660	0.687	1.650	0.417	1.347

Table 14.1 (Continued)

$2\delta_0$	g'	g	Q	y_{01}
		$R=1.15$		
0.100	234.000	43.247	5.411	7.052
0.150	70.190	19.750	3.554	4.766
0.200	30.113	11.530	2.612	3.641
0.250	15.738	7.724	2.038	2.980
0.300	9.328	5.656	1.649	2.550
0.350	6.032	4.410	1.368	2.252
0.400	4.158	3.601	1.155	2.035
0.450	3.010	3.047	0.988	1.872
0.500	2.264	2.651	0.854	1.746
0.550	1.756	2.358	0.745	1.647
0.600	1.397	2.135	0.654	1.567
0.660	1.091	1.932	0.565	1.491

$2\delta_0$	b'	g	Q	y_{01}
		$R=1.20$		
0.100	344.387	54.983	6.264	8.123
0.150	102.999	24.962	4.126	5.473
0.200	44.010	14.454	3.045	4.165
0.250	22.892	9.591	2.387	3.393
0.300	13.498	6.950	1.942	2.888
0.350	8.677	5.357	1.620	2.536
0.400	5.947	4.324	1.375	2.278
0.450	4.279	3.616	1.184	2.083
0.500	3.200	3.109	1.029	1.932
0.550	2.469	2.736	0.903	1.812
0.600	1.954	2.451	0.797	1.715
0.660	1.517	2.191	0.692	1.622

between $2\delta_0$ and $2\delta_{+1}$ for this type of network. In addition to this constraint the entries in these tables for which either $Q_L < 0.77$ or $y_{01} > 5.5$ are also not suitable for the construction of quarterwave-coupled circulators. The permissible solution for the $n=2$ network are boxed in Table 14.1. It shows, among other results, that a junction with a constant susceptance slope parameter can be used to construct circulators with widely different ripple levels and frequency responses. This makes this type of junction very versatile indeed. This is in good agreement with practical experience.

14.4.   *SYNTHESIS OF $n=3$ NETWORK*

For a double step transformer the elements of the normalized *ABCD* matrix are

$$A = \cos^2\theta - \frac{y_{02}}{y_{01}}\sin^2\theta$$

$$B = \left(\frac{1}{y_{01}} + \frac{1}{y_{02}}\right)\sin\theta\cos\theta \qquad (14.25)$$

$$C = (y_{01} + y_{02})\sin\theta\cos\theta$$

$$D = -\frac{y_{01}}{y_{02}}\sin^2\theta + \cos^2\theta$$

For $n=3$ the Chebyshev polynomial passes through zero when

$$4x^3 - 3x = 0 \qquad (14.26)$$

This occurs when the frequency variable takes the values

$$x = 0 \qquad (14.27)$$

and

$$x = \pm\sqrt{\frac{3}{4}} \qquad (14.28)$$

The Chebyshev polynomial passes through a maxima when

$$4x^3 - 3x = \pm 1 \qquad (14.29)$$

This occurs when the frequency variable takes the values

$$x = \pm \tfrac{1}{2} \tag{14.30}$$

and

$$x = \pm 1 \tag{14.31}$$

The admittances of the two transformers y_{01} and y_{02}, the susceptance slope parameter b', and the gyrator conductance g are now obtained by adjusting the zeros and maxima of the reflection coefficient to coincide with those of the Chebyshev polynomial. This gives four equations from which y_{01}, y_{02}, b' and g are obtained.

At $x=0$ the reflection coefficient $|\Gamma|$ is zero. Here, Eqs. 14.10 and 14.11 apply. This condition yields

$$y_{01}^2 = \frac{y_{02}^2}{g} \tag{14.32}$$

when Eq. 14.10 is evaluated with

$$\cos\theta = 0 \tag{14.33}$$

At $x = \pm \sqrt{\dfrac{3}{4}}$, the reflection coefficient is also zero, so that Eqs. 14.10 and 14.11 apply again. From Eqs. 14.10, one obtains

$$y_{02}^2 = \frac{g\left(1+\sqrt{g}\,\right)^2 \sin^2\theta \cos^2\theta}{1 - \left(\sqrt{g}\,\cos^2\theta - \sin^2\theta\right)^2} \tag{14.34}$$

which must be evaluated with

$$\cos\theta = \sqrt{\frac{3}{4}}\ \cos\theta_0 \tag{14.35}$$

g and y are now obtained by forming two additional equations. One of these is Eq. 14.11, which applies when $|\Gamma|=0$ at $x= \pm \sqrt{3/4}$. The other equation is obtained at $x= \pm 1/2$ when $|\Gamma|=\gamma$. Here Eq. 14.12 applies. The result is

$$y\cot\theta = \frac{CD - AB}{B^2 + D^2} \tag{14.36}$$

which must be evaluated at

$$\cos\theta = \sqrt{\frac{3}{4}}\ \cos\theta_0 \tag{14.37}$$

and

$$\gamma\left[\left(Dg+A+By\cot\theta\right)^2+\left(C+Bg-Dy\cot\theta\right)^2\right]^{1/2}$$

$$=\left[\left(Dg-A-By\cot\theta\right)^2+\left(C-Bg-Dy\cot\theta\right)^2\right]^{1/2} \tag{14.38}$$

which must be evaluated at

$$\cos\theta = \tfrac{1}{2}\cos\theta_0 \tag{14.39}$$

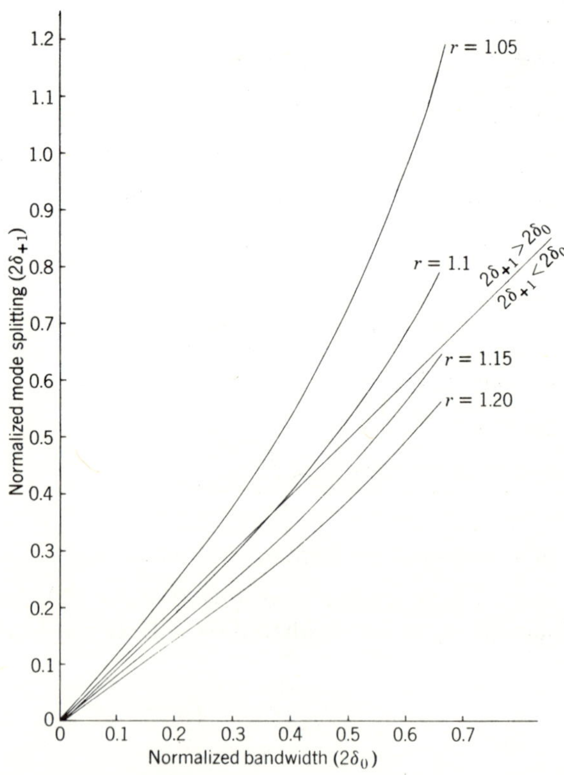

Figure 14.7.   *Relation between $2\delta_0$ and $2\delta_{+1}$ for $n=3$ and parametric values of r (Ref. 7).*

Table 14.2.   *Network Parameters for $n=3$.*

$R=1.05$

$2\delta_0$	b'	Q	g	y_1	y_2
0.100	20633.204	4.830	4271.997	6.245	408.185
0.150	2724.182	3.193	853.154	4.210	122.959
0.200	649.105	2.366	274.365	3.206	53.101
0.250	213.973	1.862	114.918	2.615	28.029
0.300	86.710	1.519	57.069	2.230	16.844
0.350	40.567	1.269	31.973	1.963	11.097
0.400	21.108	1.076	19.624	1.769	7.836
0.450	11.925	0.921	12.948	1.624	5.844
0.500	7.196	0.794	9.066	1.513	4.556
0.550	4.583	0.687	6.674	1.427	3.686
0.600	3.054	0.596	5.127	1.358	3.076
0.660	1.973	0.503	3.921	1.294	2.563

$R=1.10$

$2\delta_0$	b'	Q	g	y_1	y_2
0.100	60931.440	6.245	9757.171	8.032	793.430
0.150	8025.626	4.138	1939.352	5.391	237.430
0.200	1905.491	3.077	619.236	4.082	101.578
0.250	625.015	2.434	256.800	3.305	52.965
0.300	251.612	1.999	125.856	2.795	31.351
0.350	116.728	1.684	69.333	2.436	20.285
0.400	60.108	1.442	41.677	2.173	14.028
0.450	33.539	1.250	26.824	1.973	10.218
0.500	19.949	1.093	18.249	1.817	7.763
0.550	12.502	0.961	13.004	1.694	6.108
0.600	8.184	0.849	9.641	1.594	4.950
0.660	5.166	0.734	7.043	1.499	3.977

14.2 (*Continued*)

$2\delta_0$	b'	Q	g	y_1	y_2
		$R = 1.15$			
0.100	112327.120	7.290	15407.534	9.358	1161.559
0.150	14781.587	4.836	3056.774	6.270	346.678
0.200	3504.736	3.601	973.366	4.736	147.766
0.250	1147.428	2.853	402.149	3.824	76.676
0.300	460.787	2.349	196.125	3.221	45.115
0.350	213.102	1.985	107.371	2.797	28.982
0.400	109.314	1.707	64.048	2.483	19.875
0.450	60.713	1.487	40.841	2.244	14.340
0.500	35.914	1.307	27.482	2.056	10.779
0.550	22.365	1.157	19.338	1.906	8.382
0.600	14.537	1.026	14.135	1.784	6.707
0.660	9.090	0.897	10.131	1.666	5.302

$2\delta_0$	b'	Q	g	y_1	y_2
		$R = 1.20$			
0.100	171035.731	8.157	20967.327	10.458	1514.381
0.150	22495.848	5.413	4155.581	7.001	451.322
0.200	5329.825	4.034	1321.288	5.281	191.973
0.250	1743.184	3.200	544.790	4.256	99.347
0.300	699.110	2.638	264.990	3.579	58.261
0.350	322.783	2.232	144.591	3.100	37.279
0.400	165.238	1.924	85.897	2.746	25.446
0.450	91.547	1.680	54.503	2.474	18.262
0.500	53.996	1.481	36.462	2.260	13.645
0.550	33.512	1.315	25.483	2.088	10.540
0.600	21.698	1.174	18.483	1.948	8.374
0.660	13.497	1.030	13.109	1.811	6.557

y_{01}, y_{02}, g, and b' are given in tabular form as a function of r and $2\delta_0$ with the help of a computer program in Table 14.2. These tables indicate the precision with which circulator parameters must be satisfied in order to obtain a specified Chebyshev response. Figure 14.7 gives the relation between $2\delta_0$ and $2\delta_{+1}$ for this type of network. The useful entries for this arrangement are constrained by $2\delta_0 > 2\delta_{+1}$. Applying the constraint Q_L > 0.77 and $y_{02} < 5.5$ to these tables indicates that the freedom to specify the overall performance with an $n = 3$ quarterwave-coupled circulator is severely curtailed with this arrangement. The permissible solutions are boxed in Table 14.2.

With $2\delta_0 = 0.66$ and $n = 3$ the realizable solutions are completely constrained by the equivalent circuit of the junction to $r = 1.11$, 1.12, 1.13, and 1.14. The values for the susceptance slope parameter are, for these solutions, particularly simple to realise.

14.5.   *OCTAVE BAND STRIP LINE CIRCULATOR*

This section describes the construction of an $n = 3$ octave band stripline circulator in the frequency range 2 to 4 GHz. The permissible solutions for such a circulator are restricted to those given in Table 14.3. The normalized magnetization for the garnet material used in this circulator was about 0.72. This coincides with a minimum loaded Q factor of 0.80. The nominal values for the transformer admittances y_{01} and y_{02} were 1.43 and 3.95. The measured susceptance slope parameter for this junction in a directly coupled configuration was 6.4. Figure 14.8 shows the experimental results obtained on such 2 to 4 GHz circulator. This result was obtained by trimming each transformer with the help of a tuning screw. Although it is not possible to state exactly the final values used for y_{01} and y_{02}, this tuning arrangement increases the admittances y_{01} and y_{02} from the nominal values as given above to the ones stated in Table 14.2. The insertion loss for this circulator was below 0.50 dB. The maximum $VSWR$ was typically 1.22 and the minimum one was 1.07. Such a result is consistent with a slightly more relaxed requirement on the loaded Q factor of the junction than stated in the theoretical results given here.

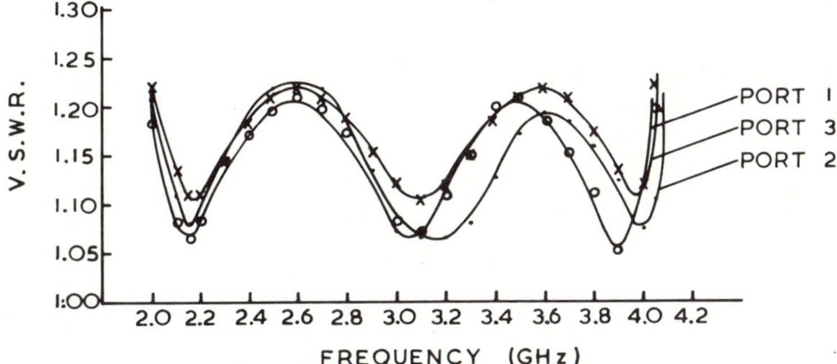

Figure 14.8.   *Frequency response of 2–4 GHz octave band circulator (Ref. 7).*

Table 14.3.   Octave band n = 3 realizable networks.

r	$2\delta_0$	$2\delta_{+1}$	b'	g	Q_L	y_1	y_2
1.11	0.660	0.750	5.903	7.667	0.770	1.534	4.249
1.12	0.660	0.719	6.665	8.288	0.804	1.569	4.517
1.13	0.660	0.690	7.452	8.906	0.837	1.602	4.781
1.14	0.660	0.666	8.261	9.521	0.868	1.634	5.043

14.6.   FUNDAMENTAL LIMITS ON EXTERNAL MATCHING

The maximum bandwidth that can be obtained with the help of external tuning can be predicted with the help of a theorem due to Bode, which establishes the maximum return loss that can be realized in a given band in the presence of "parasitic" shunt reactance. Mathematically, Bode's result can be written

$$\log \frac{1}{\Gamma} \geqslant \frac{\pi}{(\omega_2 - \omega_1)RC} \tag{14.40}$$

where Γ is the voltage reflection coefficient, C the capacitance shunting a source of resistance R, and $\omega_2 - \omega_1$ is the frequency interval of interest. The equality can hold only if the reflection coefficient has a constant value in the passband and magnitude unity elsewhere.

The result can be written in terms of the loaded Q factor of the junction circulator by first writing RC in terms of Q_L

$$\omega_0 RC = Q_L \tag{14.41}$$

By expressing the reflection coefficient in terms of the return loss (in decibels), one now obtains

$$2\delta_{\max} Q_L \leqslant \frac{27.2}{\text{Return Loss}} \tag{14.42}$$

where $2\delta_{\max}$ is the normalized passband and the return loss is expressed in decibels.

REFERENCES

1. L. K. Anderson, "An Analysis of Broadband Circulators with External Tuning Elements," *IEEE Trans. Microwave Theory Tech.*, **MTT-15**, 42–47 (1967).

2. C. E. Fay and R. L. Comstock, "Operation of the Ferrite Junction Circulator," *Trans. IEEE Microwave Theory Tech.*, **MTT-13**, 15–27 (1965).

3. R. Levy, "Explicit formulaes for Chebyshev Impedance Matching Networks, Filters and Interstages," *Proc. IEEE*, **111** (6), 1099–1106 (1964).

4. E. Schwartz, "Broadband Matching of Resonant Circuits and Circulators," *IEEE Trans. Microwave Theory Tech.*, **MTT-16**, 158–165 (1968).

5. R. Levy, "A Guide to the Practical Application of Chebyshev Functions to the Design of Microwave Components," *IEEE Monograph* No. 337E, 193–199 (1959).

6. R. E. Collin, *Foundation for Microwave Engineering*, 229–237, McGraw-Hill, New York, (1966).

7. J. Helszajn, "Synthesis of Quarter Wave Coupled Circulators with Chebyshev Characteristics," *IEEE Trans. Microwave Theory Tech.*, **MTT-20** (1972), 764–769.

8. Y. Naito and N. Tanaka, "Broadband and Changing Operating Frequency of Circulator," *IEEE Trans. Microwave Theory Tech.*, **MTT-19**, 367–372 (1971).

9. J. W. Simon, "Broadband Strip-Transmission line *Y* Junction Circulators," *IEEE Trans. Microwave Theory Tech.*, **MTT-13**, 335–345 (1965).

CHAPTER FIFTEEN

Quarterwave-Coupled Reciprocal Junctions

This chapter gives the frequency response of quarterwave-coupled reciprocal 3-port junctions. It concludes that the problem of constructing wideband circulators is essentially that of making wideband reciprocal junctions. The element values for the matching networks are, therefore, chosen to correspond to those that result in a circulator with an overall Chebyshev response when the junction is magnetized. The element values for quarterwave-coupled circulators have been given in Chapter 14. The frequency response of such junctions is obtained in this chapter for $n=1$, 2, and 3. The equivalent circuit of the reciprocal junction for $n=1$ has been derived in Chapter 5 in terms of the eigenvalues of the admittance matrix of the 3-port network. It is the same eigenvalue that characterizes the susceptance slope parameter of the magnetized junction. Once the equivalent network of the reciprocal junction is given it is possible to connect matching networks at each port and calculate the overall response in a straightforward manner using a computer. This is done for $n=2$ and 3. The frequency response of such junctions is also derived at the terminals of each network. This allows the precise experimental adjustment of each matching network to obtain the desired overall response.

15.1. *EQUIVALENT CIRCUIT OF RECIPROCAL 3-PORT JUNCTION*

The equivalent circuit is obtained by starting with the relation between the reflection coefficient S_{11} and the scattering matrix eigenvalues

$$S_{11} = \frac{s_0 + s_{+1} + s_{-1}}{3} \tag{15.1}$$

Making use of the relations between the eigenvalues of the scattering and admittance matrices, one has the result derived in Chapter 5

$$S_{11} = \frac{1 - 3y_1}{3 + 3y_1} \tag{15.2}$$

The equivalent circuit, for which S_{11} is given by the last equation, is shown in Figure 15.1.

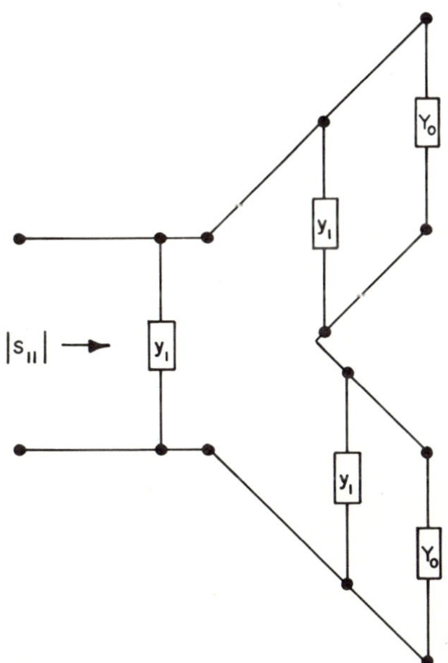

Figure 15.1.   *Equivalent circuit of reciprocal junction in terms of eigenvalues of admittance matrix ($n = 1$).*

The admittance eigenvalue y_1 has the nature of a $\lambda/4$ short-circuited transmission line:

$$y_1 = -j\left(\frac{4b' \cot\theta}{\pi}\right) \tag{15.3}$$

where

$$\theta = \frac{\pi}{2}(1 + \delta) \tag{15.4}$$

Here b' is the normalized susceptance slope parameter, and 2δ is a normalized frequency variable.

The equivalent network in terms of distributed $\lambda/4$ short-circuited transmission lines is indicated in Figure 15.2.

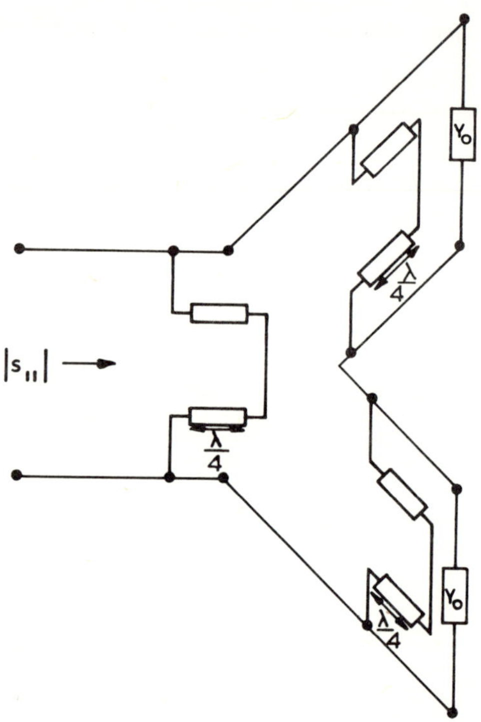

Figure 15.2.   *Equivalent circuit of reciprocal junction in terms of quarterwave stubs* ($n = 1$).

15.2.   *REFLECTION COEFFICIENT OF QUARTERWAVE RECIPROCAL 3-PORT JUNCTIONS*

The equivalent circuit of the overall network is shown in Figure 15.3. The matching network is represented in terms of an overall *ABCD* matrix. The ones studied here consist of quarterwave-long impedance transformers. The reflection coefficient of the overall network is given by straightforward calculation by

$$|\Gamma| = \left[\frac{(DE - A - BF)^2 + (C - BE - DF)^2}{(DE + A + BF)^2 + (C + BE - DF)^2}\right] \tag{15.5}$$

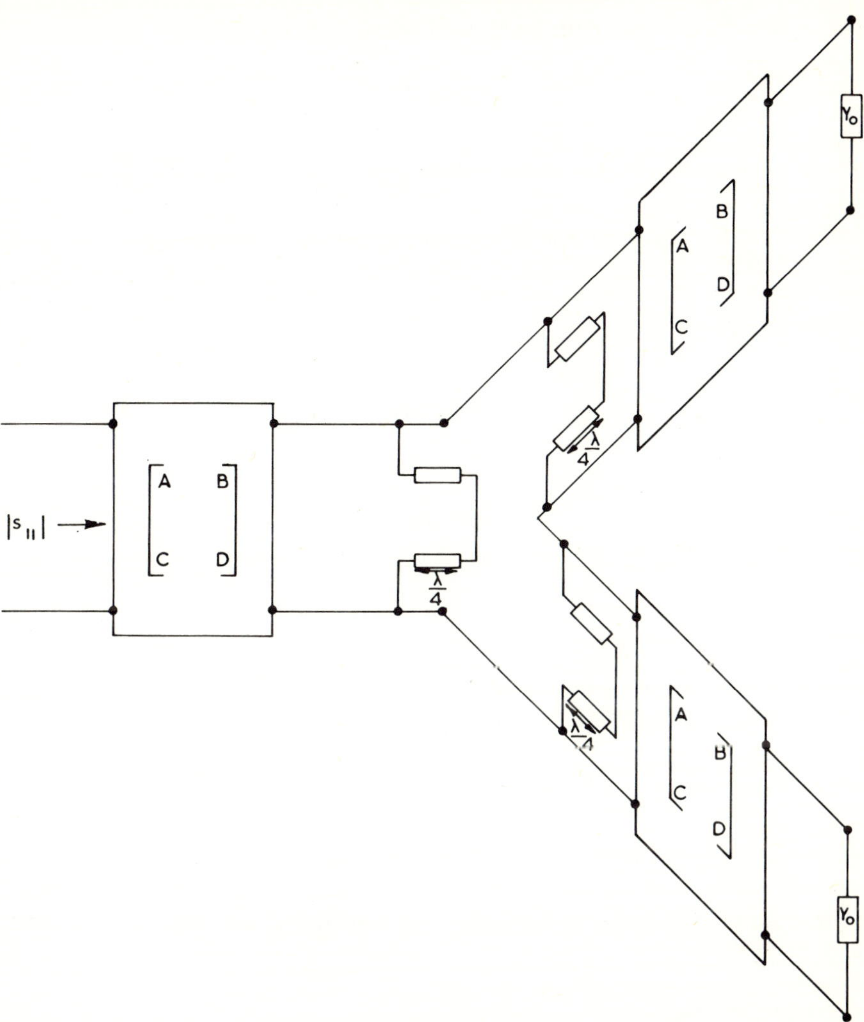

Figure 15.3.   *Equivalent circuit of wide band reciprocal junction in terms of ABCD networks.*

E and F are the real and imaginary parts of the equivalent network due to
ports 2 and 3, which load the $ABCD$ matrix at port 1, and A, B, C, and D
are real numbers. The VSWR r is given in terms of the reflection
coefficient in the usual way by

$$r = \frac{1 + |\Gamma|}{1 - |\Gamma|}$$

(15.6)

The minimum value for the reflection coefficient is

$$|\Gamma| = \frac{1}{3} \tag{15.7}$$

which corresponds with maximum power transfer through the junction.

15.3.   FREQUENCY RESPONSE OF $n=1$ NETWORK

The frequency response at the junction terminals is obtained with $ABCD$ parameters given by

$$A = \cos\theta \tag{15.8}$$

$$B = \sin\theta \tag{15.9}$$

$$C = \sin\theta \tag{15.10}$$

$$D = \cos\theta \tag{15.11}$$

and

$$E = 2 \tag{15.12}$$

$$F = 3\left(\frac{4b'\cot\theta}{\pi} \right) \tag{15.13}$$

The definition used here for θ is defined in Eq. 15.4.

It is possible to obtain an approximate relation in the neighborhood of the resonance frequency by replacing the distributed network by a lumped element one.

$$y_1 = j2\delta b' \tag{15.14}$$

This last equation is obtained directly from Eq. 15.3.

The result in terms of the VSWR r is

$$b' = \frac{\frac{2}{3}\left[(r^2 - 2.5r + 1)/2r \right]^{\frac{1}{2}}}{2\delta} \tag{15.15}$$

This result has been derived in Chapter 13.

15.4.   *FREQUENCY RESPONSE OF $n=2$ NETWORK*

The frequency response of the $n=2$ network is obtained with

$$A = \cos\theta \tag{15.16}$$

$$B = \frac{\sin\theta}{y_t} \tag{15.17}$$

$$C = y_t \sin\theta \tag{15.18}$$

$$D = \cos\theta \tag{15.19}$$

$$E = \frac{2(AD + BC)}{A^2 + B^2} \tag{15.20}$$

$$F = \frac{[3(4b'\cot\theta)/\pi](A^2 + B^2) - 2(AC - BD)}{A^2 + B^2} \tag{15.21}$$

where y_t is the transformer admittance. The arrangement used here is illustrated in Figure 15.4.

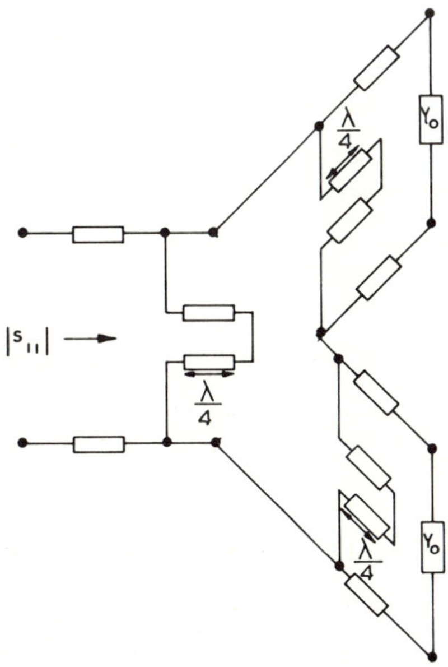

Figure 15.4.   Equivalent circuit of quarterwave coupled reciprocal junction ($n=2$).

15.5. *FREQUENCY RESPONSE OF $n=3$ NETWORK*

The frequency response of the $n=3$ network is obtained with

$$A = \cos^2\theta - \frac{y_{02}}{y_{01}}\sin^2\theta \tag{15.22}$$

$$B = \left(\frac{1}{y_{01}} + \frac{1}{y_{02}}\right)\sin\theta\cos\theta \tag{15.23}$$

$$C = (y_{01}+y_{02})\sin\theta\cos\theta \tag{15.24}$$

$$D = \frac{-y_{01}}{y_{02}}\sin^2\theta + \cos^2\theta \tag{15.25}$$

$$E = \frac{2(A_1D_1 + B_1C_1)}{A_1^2 + B_1^2} \tag{15.26}$$

$$F = \frac{[3(4b'\cot\theta)/\pi](A_1^2 + B_1^2) - 2(A_1C_1 - B_1D_1)}{A_1^2 + B_1^2} \tag{15.27}$$

where

$$A_1 = \cos^2\theta - \frac{y_{01}}{y_{02}}\sin^2\theta \tag{15.28}$$

$$B_1 = \left(\frac{1}{y_{01}} + \frac{1}{y_{02}}\right)\sin\theta\cos\theta \tag{15.29}$$

$$C_1 = (y_{01}+y_{02})\sin\theta\cos\theta \tag{15.30}$$

$$D_1 = \frac{-y_{02}}{y_{01}}\sin^2\theta + \cos^2\theta \tag{15.31}$$

Here y_{01} and y_{02} are the normalized admittances of the transformers. The schematic diagram for this situation is depicted in Figure 15.5.

15.6. *COMPUTATIONS*

This section describes some computations on the frequency responses of the $n=2$ and 3 networks. The examples treated in this section apply to networks that when magnetized yield circulators with $r=1.15$ and $2\delta_0 = 0.35$, 0.50, and 0.66. The element values for the networks are given in

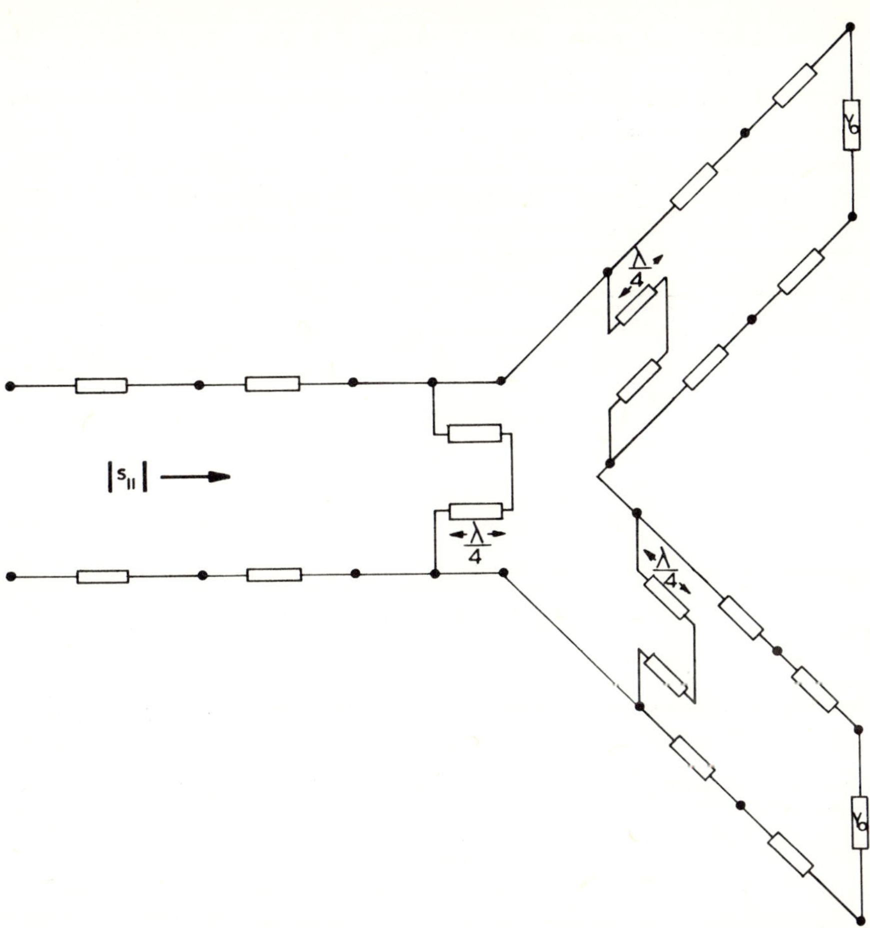

Figure 15.5.   *Equivalent circuit of two section quarterwave coupled reciprocal junction ($n=3$).*

Chapter 14. The results are shown in Figures 15.6 and 15.7 in the case of $n=2$ and in Figures 15.8, 15.9, and 15.10 in the case of $n=3$. The frequency responses are given in each case at the terminals of each network. The results for the frequency response at the intermediate transformer terminals in the case of $n=3$ is obtained from the results stated in Section 15.4 with y_t replaced by y_{02}. This allows the precise experimental adjustment of each matching network to obtain the overall response. It is seen from these illustrations that the bandwidth of the overall reciprocal network is comparable with that of the junction magnetized to form a circulator. It is also noticed that the frequency responses of the $n=2$ and 3 reciprocal junctions, associated with similar circulator responses, are similar. The only difference being that the slope at the band edges for the $n=2$ network is less than the one for the $n=3$ one. Here, the bandwidth of the reciprocal junction refers to that which coincides with maximum power transfer through the junction. This means that the construction of wideband circulators is closely related to that of wideband reciprocal 3-port junctions. Figure 15.11 illustrates one experimental result.

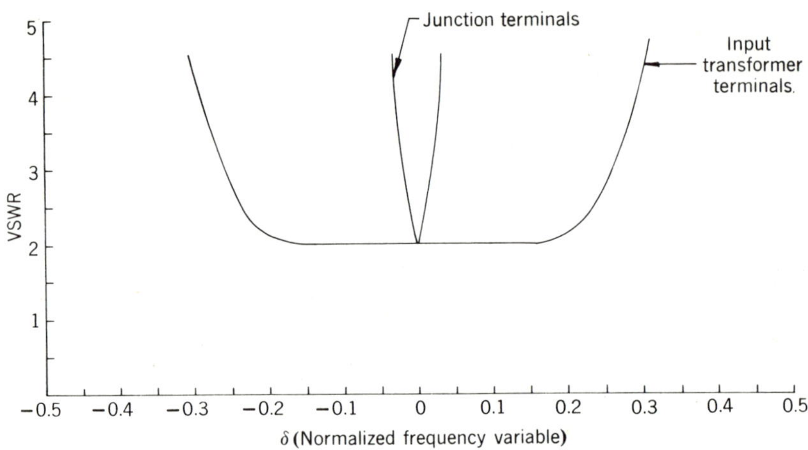

Figure 15.6.   *Frequency response for n = 2 junction with r = 1.15 and $2\delta_0 = 0.35$ at each pair of terminals.*

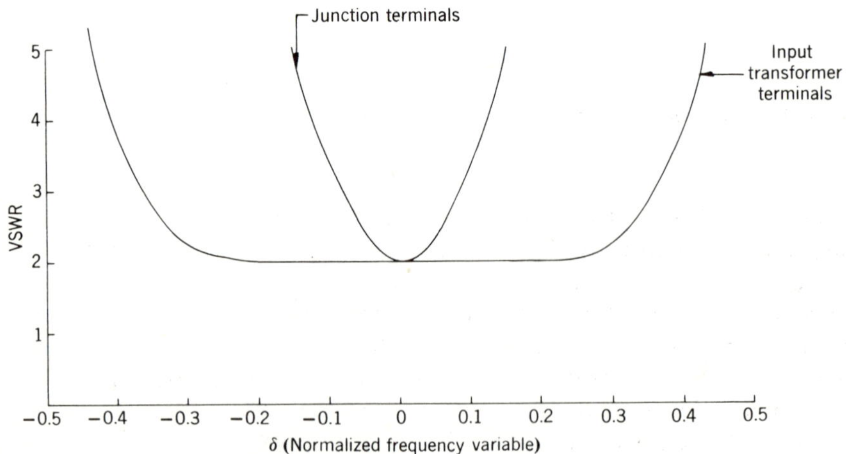

Figure 15.7.   *Frequency response for n = 2 junction with r = 1.15 and $2\delta_0 = 0.50$ at each pair of terminals.*

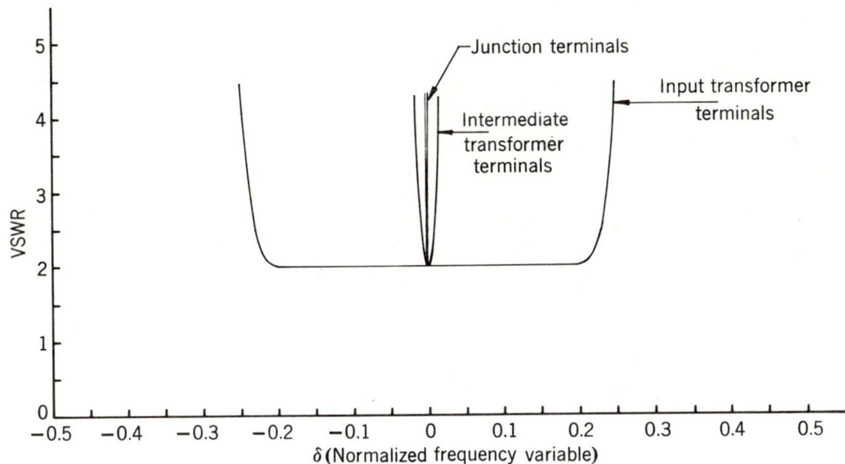

Figure 15.8.   *Frequency response for n = 3 junction with r = 1.15 and 2δ₀ = 0.35 at each pair of* *terminals*.

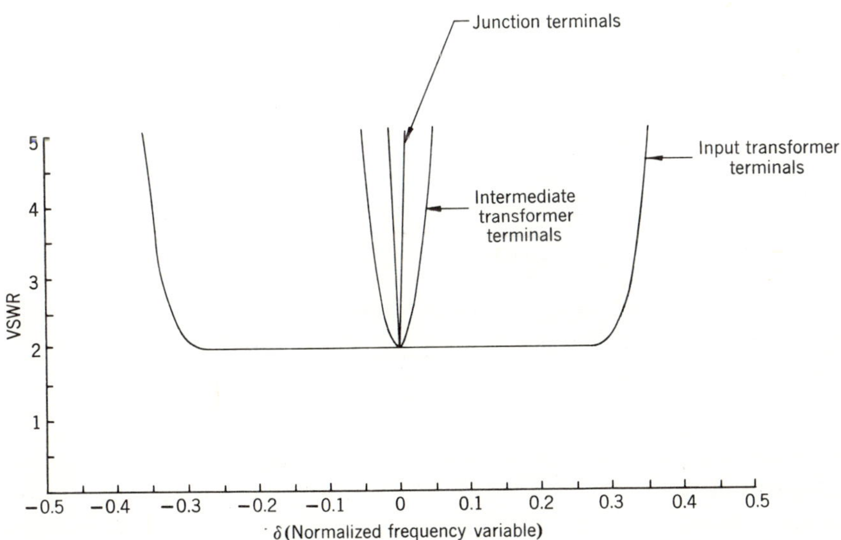

Figure 15.9.   *Frequency response for n = 3 junction with r = 1.15 and 2δ₀ = 0.50 at each pair of* *terminals*.

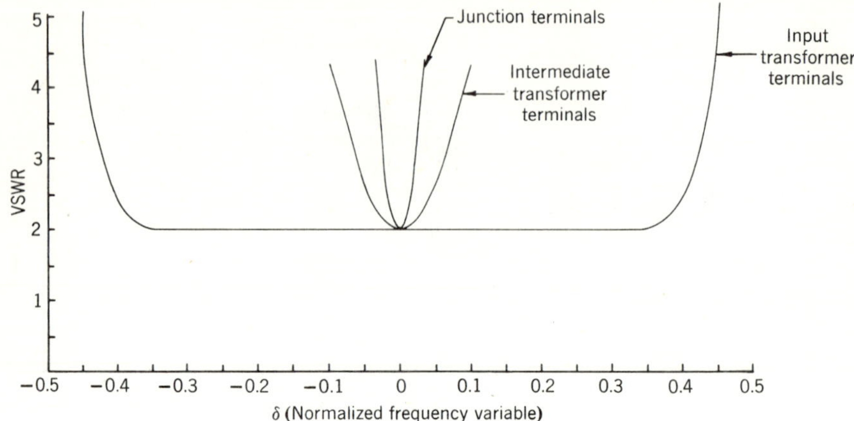

Figure 15.10. *Frequency response for $n = 3$ junction with $r = 1.15$ and $2\delta_0 = 0.65$ at each pair of terminals.*

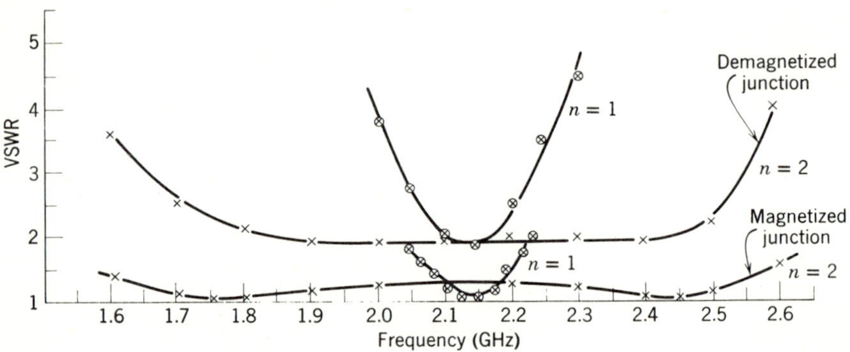

Figure 15.11. *Experimental frequency responses for $n = 1$ and 2 networks.*

272

CHAPTER SIXTEEN

Scattering Matrices of Junction Circulators with Chebyshev Characteristics

The theory of wideband circulators using external matching networks usually starts by assuming that the equivalent circuit of the device is a 1-port network. This 1-port circuit consists of the gyrator conductance in parallel with either a lumped or distributed resonator. It assumes that the frequency behaviour of the in-phase eigennetwork at the gyrator terminals may be omitted compared with that of the two counterrotating ones. The bandwidth over which this approximation applies has been discussed in Chapter 12 in terms of the resonant frequencies of the counterrotating eigennetworks, but a fuller investigation of the omission of the frequency variation of the in-phase eigennetwork on the quality of this equivalent circuit appears desirable.

The most general representation of the three port circulator is in terms of the eigenvalues of the scattering matrix. The eigenvalues of this matrix are reflection coefficients associated with the different ways of exciting the junction. The entries of the scattering matrix are constructed by taking linear combinations of these eigenvalues. This method therefore yields not only the reflection coefficient at the input port but also the transmission coefficients of the junction. The approach is quite general and applies to the *m*-port junction also. It starts by representing the matching network at each port by its A B C D matrix. The eigenvalues at the input terminals of the junction are then obtained one at a time in terms of the A B C D coefficients and the initial set of eigenvalues at the gyrator terminals.

In this chapter the boundary condition for circulators with Chebyshev frequency characteristics is first established at the terminals of the matching network in terms of the eigenvalues of the scattering matrix by omitting the frequency variation of the in-phase eigennetwork at the gyrator terminals and subsequently reintroducing it to study its influence on the overall frequency response. It is found that the former results are in excellent agreement with those obtained in Chapter 14 by connecting the matching network directly to the 1-port circuit.

The influence of the in-phase eigenvalue on the overall frequency response of the circulator is studied separately in the case of the stripline circulator by combining the electromagnetic and network problems. This result indicates that the in-phase eigennetwork cannot be neglected for high quality communication circulators. However, the one port results can be used provided the in-phase eigennetwork is tuned by an additional independent variable such as a thin metal post at the center of the junction or thin metal posts at the input terminals of the quarter wave transformers.

16.1. EIGENVALUES OF SCATTERING MATRIX

The entries of the scattering matrix of junctions are usually directly constructed in terms of their eigenvalues. In the case of the 3-port junction, the following standard equations obtained in Chapter 2 apply:

$$S_{11} = \frac{s_0 + s_{+1} + s_{-1}}{3} \tag{16.1}$$

$$S_{12} = \frac{s_0 + s_{+1}e^{j2\pi/3} + s_{-1}e^{-j2\pi/3}}{3} \tag{16.2}$$

$$S_{13} = \frac{s_0 + s_{+1}e^{-j2\pi/3} + s_{-1}e^{j2\pi/3}}{3} \tag{16.3}$$

The schematic of the junction is depicted in Figure 16.1.

The eigenvalues of the scattering matrix are the reflection coefficients associated with each possible way of exciting the junction

$$s_0 = e^{-j2\theta_0} \tag{16.4}$$

$$s_{+1} = e^{-j2(\theta_1 + \theta_{+1} + \pi/2)} \tag{16.5}$$

$$s_{-1} = e^{-j2(\theta_1 + \theta_{-1} + \pi/2)} \tag{16.6}$$

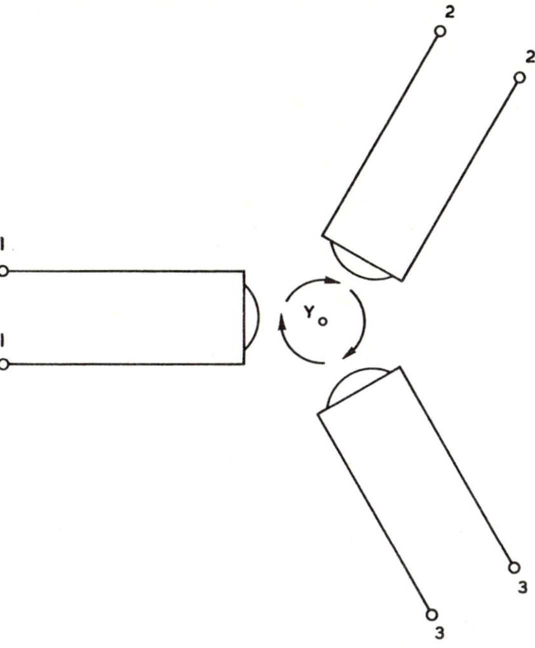

Figure 16.1.   Schematic of 3-port junction circulator.

The angles of the eigenvalues are defined below. The admittance eigenvalues are related to those of the scattering matrix by

$$y_0 = \frac{1 - s_0}{1 + s_0} = j \tan \theta_0 \tag{16.7}$$

$$y_{+1} = \frac{1 - s_{+1}}{1 + s_{+1}} = j \tan\left(\theta_1 + \theta_{+1} + \frac{\pi}{2}\right) \tag{16.8}$$

$$y_{-1} = \frac{1 - s_{-1}}{1 + s_{-1}} = j \tan\left(\theta_1 + \theta_{-1} + \frac{\pi}{2}\right) \tag{16.9}$$

The equivalent circuits of the admittance eigenvalues y_{+1} and y_{-1} are short-circuited transmission lines of electrical length $\theta_1 + \theta_{\pm 1}$ as shown in Figure 16.2. For a 3-port circulator for which $S_{12} = -1$, the equivalent circuit for the admittance eigenvalue y_0 is a quarterwave-long open-circuited transmission line of length θ_0. The admittance eigenvalues defined by equations 16.7 through 16.9 are in terms of transmission lines of characteristic impedance equal to that of the input lines. A more suitable

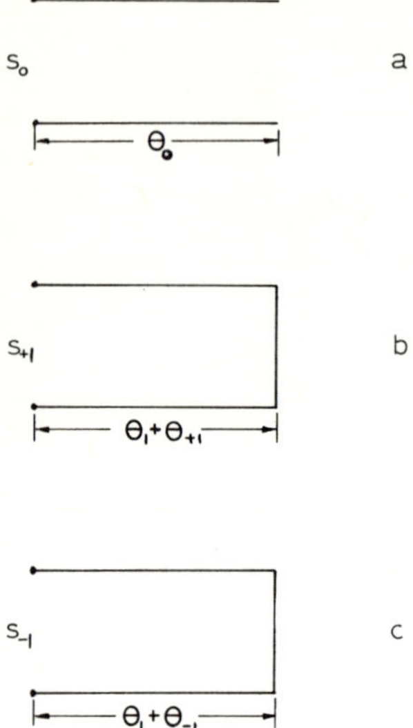

Figure 16.2.   *Eigennetworks of 3-port junction circulator.*

set for stripline circulators is

$$y_0 = jy_0 \tan\theta \tag{16.10}$$

$$y_{+1} = -jy\cot\theta + jy'_{+1}\tan\theta'_{+1} \tag{16.11}$$

$$y_{-1} = -jy\cot\theta + jy'_{-1}\tan\theta'_{-1} = -jy\cot - jy'_{+1}\tan\theta'_{+1} \tag{16.12}$$

The form of these equations is obtained by forming the input admittance of the networks in terms of the *ABCD* matrices.

The equivalent circuit for the admittance y_{+1} is a transmission line of admittance y'_{+1} and electrical length θ'_{+1} in cascade with a short-circuited transmission line of admittance y of electrical length θ. A similar statement applies to the admittance y_{-1}. These transmission lines are shown in Figure 16.3. The equivalent circuit for the admittance eigenvalue y_0 remains unchanged since it is unaffected by the direct magnetic field.

<h3 style="text-align:center">16.2.   EIGENVALUES OF AUGMENTED SCATTERING
MATRIX</h3>

If matching networks are now connected to each of the ports of the junction, one obtains the schematic diagram shown in Figure 16.4. The eigennetworks for this circuit are shown in Figure 16.5. The entries of the

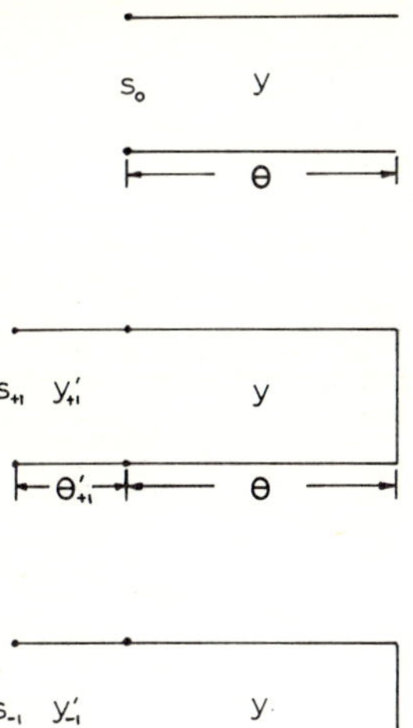

Figure 16.3. *Eigennetworks of magnetized 3-port junction circulator in terms of frequency and magnetic variables.*

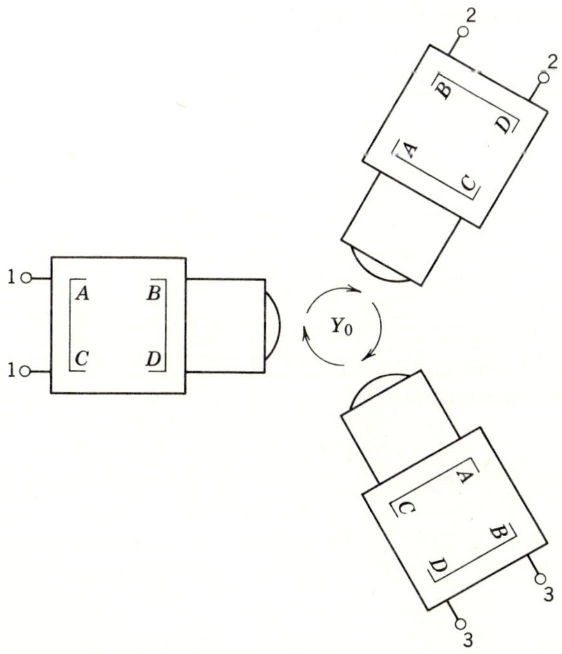

Figure 16.4. *Schematic of 3-port junction circulator with matching networks.*

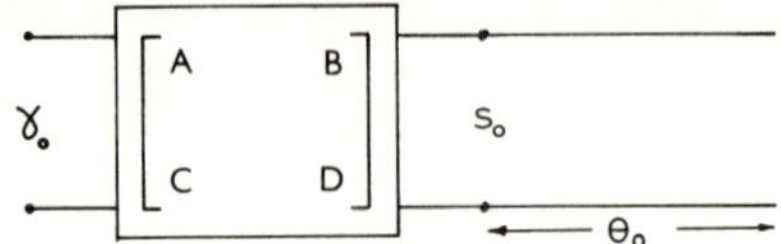

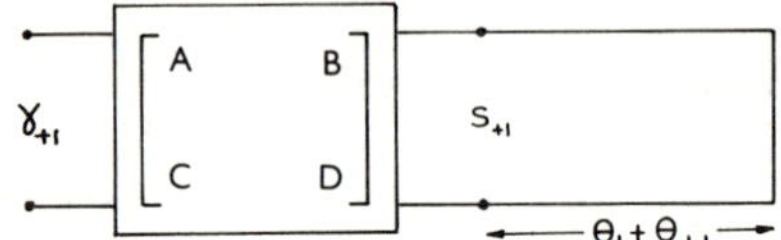

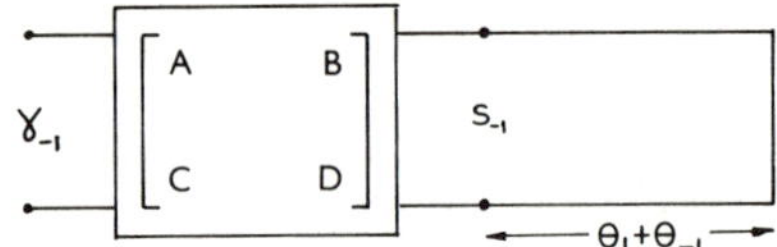

Figure 16.5.   *Eigennetworks of 3-port circulator with matching networks.*

scattering matrix at the new terminals are now given by

$$\Gamma_{11} = \frac{\gamma_0 + \gamma_{+1} + \gamma_{-1}}{3} \tag{16.13}$$

$$\Gamma_{12} = \frac{\gamma_0 + \gamma_{+1}e^{j2\pi/3} + \gamma_{-1}e^{-j2\pi/3}}{3} \tag{16.14}$$

$$\Gamma_{13} = \frac{\gamma_0 + \gamma_{+1}e^{-j2\pi/3} + \gamma_{-1}e^{j2\pi/3}}{3} \tag{16.15}$$

The eigenvalues at the input terminals of the *ABCD* networks are given by straightforward calculation by

$$\gamma_0 = e^{-j2\psi_0} \tag{16.16}$$

$$\gamma_{+1} = e^{-j2(\psi_{+1} + \pi/2)} \tag{16.17}$$

$$\gamma_{-1} = e^{-j2(\psi_{-1} + \pi/2)} \tag{16.18}$$

where

$$j\tan(\psi_0) = \frac{jC + Dy_0}{A + jBy_0} \tag{16.19}$$

$$j\tan\left(\psi_{+1}+\frac{\pi}{2}\right)=\frac{jC+Dy_{+1}}{A+jBy_{+1}} \qquad (16.20)$$

$$j\tan\left(\psi_{-1}+\frac{\pi}{2}\right)=\frac{jC+Dy_{-1}}{A+jBy_{-1}} \qquad (16.21)$$

It is also observed that $\gamma\gamma^*=1$. Hence, the new eigenvalues lie on the unit circle also.

For a single quarterwave transformer, the $ABCD$ parameters are

$$A=\cos\theta \qquad (16.22)$$

$$B=\frac{\sin\theta}{y_t} \qquad (16.23)$$

$$C=y_t\sin\theta \qquad (16.24)$$

$$D=\cos\theta \qquad (16.25)$$

The frequency variable θ is

$$\theta=\frac{\pi}{2}(1+\delta) \qquad (16.26)$$

where

$$2\delta=2\left(\frac{\omega-\omega_0}{\omega_0}\right) \qquad (16.27)$$

At the band edges the frequency variable becomes

$$2\delta_0=2\left(\frac{\omega_1-\omega_0}{\omega_0}\right) \qquad (16.28)$$

The Chebyshev polynomial passes through zero for $n=2$ when

$$\sqrt{2}\,\cos\theta=\cos\theta_0 \qquad (16.29)$$

Fig. 16.6 depicts the frequency response considered in this section.

16.3. CIRCULATION ADJUSTMENT

The design proceeds by having the frequency at which the reflection coefficient passes through its zeros and maxima coincide with those of a Chebyshev polynomial.

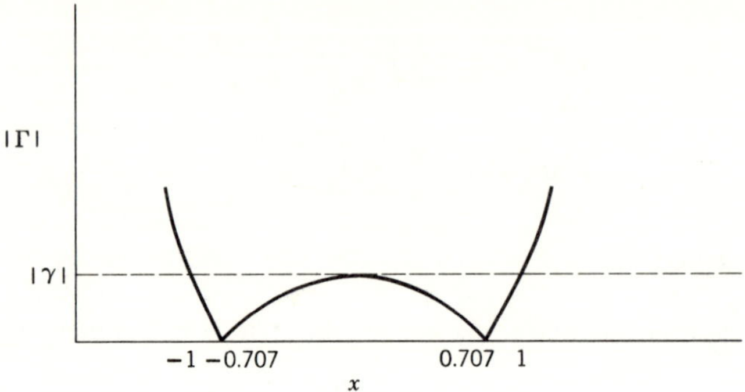

Figure 16.6.   *Chebyshev response of n = 2 quarterwave coupled circulator.*

When the reflection coefficient passes through zero, the eigenvalues lie equally spaced on the unit circle:

$$\psi_{+1} + \frac{\pi}{2} = \psi_0 + \frac{\pi}{3} \tag{16.30}$$

$$\psi_{-1} + \frac{\pi}{2} = \psi_0 - \frac{\pi}{3} \tag{16.31}$$

Substituting the last two equations into Eqs. 16.19 through 16.21 gives

$$j \tan \psi_0 = -\frac{jD}{B} \tag{16.32}$$

$$j \tan\left(\psi_0 + \frac{\pi}{3}\right) = \frac{jC + Dy_{+1}}{A + jBy_{+1}} \tag{16.33}$$

$$j \tan\left(\psi_0 - \frac{\pi}{3}\right) = \frac{jC + Dy_{-1}}{A + jBy_{-1}} \tag{16.34}$$

The simplified form for y_0 comes about because $y_0 = \infty$ for a circulator for which $s_0 = -1$. Expanding the above equations in terms of $t = \tan \psi_0$ and putting $y_{\pm 1} = j(\lambda \mp \mu)$ gives

$$t = \frac{D}{-B} \tag{16.35}$$

$$\frac{t - \sqrt{3}}{1 + t\sqrt{3}} = \frac{C - \lambda D + \mu D}{A + \lambda B - \mu B} = \frac{t - \dfrac{(C - D\lambda)}{\mu B}}{1 + t\dfrac{(A + B\lambda)}{\mu D}} \tag{16.36}$$

$$\frac{t+\sqrt{3}}{1-t\sqrt{3}} = \frac{C-\lambda D-\mu D}{A+\lambda B+\mu B} = \frac{t+\dfrac{C-\lambda D}{\mu B}}{1-t\dfrac{A+\lambda B}{\mu D}} \tag{16.37}$$

These equations are consistent provided

$$y\cot\theta = \frac{CD-AB}{B^2+D^2} \tag{16.38}$$

and

$$\sqrt{3}\,y'_{+1}\tan\theta'_{+1} = \frac{AD+BC}{B^2+D^2} \tag{16.39}$$

which must be evaluated at $\sqrt{2}\ \cos\theta = \cos\theta_0$. These are the equations previously derived in Chapter 14.

To obtain y_t it is necessary to directly evaluate Γ_{11} at the center frequency in terms of the original variables

$$\gamma_0 = 1$$

$$\gamma_{+1} = \frac{\left(y'_{+1}\tan\theta'_{+1}/y_t\right)^2 - y_t^2 + 2j\left(y'_{+1}\tan\theta'_{+1}/y_t\right)y_t}{\left(y'_{+1}\tan\theta'_{+1}/y_t\right)^2 + y_t^2} \tag{16.40}$$

$$\gamma_{-1} = \frac{\left(y'_{-1}\tan\theta'_{-1}/y_t\right)^2 - y_t^2 + 2j\left(y'_{-1}\tan\theta'_{-1}/y_t\right)y_t}{\left(y'_{-1}\tan\theta'_{-1}/y_t\right)^2 + y_t^2} \tag{16.41}$$

The imaginary part of Γ_{11} is zero provided

$$y'_{-1}\tan\theta'_{-1} = -y'_{+1}\tan\theta'_{+1} \tag{16.42}$$

The real part of Γ_{11} is given in terms of the VSWR by

$$|\Gamma_{11}| = \frac{r-1}{r+1} = \frac{\left(\sqrt{3}\,y'_{+1}\tan\theta'_{+1}\right)^2 - y_t^4}{\left(\sqrt{3}\,y'_{+1}\tan\theta'_{+1}\right)^2 + 3y_t^4} \tag{16.43}$$

For $g < y_t^2$ the result is

$$\sqrt{3}\,y'_{+1}\tan\theta'_{+1} = y_t^2\sqrt{\frac{2-r}{r}} \tag{16.44}$$

In terms of the original variables Eqs. 16.38, 16.39, and 16.44 become

$$g = \frac{\sqrt{r/(2-r)} - \sin^2\theta_0}{\sqrt{r/(2-r)}\ \cos^2\theta_0} \tag{16.45}$$

$$y = \left[g\sqrt{\frac{2-r}{r}} \right]^{1/2} \left[g\sqrt{\frac{r}{2-r}} - 1 \right] \sin^2\theta_0 \tag{16.46}$$

$$y_t^2 = \sqrt{\frac{r}{2-r}}\ g \tag{16.47}$$

The present results are identical with equations 14.20, 14.22 and 14.23 provided

$$r \approx \sqrt{\frac{r}{2-r}} \tag{16.48}$$

which is a good approximation for the values of r usually encountered in circulator design.

16.4. *QUARTERWAVE COUPLED BELOW RESONANCE STRIPLINE CIRCULATOR*

The theory developed so far will now be combined with the e.m. problem in the case of the stripline circulator.

$$y_{+1} = \frac{-j\pi Y_e}{3\sin\psi} \cdot \left[J_1'(kR) - \frac{K}{\mu}\frac{J_1(kR)}{kR} \right] [J_1(kR)]^{-1} \tag{16.49}$$

$$y_{-1} = \frac{-j\pi Y_e}{3\sin\psi} \cdot \left[J_1'(kR) + \frac{K}{\mu}\frac{J_1(kR)}{kR} \right] [J_1(kR)]^{-1} \tag{16.50}$$

$$y_0 = \frac{j\pi Y_e}{3\psi} \cdot \left[\frac{J_0'(kR)}{J_0(kR)} \right] \tag{16.51}$$

Figure 16.7 depicts the above eigenadmittances as a function of kR for a circulator which when magnetized gives $r = 1.07$. The phase angles of the eigenreflection coefficients are shown in Figure 16.8. These results show that the equivalent circuits for y_{+1} and y_{-1} are short-circuited radial transmission lines, while that of the eigennetwork y_0 is an open-circuited transmission line. If it assumed from the e.m. problem, in the usual way, that the frequency variation of the y_0 network can be omitted compared to the $y_{\mp1}$ networks as was done in the network problem the two are related. The result is

$$g = \sqrt{3}\, y_{+1} \tan\theta_{+1} = \frac{\pi Y_e}{\sqrt{3}\, Y_0 \sin\psi kR} \cdot \frac{K}{\mu} \tag{16.52}$$

$$b' = \frac{\pi y}{4} = \frac{\pi Y_e}{3 Y_0 \sin\psi}\left[\frac{(kR)^2 - 1}{2(kR)}\right] \tag{16.53}$$

where it has been assumed that the susceptance slope parameters for the two equivalent reciprocal circuits are the same.

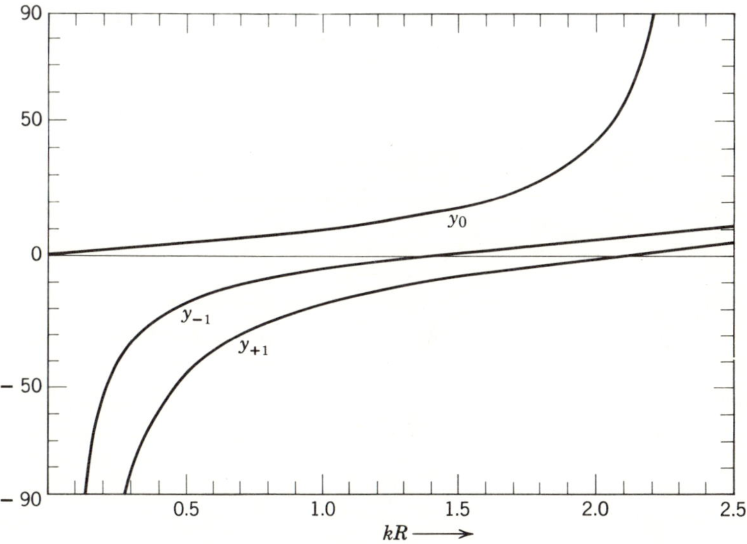

Figure 16.7.   *Eigenadmittances of stripline circulator with* $r = 1.07, 2\delta_0 = 20\%$.

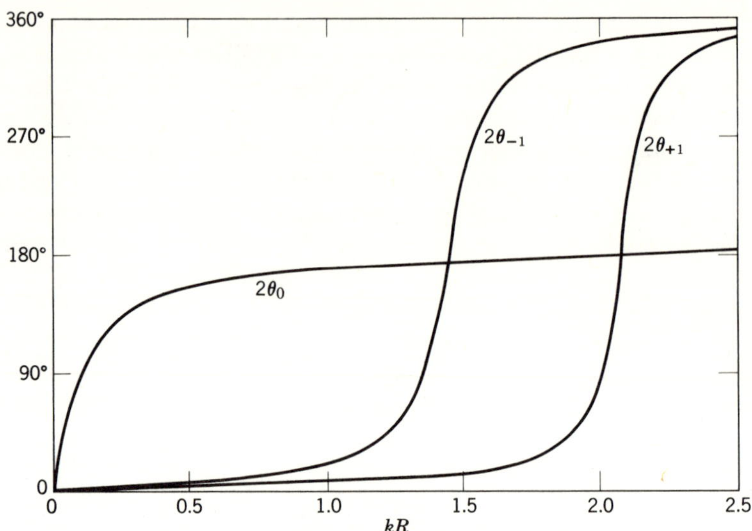

Figure 16.8.   *Electrical lengths of eigennetworks for stripline circulator with* $r = 1.07, 2\delta_0 = 20\%$.

Here g is the normalized gyrator admittance of the junction, and b' is the normalized susceptance slope parameter of the reciprocal eigennetworks. The electromagnetic problem that must be satisfied is therefore

$$\frac{\pi Y_e}{3 Y_0 \sin\psi} \cdot \left[\frac{(kR)^2 - 1}{2kR} \right] = \frac{\pi}{4} \left[\frac{r - \sin^2\theta_0}{r^2 \cos^2\theta_0} \right]^{1/2} (r-1)\tan^2\theta_0 \quad (16.54)$$

$$\frac{\pi Y_e}{\sqrt{3}\, Y_0 \sin\psi kR} \cdot \frac{K}{\mu} = \frac{r - \sin^2\theta_0}{r \cos^2\theta_0} \quad (16.55)$$

The first equation determines $\sin\psi$ and the second gives K/μ. In the above eqs.

$$Y_e = 4\sqrt{\frac{\epsilon_r \epsilon_0}{\mu_e \mu_0}} \left[\ln\left(\frac{W+b}{W+t} \right) \right]^{-1} \quad (16.56)$$

$$Y_0 = 4\sqrt{\frac{\epsilon_0}{\mu_0}} \left[\ln\left(\frac{W+b}{W+t} \right) \right]^{-1} \quad (16.57)$$

$$\sin\psi = \frac{W}{2R} \quad (16.58)$$

where W is the width of the stripline, b is the ground plane spacing, t is the thickness of the center conductor, and the other quantities have the usual meanings.

The width of the center conductor is now obtained from Eq. 16.58 and for a 50 Ω line the ground plane spacing and center conductor thickness are obtained from Eq. 16.57.

16.5. *FREQUENCY VARIATION OF QUARTERWAVE COUPLED CIRCULATOR*

The assumption used throughout this text is that the frequency variation of the s_0 eigenvalue may be omitted compared to that of the $s_{\pm 1}$ eigenvalues. This assumption will now be tested in the case of stripline circulators with $2\delta_0 = 20\%$ and $r = 1.10$, 1.15, and 1.22. The physical variables $\pi Y_e /$ $3\sin\psi$ and K/μ in Eqs. 16.49 through 16.51 are given by Eqs. 16.54 and 16.55 once the bandwidth and VSWR are stated.

Figures 16.9, 16.10, and 16.11 give the frequency behavior of the overall quarterwave-coupled circulator for the ideal and actual cases. These results have been obtained by assuming that the center frequency lies midway between the two split frequencies rather than at $kR = 1.84$.

The above results show that although the frequency variation of the in phase reflection coefficient is small compared to those of the counter-rotating ones it cannot be ignored for devices with relatively widebands and low VSWR's.

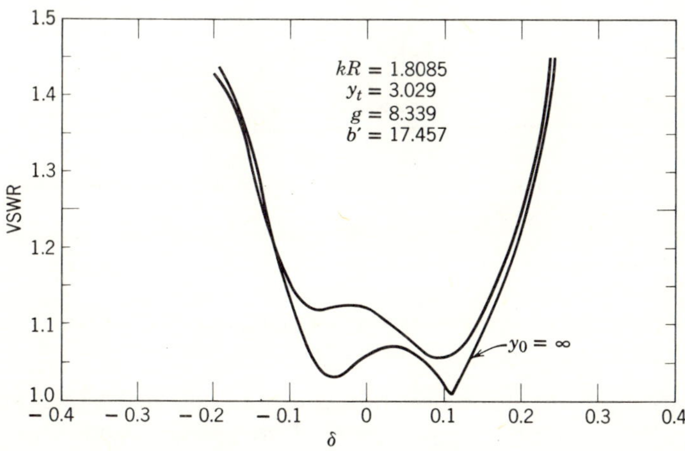

Figure 16.9. *Frequency response of quarterwave coupled circulator with and without third eigennetwork for* $r = 1.10, 2\delta_0 = 20\%$.

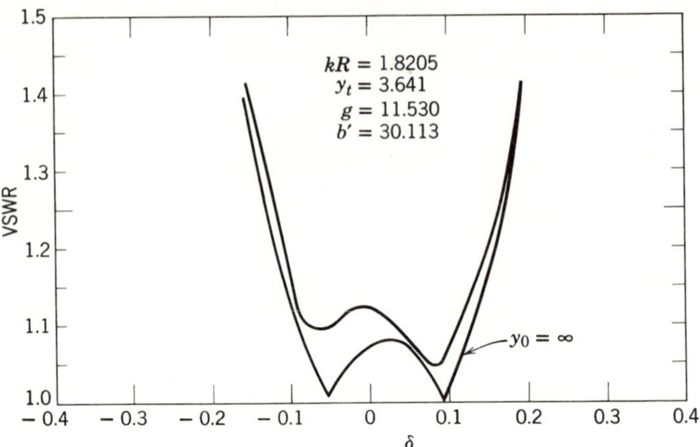

Figure 16.10.　*Frequency response of quarterwave coupled circulator with and without third eigennetwork for* $r = 1.15, 2\delta_0 = 20\%$.

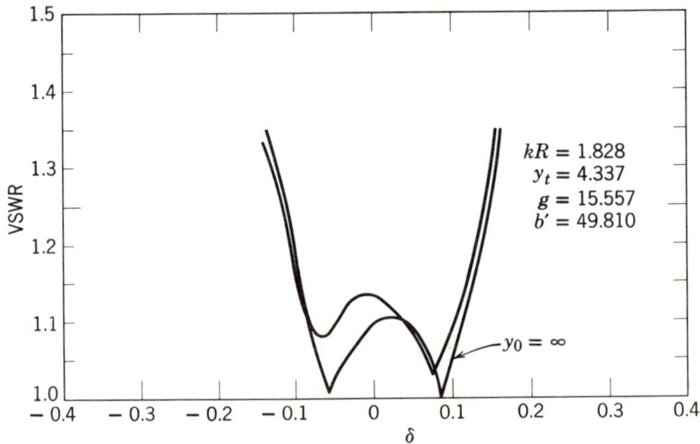

Figure 16.11.　*Frequency response of quarterwave coupled circulator with and without third eigennetwork for* $r = 1.22, 2\delta_0 = 20\%$.

16.6.　FREQUENCY RESPONSE OF QUARTERWAVE COUPLED CIRCULATOR WITH CAPACITIVE TUNING

One way in which the theoretical results can be obtained in practice is to introduce a thin metal post in the center of the junction. Equivalently, thin metal posts may be introduced at the input transformer terminals, which is the normal approach in stripline devices. This last statement comes about because the in phase eigennetwork is an opencircuited half wave resonator which can be tuned at either input or output terminals.

The *ABCD* matrix for a single shunt capacitor is

$$A = 1 \tag{16.59}$$

$$B = 0 \tag{16.60}$$

$$C = \omega C \tag{16.61}$$

$$D = 1 \tag{16.62}$$

Cascading this *ABCD* network with that for the single quarterwave transformer gives the following *ABCD* parameters for the overall matching network:

$$A = \cos \theta \tag{16.63}$$

$$B = \frac{\sin \theta}{Y_t} \tag{16.64}$$

$$C = Y_t \sin \theta + \omega C \cos \theta \tag{16.65}$$

$$D = \cos \theta - \omega C \frac{\sin \theta}{Y_t} \tag{16.66}$$

The arrangement used here is shown in Figure 16.12.

Figures 16.13, 16.14, and 16.15 indicate the influence of the capacitor C on the overall frequency response of the device for $r = 1.10$, 1.15, and 1.22 and $2\delta_0 = 20\%$. They show that such a capacitor can indeed be used to improve the correlation between the two and three eigennetworks models of the stripline circulator.

16.7.   *FREQUENCY RESPONSE OF RECIPROCAL 3-PORT JUNCTION*

It is also possible to obtain the frequency response of the reciprocal junction by taking a linear combination of the eigenreflection coefficients of the junction. The result is

$$3\Gamma_{11} = \gamma_0 + \gamma_{+1} + \gamma_{-1} \tag{16.67}$$

$$= \gamma_0 + 2\gamma_1 \tag{16.68}$$

where

$$\gamma_0 = e^{-j2\psi_0} \tag{16.69}$$

$$\gamma_1 = e^{-j2(\psi_1 + \pi/2)} \tag{16.70}$$

and

$$j\tan(\psi_0 + \pi/2) = \frac{jC + Dy_0}{A + jBy_0} \tag{16.71}$$

$$j\tan(\psi_1 + \pi/2) = \frac{jC + Dy_1}{A + jBy_1} \tag{16.72}$$

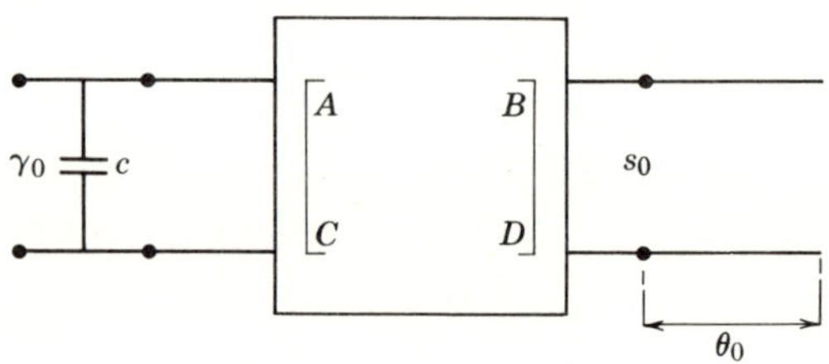

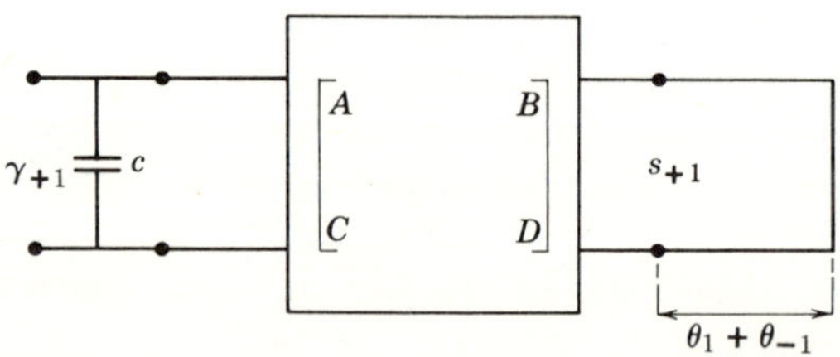

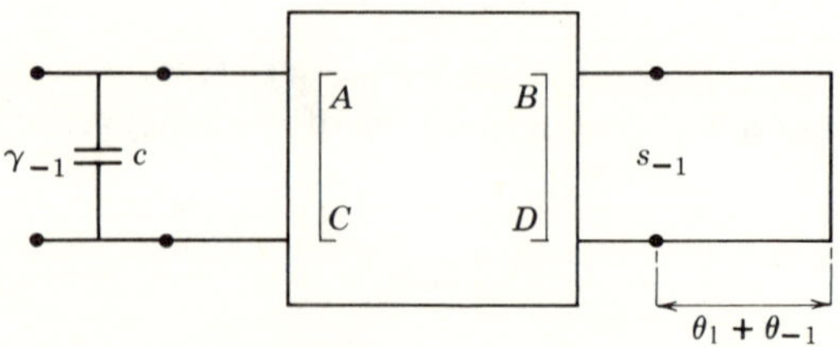

Figure 16.12.   *Eigennetworks of 3-port circulator with capacitive tuning.*

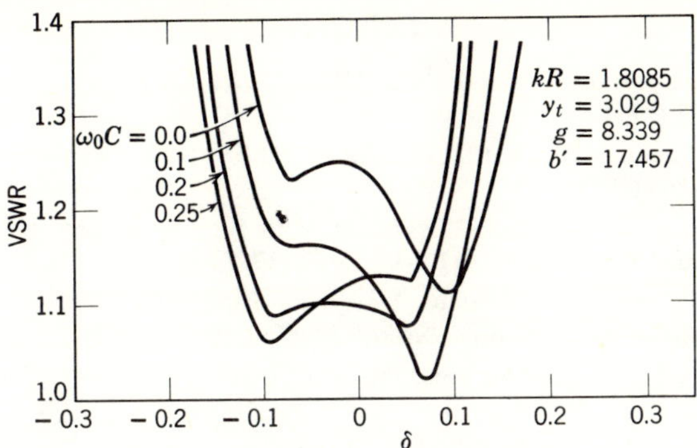

Figure 16.13. *Frequency response of quarterwave coupled circulator with capacitive tuning for* $r = 1.1$, $2\delta_0 = 20\%$.

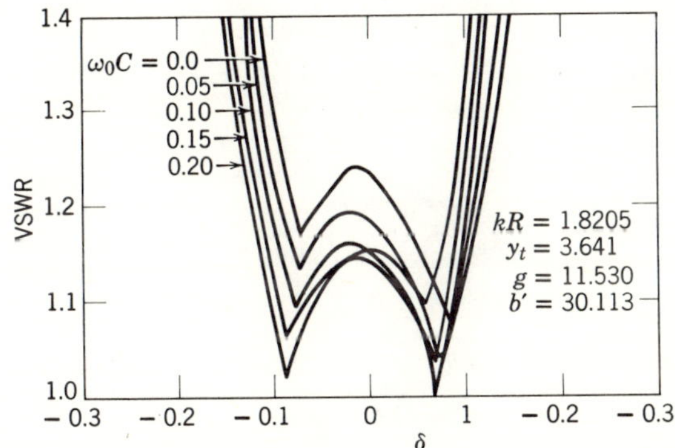

Figure 16.14. *Frequency response of quarterwave coupled circulator with capacitive tuning for* $r = 1.15$, $2\delta_0 = 20\%$.

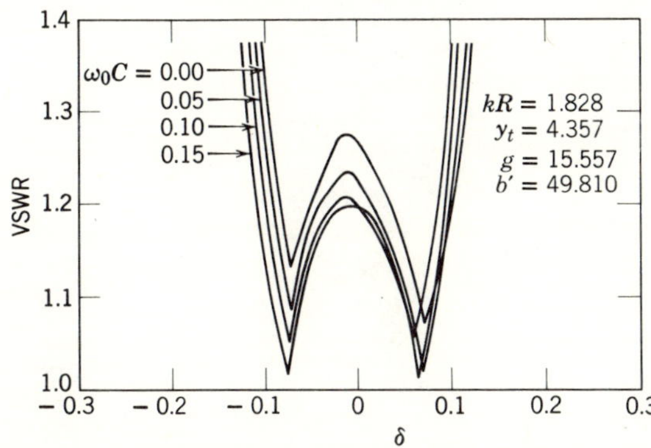

Figure 16.15. *Frequency response of quarterwave coupled circulator with capacitive tuning for* $r = 1.22$, $2\delta_0 = 20\%$.

REFERENCES

1. H. Bosma, "On Stripline *Y*-circulator at UHF," *IEEE Trans. Microwave Theory Tech.*, **MTT-12**, 61–72 (1964).

2. Y. Konishi, "Lumped Element *Y* Circulator," *IEEE Trans. Microwave Theory Tech.*, **MTT-13**, 852–864 (1965).

3. C. E. Fay and R. L. Comstock, "Operation of the Ferrite Junction Circulator," *IEEE Trans. Microwave Theory Tech.*, **MTT-13**, 15–27 (1965).

4. C. E. Barnes, "Integrated Circulator Design for Parametric Amplifier Design," IEEE G-MTT, *Int. Symp. Tech. Program Digest*, 170–175 (1964).

5. J. Helszajn, "A Ferrite Ring Stripline Junction Circulator," *Radio Electron. Eng.*, **32** (1), 55–60 (1966).

6. L. K. Anderson, "An Analysis of Broadband Circulators with External Tuning Elements," *IEEE Trans. Microwave Theory Tech.*, **MTT-15**, 42–47 (1967).

7. Y. Neuto and N. Zanaka, "Broadbanding and Changing Operation Frequency of Circulator," *IEEE Trans. Microwave Theory Tech.*, **MTT-19**, 367–372 (1971).

8. J. Helszajn, "The Synthesis of Quarter-Wave Coupled Circulators with Chebyshev Characteristics," *IEEE Trans.*, **MTT-20**, 764–769 (1972).

9. B. A. Auld, "The Synthesis of Symmetrical Waveguide Circulators," *Trans. I.R.E.* **MTT-7**, 238–246 (1959).

CHAPTER SEVENTEEN

Switchable Junction Circulators

It has been observed in Chapter 1 that μ is an even function of the direct magnetic bias field and that K is an odd function of the bias field. This suggests that reversing the direction of the bias magnetic field reverses the direction of circulation. Hence, junction circulators can be used to construct microwave switches. There are in general two ways in which this can be done.

In the conventional approach the device is usually operated by an electromagnet. A shortcoming of this technique is that the switching power is moderately large, and that the switching speed is limited to about 10 μsec because of demagnetizing fields and eddy currents. The demagnetizing fields are determined by the shape of the ferrite. The eddy currents are induced in the waveguide walls due to the changing magnetic field, which results in a certain degree of shielding. One way of removing both these limitations is to use a ferrite shape in the microwave circuit that can be biased by a single wire loop and that preferably has a square hysteresis loop. In this way the demagnetizing fields of the ferrite shape and the eddy currents in the junction housing are eliminated. The magnetic energy is here determined by the inductance of a single wire turn and that required to move between the two states of the hysteresis loop and is extremely small. Switching speeds of less than $1/2$ μsec are possible with this type of operation because the switching wire is inside the waveguide. A typical hysteresis loop is shown in Figure 17.1. The material can be biased to either of the two remanence states by applying a positive or negative current pulse sufficiently large to overcome the coercive field.

One advantage of this technique is, of course, that no holding current is required. The most useful parameters related to the loop is the saturated remanence ratio M_r/M_s and the coercive field H_c.

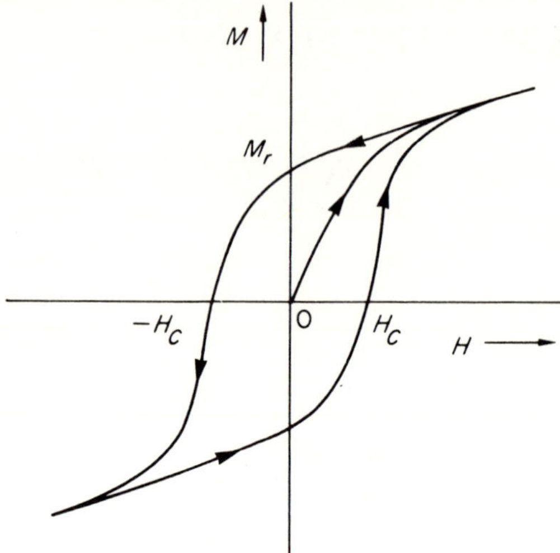

Figure 17.1.   *Hysteresis loop showing parameters of loop.*

If the magnitude of the magnetic field is changed, energy must be provided to, or removed from, the magnetic field. Hence the magnetic energy stored in the circuit is an important quantity in rating different switches. This energy, divided by the switching time, is related to the instantaneous power required from the driver. This means that the switching power is inversely proportional to the switching time.

There is also a class of 3- and 4-port circulators that only use a single gyrator as discussed in Chapter 4. By using tunable 2-port gyrators, it is possible to tune such circulators over very wide bands.

17.1.   *IMPEDANCE MATRIX OF LATCHING CIRCULATOR*

One important latching geometry is the 3-port TEM switch, illustrated in Figure 17.2, which is biased by a single wire loop embedded in each of the two ferrite disks of the junction. The arrangement of the wire loop is shown in Figure 17.3 for one of the ferrite disks. The behavior of the junction can be understood with the help of Figure 17.3 which depicts the biasing magnetic field. The perpendicular component of the magnetic field links downwards that part of the ferrite material that is outside the wire loop, and above that part of the material that is inside the wire loop. Hence, the center of the ferrite disk is biased in one sense, and the outer part is biased in the opposite sense. A preferred direction of circulation results because the r.f. energy is not uniformly distributed over the whole disk cross section. In fact, it is circularly polarized at the center of the disk,

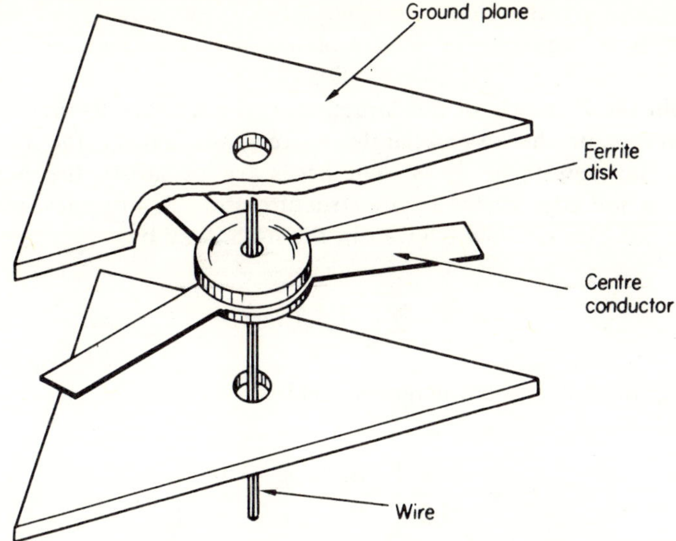

Figure 17.2.   *Latching TEM junction circulator* (*Ref.* 17.18).

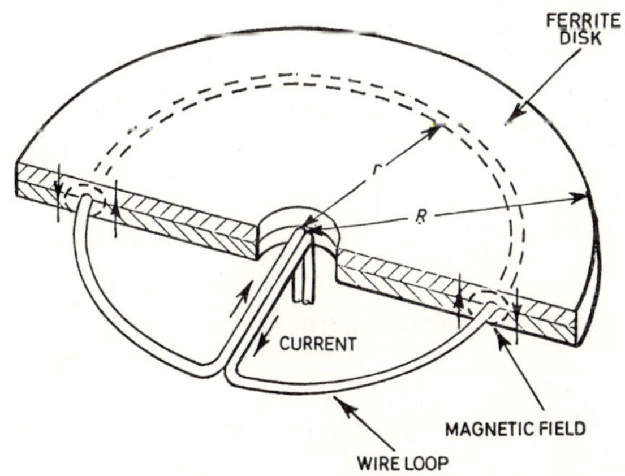

Figure 17.3.   *Current and magnetic field of ferrite disk.*

and becomes more elliptically polarized as the radius increases. At the edge of the disk it is linearly polarized.

The derivation of the impedance matrix of the latching circulator follows closely that of the conventional one discussed in Chapter 10. The configuration considered in this section is that in Figure 17.3. It consists of two regions in which the applied direct magnetic field is directed along either the $+z$ or $-z$ directions. It is assumed, as an approximation, that the magnitudes of the direct magnetic fields are equal and constant throughout both regions. The effects of d.c. magnetization are also neglected.

To obtain the $\overline{Z}$-matrix of this structure, it is necessary to first form the field equations in the two separate regions and satisfy the boundary conditions between them. It is then necessary to satisfy the boundary conditions at the edge of the ferrite structure and the striplines.Inside the ferrite disk the electric field is given in the usual way by

$$E_z = \sum_{n=-\infty}^{\infty} A_n J_n(x) e^{jn\phi} \tag{17.1}$$

The ϕ component H_ϕ of the magnetic field is

$$H_\phi = \frac{-j}{\omega\mu_0\mu_e} \left[\frac{\partial E_z}{\partial r} + j\frac{K}{\mu}\frac{1}{r}\frac{\partial E_z}{\partial\phi} \right] \tag{17.2}$$

from which one obtains

$$H_\phi = \sum_{n=-\infty}^{\infty} -jA_n\zeta_e\left[J_n'(x) - \frac{K}{\mu}\frac{nJ_n(x)}{x} \right] e^{jn\phi} \tag{17.3}$$

where

$$x = \omega\sqrt{\epsilon_0\epsilon_r\mu_0\mu_e}\, r \tag{17.4}$$

The solution for the electric field E_z in the ferrite ring is given by the following expression

$$E_z = \sum_{n=-\infty}^{\infty} A_n[B_n J_n(x) + C_n Y_n(x)] e^{jn\phi} \tag{17.5}$$

where B_n and C_n are constants, Y_n and J_n denote Bessel functions of order n of the first and second kinds, respectively.

The ϕ component H_ϕ of the magnetic field in the ferrite ring can be obtained from Eq. 17.2 by changing the sign of K, as required by reversal of the direct magnetic field. Substituting Eq. 17.5 into Eq. 17.2 and reversing the sign of K gives

$$H_\phi = \sum_{n=-\infty}^{\infty} -jA_n \zeta_e \left\{ \left[J_n'(x) + \frac{K}{\mu} \frac{nJ_n(x)}{x} \right] B_n \right.$$

$$\left. + \left[Y_n'(x) + \frac{K}{\mu} \frac{nY_n(x)}{x} \right] C_n \right\} e^{jn\phi} \qquad (17.6)$$

The constants B_n and C_n can be expressed in terms of the physical variables by using the conditions that E_z and H_ϕ are continuous at $x = x_1$:

$$B_n J_n(x_1) + C_n Y_n(x_1) = J_n(x_1) \qquad (17.7)$$

$$B_n \left[J_n'(x_1) + \frac{K}{\mu} \frac{nJ_n(x_1)}{x_1} \right] + C_n \left[Y_n'(x_1) + \frac{K}{\mu} \frac{nY_n(x_1)}{x_1} \right]$$

$$= \left[J_n'(x_1) - \frac{K}{\mu} \frac{nJ_n(x_1)}{x_1} \right] \qquad (17.8)$$

The result is

$$B_n = 1 - \frac{\pi K}{\mu} nJ_n(x_1) Y_n(x_1) \qquad (17.9)$$

$$C_n = \frac{\pi K}{\mu} nJ_n^2(x_1) \qquad (17.10)$$

The two roots for the uncoupled resonators are given in the usual way by

$$H_\phi(x = x_2) = 0 \qquad (17.11)$$

Another solution for H_ϕ that satisfies the boundary conditions over the width of the striplines is given in Chapter 10 by

$$H_\phi = \sum_{n=-\infty}^{\infty} b_n e^{jn\phi} \qquad (17.12)$$

where

$$b_n = \frac{\sin n\psi}{\pi n} [H_1 + H_2 e^{j2\pi n/3} + H_3 e^{-j2\pi n/3}] \qquad (17.13)$$

Comparing Eq. 17.6 and 17.12 at $x = x_2$ gives for the amplitude constant A_n

$$A_n = \frac{-\eta_e \sin n\psi}{j\pi n} \left[H_1 + H_2 e^{j2\pi n/3} + H_3 e^{-j2\pi n/3} \right]$$

$$\times \left\{ B_n \left[J_n'(x_2) + \frac{K}{\mu} n \frac{J_n(x_2)}{x_2} \right] + C_n \left[Y_n'(x_2) + \frac{K}{\mu} n \frac{Y_n(x_2)}{x_2} \right] \right\}^{-1} \tag{17.14}$$

The average electric fields at the three ports are now given with the help of Eq. 17.5 and 17.14 by

$$E_1 = \left(\frac{1}{2\psi} \right) \int_{-\psi}^{\psi} E_z \, d\phi = \eta_{11} H_1 + \eta_{12} H_2 + \eta_{21} H_3 \tag{17.15}$$

$$E_2 = \left(\frac{1}{2\psi} \right) \int_{-2\pi/3 - \psi}^{-2\pi/3 + \psi} E_z \, d\phi = \eta_{21} H_1 + \eta_{11} H_2 + \eta_{12} H_3 \tag{17.16}$$

$$E_3 = \left(\frac{1}{2\psi} \right) \int_{2\pi/3 - \psi}^{2\pi/3 + \psi} E_z \, d\phi = \eta_{12} H_1 + \eta_{21} H_2 + \eta_{11} H_3 \tag{17.17}$$

These last equations define the wave impedance matrix. Using the relations between E and V and H and I gives the following entries for the characteristic impedance matrix

$$R_{11} = \sum_{n=-\infty}^{\infty} \frac{r_n}{3} \tag{17.18}$$

$$R_{12} = \sum_{n=-\infty}^{\infty} \frac{r_n e^{j2\pi n/3}}{3} \tag{17.19}$$

$$R_{21} = -R_{12}^* = \sum_{n=-\infty}^{\infty} \frac{r_n e^{-j2\pi n/3}}{3} \tag{17.20}$$

and

$$r_n = \frac{3 R_e \sin^2 n\psi}{-j\pi n^2 \psi} \left\{ [B_n J_n(x_2) + C_n Y_n(x_2)] \right.$$

$$\times \left(B_n \left[J_n'(x_2) + \frac{K}{\mu} \frac{n J_n(x_2)}{x_2} \right] + C_n \left[Y_n'(x_2) + \frac{K}{\mu} \frac{n Y_n(x_2)}{x_2} \right] \right)^{-1} \right\} \tag{17.21}$$

where R_e has the form of Eq. 10.25.

The two circulation conditions are now obtained by comparing the coefficients of the above impedance matrix with those of an ideal circulator.

$$R_{11}=0 \qquad (17.22)$$

$$R_{12}=-R_0 \qquad (17.23)$$

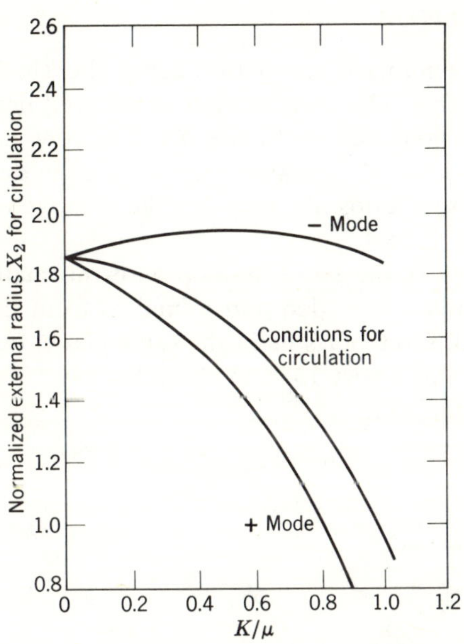

Figure 17.4.   Normalized outside radius for circulation of oppositely magnetized ferrite elements for the $n=1$ mode as function of K/μ for $x_2=\sqrt{2}\,x_1$ (Ref. 17.19).

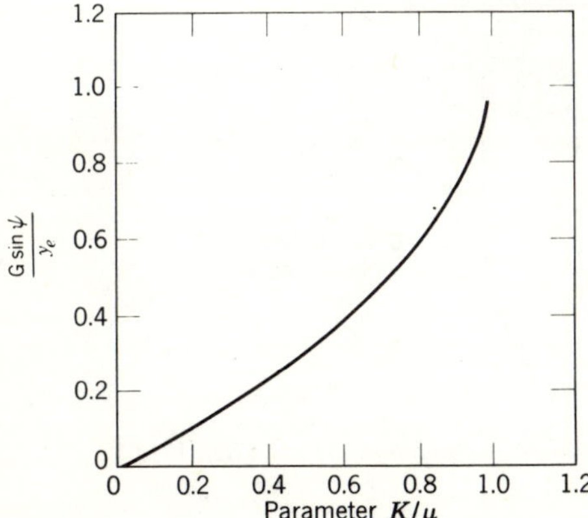

Figure 17.5.   Dependence of admittance function for oppositely magnetized ferrite elements on K/μ for $x_2=\sqrt{2}\,x_1$ (Ref. 17.19).

Figure 17.4 gives the roots of x_2 for $n = \pm 1$ as a function of K/μ for $x_2 = \sqrt{2}\, x_1$. It indicates that for a given value of K/μ the splitting between the resonant modes of the latched configuration is here approximately reduced by a factor of 0.70. The operating frequency is also now a function of K/μ. Figure 17.5 depicts the dependence of the admittance function under the same condition.

17.2. 3-PORT CIRCULATOR USING SINGLE 2-PORT GYRATOR

This section describes a 3-port assymmetric circulator using the single gyrator network described in Chapter 4. The construction considered here corresponds to the series network connection in Figure 4.7. The physical arrangement is illustrated in Figure 17.6. The gyrator network used is that described in Chapter 4 except that semiloops are used for the orthogonal circuits instead of the full loops in Figure 4.5.

Taking this basic YIG filter structure the two semiloop wires normally terminated to earth beyond the sphere are folded and joined to form the inner conductor of the third port. All three ports lie in the same plane, two ports at 90° to each other and the third port 135° to the other two. The device has only one plane of rotational symmetry.

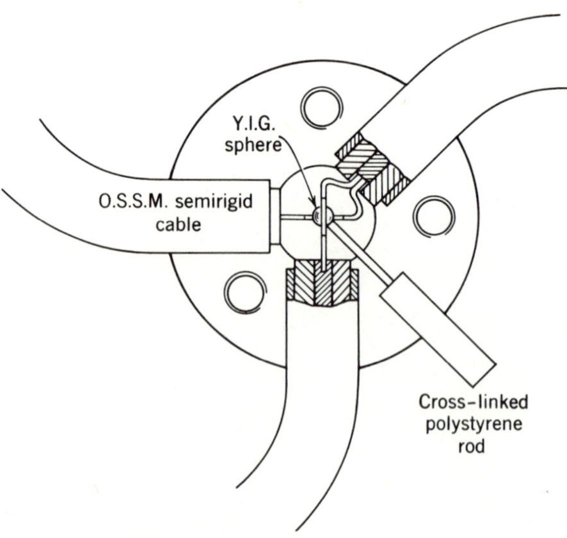

Figure 17.6. *Assymmetric 3-port tunable circulator using 2-port gyrator* (*Ref.* 17.20).

The semiloops structure was constructed in 0.15 mm diameter copper wire with semiloop diameter of approximately 1 mm. The wire semiloops were soldered into 50 Ω O.S.S.M. semirigid cable which forms the 50 Ω cables into a brass ring. The ring thickness was 3 mm and was positioned between the metallic poles of an electromagnet. The inner space of the ring containing the centrally positioned YIG sphere and semiloops was 3 mm diameter. The 3-port circulator can be tuned in frequency by varying the direct magnetic field.

17.3.   4-PORT CIRCULATOR USING 2-PORT GYRATOR

Chapter 4 also suggests that it is possible to construct an assymmetrical 4-port circulator using a single gyrator network. A possible waveguide arrangement is that in Figure 17.7. The gyrator network used here is the dual of the one in Figure 4.5 with the wire loops replaced by orthogonal waveguides. The physical arrangement consists of two waveguides separated by a thin wall at the junction. The ferrite sphere is placed in a small hole in the wall at the coordinates x,y. With matched terminations at all four ports, the equivalent circuit in Figure 17.8 describes the coupling at ferrimagnetic resonance. The scattering matrix for this device is

$$\bar{S} = \begin{bmatrix} \pm\dfrac{2Q_l}{\sqrt{Q_{ex1}Q_{ex4}}}, & \mp j\dfrac{2Q_l}{\sqrt{Q_{ex2}Q_{ex4}}}, & \pm j\dfrac{2Q_l}{\sqrt{Q_{ex3}Q_{ex4}}}, & 1-\dfrac{2Q_l}{Q_{ex4}} \\[3ex] \pm j\dfrac{2Q_l}{\sqrt{Q_{ex1}Q_{ex3}}}, & \pm\dfrac{2Q_l}{\sqrt{Q_{ex2}Q_{ex3}}}, & 1-\dfrac{2Q_l}{Q_{ex3}}, & \mp j\dfrac{2Q_l}{\sqrt{Q_{ex3}Q_{ex4}}} \\[3ex] \mp j\dfrac{2Q_l}{\sqrt{Q_{ex1}Q_{ex2}}}, & 1-\dfrac{2Q_l}{Q_{ex2}}, & \pm\dfrac{2Q_l}{\sqrt{Q_{ex2}Q_{ex3}}}, & \pm j\dfrac{2Q_l}{\sqrt{Q_{ex2}Q_{ex4}}} \\[3ex] 1-\dfrac{2Q_1}{Q_{ex1}}, & \pm j\dfrac{2Q_l}{\sqrt{Q_{ex1}Q_{ex2}}}, & \mp j\dfrac{2Q_l}{\sqrt{Q_{ex1}Q_{ex3}}}, & \pm\dfrac{2Q_l}{\sqrt{Q_{ex1}Q_{ex4}}} \end{bmatrix}$$

$$(17.24)$$

Q_l is the total loaded Q. (The sign $\pm$ depends on the position of the sphere.)

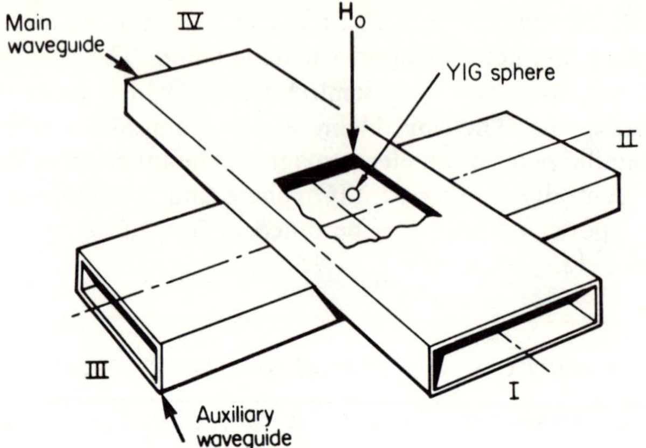

Figure 17.7. *Assymmetric 4-port tunable circulator using 2-port gyrator* (*Ref.* 17.17).

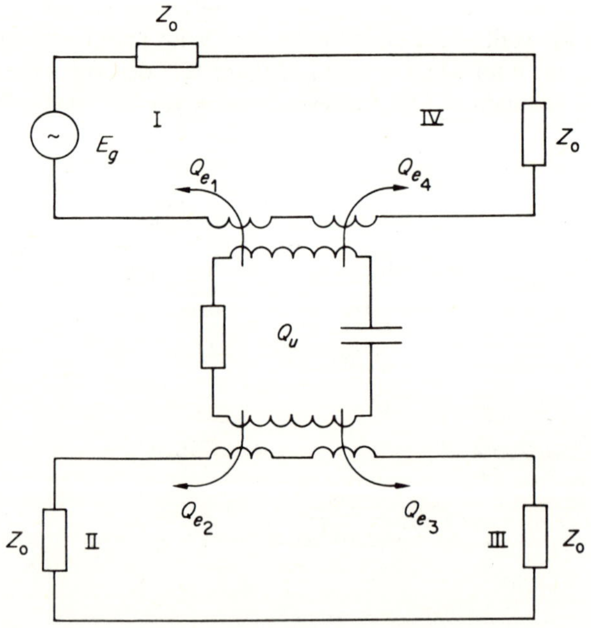

Figure 17.8. *Equivalent circuit of assymmetric 4-port tunable circulator using 2-port gyrator.* (*Ref.* 17.17).

300

If the sphere is placed at the point of circularly polarized fields for both guides ($\pi x/a = \pi y/a = \alpha$), a resonant circulator is obtained. The scattering matrix will be

$$\bar{S} = \begin{bmatrix} 0, & -j\dfrac{2Q_l}{\sqrt{Q_{ex2}Q_{ex4}}}, & 0, & 1-\dfrac{2Q_l}{Q_{ex4}} \\[4mm] 0, & 0, & 1, & 0 \\[4mm] 0, & 1-\dfrac{2Q_l}{Q_{ex2}}, & 0, & j\dfrac{2Q_l}{\sqrt{Q_{ex2}Q_{ex4}}} \\[4mm] 1, & 0, & 0, & 0 \end{bmatrix} \qquad (17.25)$$

Since the field at the center of a small hole in an infinitely thin wall must be half the field at the wall outside the hole, the Q values are readily found to be

$$Q_{ex1} = 4Q_{ex0}\frac{\sin^2\alpha}{\sin^2\left(\alpha - \dfrac{\pi y}{a}\right)} \qquad (17.26)$$

$$Q_{ex2} = 4Q_{ex0}\frac{\sin^2\alpha}{\sin^2\left(\alpha + \dfrac{\pi x}{a}\right)} \qquad (17.27)$$

$$Q_{ex3} = 4Q_{ex0}\frac{\sin^2\alpha}{\sin^2\left(\alpha - \dfrac{\pi x}{a}\right)} \qquad (17.28)$$

$$Q_{ex4} = 4Q_{ex0}\frac{\sin^2\alpha}{\sin^2\left(\alpha + \dfrac{\pi y}{a}\right)} \qquad (17.29)$$

where

$$Q_{ex0} = \frac{2Z_0 ab}{\mu_0 \omega_m v_f}\frac{\lambda_g}{\lambda}, \qquad (17.30)$$

when it is assumed that the two guides have equal dimensions.

An experimental circulator of this kind was constructed of X-band using a 1.95 mm diameter YIG sphere. The waveguide junction construction is shown in Figure 17.7. The insertion wall is 0.1 mm thick and the hole has a diameter of 3.5 mm. The hole is placed half way between the center and the side wall of each guide, which yields circularly polarized fields for $\alpha = 45°$, or $\lambda_s = \lambda_e$, that is, at approximately 9250 MHz. The theoretical scattering matrix, based on an unloaded Q of the sphere of 8100 was calculated for 9250 MHz.

$$S_{\text{theoret}} = \begin{bmatrix} 0 & -j0.835 & 0 & 0.165 \\ 0 & 0 & 1 & 0 \\ 0 & 0.165 & 0 & j0.835 \\ 1 & 0 & 0 & 0 \end{bmatrix} \quad (17.31)$$

Coupling through the hole without ferrite is neglected.

The following scattering matrix was measured:

$$S_{\text{exp}} = \begin{bmatrix} 0.05 & -j0.84 & -j0.08 & 0.156 \\ (j0.06) & (-0.025) & (0.96) & (0.07) \\ j0.08 & 0.18 & 0.05 & j0.835 \\ (0.96) & (-j0.05) & (-j0.05) & (-0.025) \end{bmatrix} \quad (17.32)$$

REFERENCES

1. L. Levy and L. M. Silber, "A Fast Switching X-Band Circulator Utilizing Ferritte Toroids," *I.R.E. Wescon Convention Rec.*, Part 1 (1960).

2. J Helszajn, "Switching Criteria for Waveguide Ferrite Devices," *Radio Electron. Eng.*, **30** (1965), 289–296.

3. P. C. Goodman, "A Latching Ferrite Junction Circulator for Phased Array-Switching Applications," 1965-G-MTT Symposium.

4. L. Freiberg, "Pulse Operated Circulator Switch," *I.R.E. Trans. Microwave Theory Tech.* (1961).

5. Van Der Heide *et al. Philips Tech. Rev.*, 18, 339 (1956).

6   G. S. Uebele, "High Speed Ferrite Microwave Switch," *IRE Nat. Convention Rec.*, **5**, Part 1, 227–234 (1957).

7. L. R. Whicker and R. R. Jones, "A Digital Current Controlled Latching Ferrite Phase Shifter," presented at the 1965 IEEE Convention, New York, N. Y.

8. D. R. Taft and L. R. Hodges, Jr., "Square Loop Materials for Digital Phase Shifter Applications,': *J. Appl. Phys.*, **36**, 1263–1264 (1965).

9. J. E. Woermbke and J. A. Myers, "Latching Ferrite Devices," *Microwaves*, **3**, 20–27 (1964).

10. R. W. Damon, "Magnetically Controlled Microwave Directional Coupler," *J. Appl. Phys.*, **26**, 1281 (1955).

11. A. D. Berk and E. Strumwasser, "Ferrite directional coluplers," *Proc. IRE*, **44**, 1439–1445 (1956).

12. R. W. DeGrasse, "Low-Loss Gyromagnetic Ouplings Through Single Crystal Garents," *J. Appl. Phys. Suppl.*, **30**, 155S (1959).

13. P. S. Carter, Jr., "Magnetically-Tunable Microwave Filters Using Singel-Crystal Yttrium–Iron–Garnet Resonators," *IRE Trans. Microwave Theory Tech.*, **MTT-9**, 252 (1962).

14. K. L. Kotzebue, "Broadband Electronically-Tuned Microwave Filters," *Wescon Convention Rec.* **4**, Part 1, 21–27 (1963).

15. G. L. Matthaei, "Magnetically-Tunable Band-Stop Filters," *IEEE Trans. Microwave Theory Tech.*, **MTT-13**, 203–212 (1965).

16. C. E. Fay, "Ferrite-Tuned Resonant Cavities," *Proc. IRE,* **44**, 1446–1449 (1956).

17. H. Skeil, "Non-Recriprocal Couplings with Single Crystal Ferrites," *IEEE Trans. Microwave Theory Tech.*, **MTT-12**, 587–594 (1964).

18. J. Helszajn and M. L. Hines, "A High Speed TEM Junction Ferrite Modulator Using a Wire Loop," *Radio Electron. Eng.*, **35**, 81–82 (1968).

19. W. W. Siekanowicz and W. A. Schilling, "A New Tyoe of Latching Switchable Ferrite Junction Circulator," *IEEE Trans. Microwave Theory Tech.*, March 177–183 (1968).

20. S. R. Longley, "Multioctave Tunable 3-Port Circulator Using a YIG Sphere," *Electro. Lett.*, **6**, (13) 406–408 (1970).

21. A. Clavin, "Reciprocal and Nonreciprocal Switches utilizing Ferrite Junction Circula tors," *IEEE Trans. Micorwave Theory Tech.*, **MTT-11**, 217–218 (1963).

22. R. L. Comstock, "A Crygenic Four-Port Circulator Using Single Crystal Yttrium Iron Garnet," *Proc. IEEE*, **51**, 1768–1769 (1963).

CHAPTER EIGHTEEN

The 4-Port Single Junction Circulator

This chapter describes a number of 4-port single junction circulators in both waveguide and stripline. In common with 3-port devices such junctions exhibit some of the properties of transmission line cavity resonators between ports 1 and 2, and a definite standing wave pattern exists within the junction with nulls at ports 3 and 4 also. However, an important difference between the two is that the 4-port device cannot be adjusted with external tuning elements only. This is because a 4-port device can be matched without being a circulator. To adjust this component requires three independent variables that may be established in a systematic way by perturbing the scattering matrix eigenvalues one at a time on the unit circle until they coincide with those of an ideal circulator. This approach has been fully discussed in Chapter 8. It is facilitated provided the nature of the field patterns used to construct the device are known. The present chapter describes three H-plane waveguide devices and a stripline component. The stripline junction is an example of the use of higher order radial modes in the construction of circulators.

The first 4-port H-plane waveguide circulator described consists of a single magnetized ferrite disk on one side of the waveguide and a metal post on the other wall. The three independent variables used in this component are a pair of $\mathrm{HE}_{\pm 1,1,1}$ open dielectric resonances in an unmagnetized ferrite disk, a $\mathrm{TM}_{0,1,1}$ resonance on a metal post, and the magnitude of a direct field to remove the degeneracy between the $\mathrm{HE}_{\pm 1,1,1}$ modes. Since the overall length of the ferrite disk and the quarterwave-long cylinder is approximately equal to the height of the waveguide, the two are allowed to make firm flat contact with each other and with the waveguide walls. These two conditions satisfy two of the three independent variables of the junction. The third and last one is satisfied by magnetizing the junction with an appropriate direct magnetic field thereby rotating the standing wave formed by the hybrid $HE_{\pm 1,1,1}$ modes by the required $45°$. The procedure used to adjust this circulator is described in Chapter 8. It allows each of the three independent variables for this type of junction to be established one at a time in a systematic way.

The second H-plane junction described in this chapter uses $n = \pm 1$ radial field patterns for the counter-rotating junction modes and again uses the $TM_{0,1,1}$ mode associated with a quarterwave-long metal post for the symmetrical mode.

The third waveguide geometry relies altogether on radial modes for both the counter-rotating and symmetrical modes of the junction. This example uses the $n = 0$ and $n = \pm 1$ modes with $H_\phi = 0$ at $r = R$ in a simple ferrite post with no variation of the electric field along the post axis.

The variables employed in the construction of the stripline junction are a pair of radial $n = \pm 3$ degenerate resonances, a radial $n = 0$ resonance, and the amplitude of a direct field to split the degeneracy between the former modes. Since the modes in this construction have nearly equal radial wave numbers, a simple ferrite disk immediately satisfies the first two of the three circulation conditions. The third and last one is obtained in the usual way by removing the degeneracy between the degenerate modes in the usual manner by biasing the junction with a direct magnetic field.

18.1.   4-PORTS SINGLE JUNCTION CIRCULATOR USING AXIAL MODES

The 4-port single junction circulator described in this chapter relies on a linear combination of $HE_{\pm 1,1,1}$, and $TM_{0,1,1}$ axial modes for its operation. Once these modes are resonant within the junction, the $HE_{\pm 1,1,1}$ one is rotated by the application of a direct magnetic field to form a circulator. The geometry considered is shown in Figure 18.1.

The electric fields employed are given in Figures 18.2 a through d. The illustrations in Figures 18.2a and b show the $HE_{\pm 1,1,1}$ and $TM_{0,1,1}$ modes of the unmagnetized junction. Figures 18.2c and d give the same field patterns with the $HE_{\pm 1,1,1}$ mode rotated through $45°$. Adding the amplitudes of the E-fields at the four ports indicates that circulation takes place between ports 1 and 2 while ports 3 and 4 are decoupled.

The angle through which the $HE_{\pm 1,1,1}$ hybrid mode is rotated is obtained by taking a linear combination of the electric fields around the

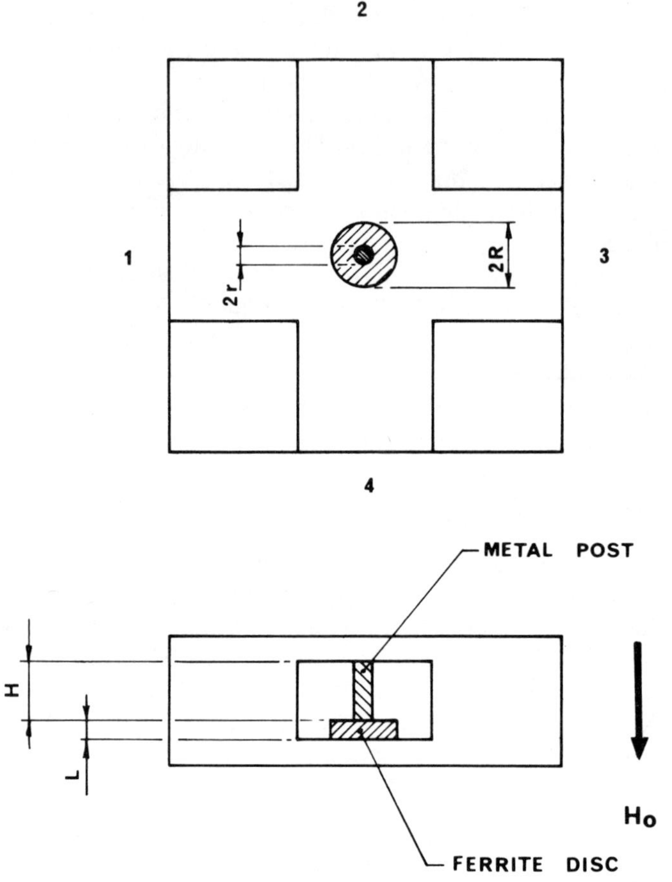

Figure 18.1.　*Schematic of single junction 4-port waveguide circulator using axial modes.*

periphery of the ferrite disk.

$$E(R,\phi) = a_{01} + a_{11}\cos(\tau_{11} + \phi) \qquad (18.1)$$

where a_{01} and a_{11} are arbitrary constants and τ_{11} is the phase angle through which the $HE_{\pm 1,1,1}$ mode is rotated. Applying the boundary conditions of an ideal circulator at the four ports gives

$$E(R,0) = a_{01} + a_{11}\cos\tau_{11} = +1 \qquad (18.2)$$

$$E\left(R,\frac{\pi}{2}\right) = a_{01} + a_{11}\cos\left(\tau_{11} + \frac{\pi}{2}\right) = +1 \qquad (18.3)$$

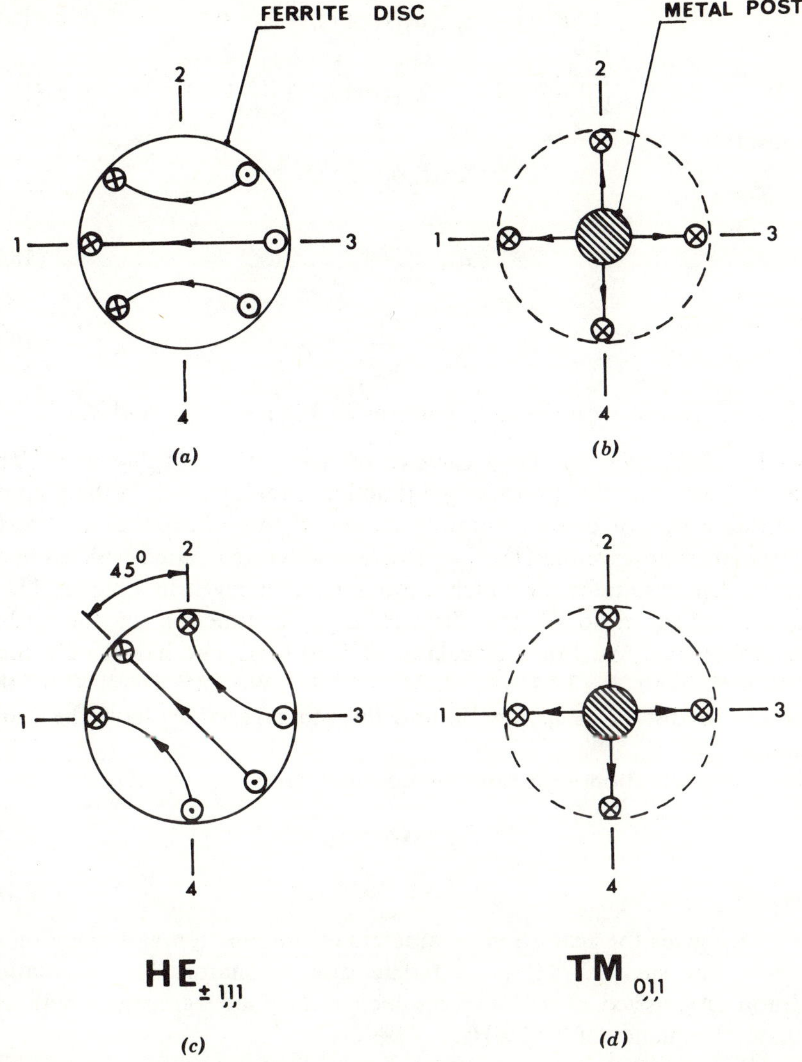

Figure 18.2.   (*a*) $HE^{\pm}_{1,1,1}$ *E-field pattern in unmagnetized disk* (*Ref.* 18). (*b*) $TM_{0,1,1}$ *E-field pattern on metal post* (*Ref.* 18). (*c*) $HE^{\pm}_{1,1,1}$ *E-field pattern in magnetized disk* (*Ref.* 18). (*d*) $TM_{0,1,1}$ *E-field pattern on metal post* (*Ref.* 18).

$$E(R,\pi) = a_{01} + a_{11}\cos(\tau_{11} + \pi) = 0 \tag{18.4}$$

$$E\left(R, \frac{3\pi}{2}\right) = a_{01} + a_{11}\cos\left(\tau_{11} + \frac{3\pi}{2}\right) = 0 \tag{18.5}$$

The result is

$$\tau_{11} = 45° \tag{18.6}$$

$$a_{01} = \frac{1}{2} \tag{18.7}$$

$$a_{11} = \frac{1}{\sqrt{2}} \tag{18.8}$$

which is consistent with the field patterns in Figures 18.2c and d.

18.1.1. *Adjustment of Axial Counter-rotating Modes of Junction.* The center frequency of the 4-port single junction circulator lies in the vicinity of a suitable pair of counter-rotating modes of the junction geometry. In this circulator they are the $HE_{\pm 1,1,1}$ hybrid modes associated with an open dielectric disk resonator for which a mode chart is given in Chapter 11. It gives the aspect ratio of the disk R/L as a function of the radial wavenumber $2\pi R/\lambda_0$. For a circulator at 9.30 GHz one has $R = 5.0$ mm, $L = 3.0$ mm, with $\varepsilon_r = 12.4$ and $4\pi M_0 = 0.2400$ Wb/m^2. Here, R is the radius of the disk, L is its length, and the other variables have the usual meaning.

The first circulation adjustment is now met with

$$S_{11} = S_{12} = S_{14} = 0 \tag{18.9}$$

$$S_{13} = 1 \tag{18.10}$$

Figure 18.3 gives the scattering parameters of the junction as a function of frequency for such an $HE_{\pm 1,1,1}$ ferrite disk resonator. The circulation condition is satisfied at 9.10 GHz, which is in good agreement with the calculated frequency of 9.30 GHz.

18.1.2. *Adjustment of Symmetrical Mode of Junction.* The second circulation adjustment is obtained by establishing a suitable symmetrical mode within the junction. The one chosen here is the axial $TM_{0,1,1}$ mode which exists along a quarterwave-long thin metal post. The condition for resonance on such a post is approximately given by

$$l \approx \frac{\lambda_0}{4} - \frac{(a-r)}{2\log_e\left(\dfrac{a}{r}\right)}$$

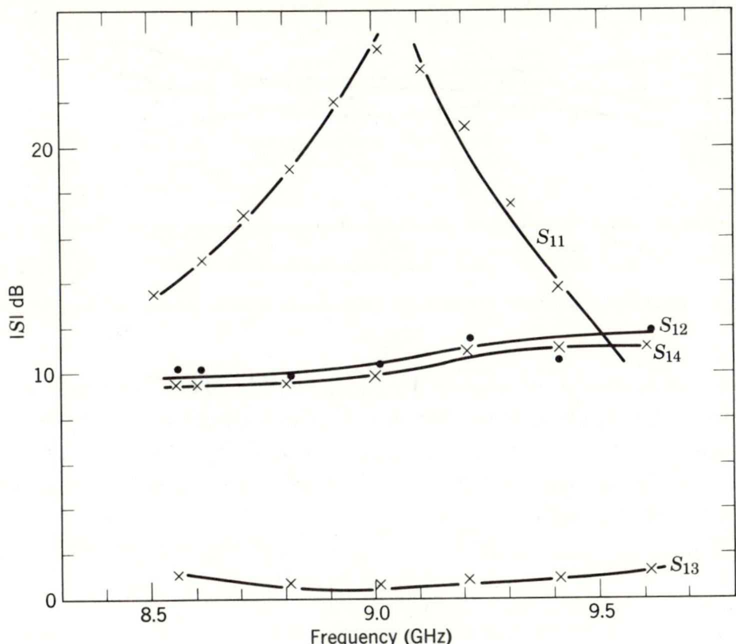

Figure 18.3.   *Scattering parameters of junction after first circulation adjustment* (*Ref.* 18).

which is asymptotic to $\lambda_0/4$ for an infinitely thin post. This equation is obtained by having the incident tangential electric field set up a current in the post the net effect of which is to cancel the incident field at the post. The current along the wire has the form of a standing wave

$$I_x = \sin \frac{2\pi(l-x)}{\lambda_0}$$

where x lies between 0 and l.

The circulation condition is satisfied with the scattering coefficients given by

$$S_{11} = S_{12} = S_{13} = S_{14} = \tfrac{1}{2}$$

Figure 18.4 gives the scattering variables of the junction for a metal post of radius 1 mm. Since the electrical length of the post is $0.22\lambda_0$ at 9.10 GHz it has been allowed to make contact with the ferrite disk to maximize its length. No effort was made to improve on the geometry used here.

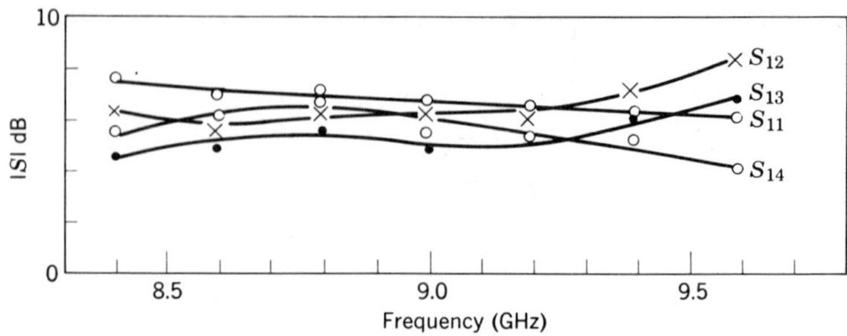

Figure 18.4.   *Scattering parameters of junction after second circulation adjustment (Ref. 18).*

18.1.3.   *Adjustment of Direct Magnetic Field.*   The third and last adjustment of the junction is the amplitude of the direct magnetic field. This adjustment rotates the standing wave formed by the counter-rotating $HE_{\pm1,1,1}$ modes by 45° to form an ideal circulator. For an ideal lossless junction the scattering parameters are

$$S_{11} = S_{13} = S_{14} = 0$$

$$S_{12} = 1$$

Figure 18.5 shows the various scattering coefficients of the junction as a function of frequency.

Figure 18.5 also indicates a property of the 4-port single junction circulator, which is that S_{11} and S_{13} are interdependent. These two parameters may be wide-banded with the help of external matching but S_{14} is a property of the basic junction alone.

The overall bandwidth of the device is of course determined by the Q factors of the individual modes used to construct the device. Since the resonant frequency of an axially resonating mode does not require a unique shape, some adjustment of the Q factor is possible in such resonators by varying the geometry. Another way of widebanding the junction is by means of a radial line transformer as in the case of the 3-port junction.

18.2.   4-PORT SINGLE JUNCTION CIRCULATOR USING MIXED RADIAL AND AXIAL MODES

This section describes the adjustment of the 4-port single junction circulator in Figure 18.6. This geometry consists of a full-height ferrite post with a partial-height metal post through its center at the junction of four waveguides. The modes used in this arrangement are the $n = \pm 1$ radial ones associated with a full-height post with no variation of the fields along

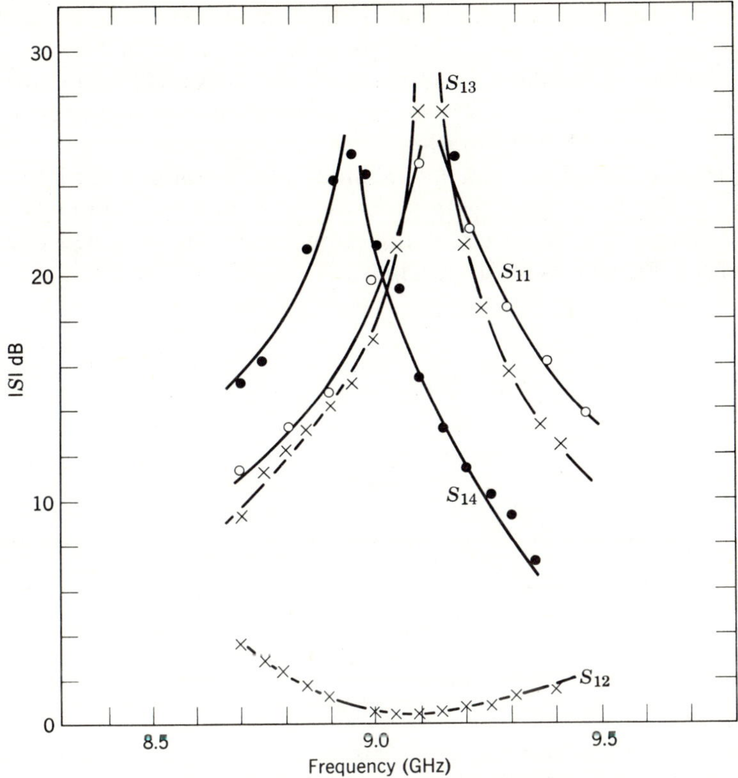

Figure 18.5.   *Frequency variation of scattering parameters of junction after third circulation adjustment (Ref. 18).*

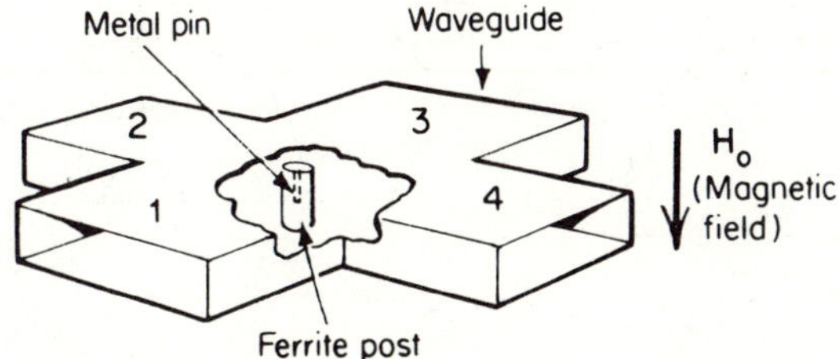

Figure 18.6.   *Schematic of 4-port single junction circulator using mixed radial and axial modes.*

the axis of the post and the axial $TM_{0,1,1}$ mode which exists along a quarterwave-long metal post. These patterns are shown in Figures 18.7a and b. The first shows the $n \pm 1$ radial modes of the simple ferrite post and the $TM_{0,1,1}$ mode associated with a quarterwave-long metal post in a waveguide. The second illustration gives the same patterns but with the $n = \pm 1$ modes rotated through 45°. A linear combination of the latter field patterns at the four ports again satisfies the boundary conditions of an ideal circulator. The three circulation adjustments are obtained one at a time in the same way as in the previous case as will now be described.

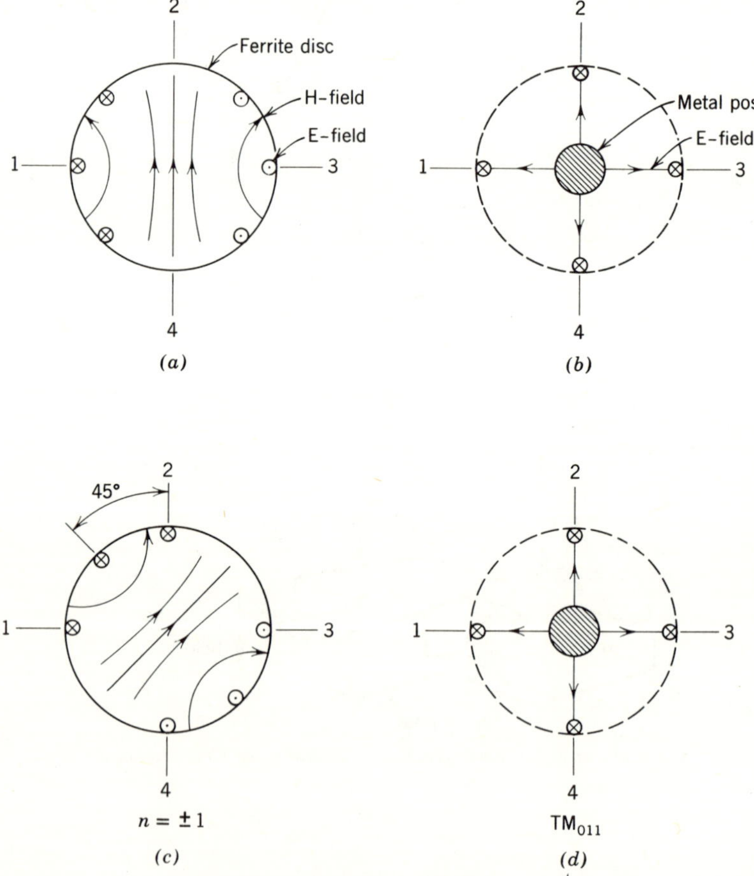

Figure 18.7.   (a) $n = \pm 1$ field patterns for unmagnetized ferrite post. (b) $TM_{0,1,1}$ E-field pattern on metal post. (c) $n = \pm 1$ modes in magnetized ferrite post. (d) $TM_{0,1,1}$ E-field on metal post.

18.2.1.   *Adjustment of Counter-rotating Radial Modes.*   The counter-rotating radial modes used for the construction of this geometry are the $n = \pm 1$ modes with $kR = 1.84$. This circulation condition is satisfied as before with

$$S_{11} = S_{11} = S_{14} = 0$$

$$S_{13} = 1$$

Figure 18.8 gives S_{13} as a function of frequency for a ferrite post 19.6 mm in diameter having a magnetization of 0.1200 Wb/m^2 and a dielectric constant of 14.4.

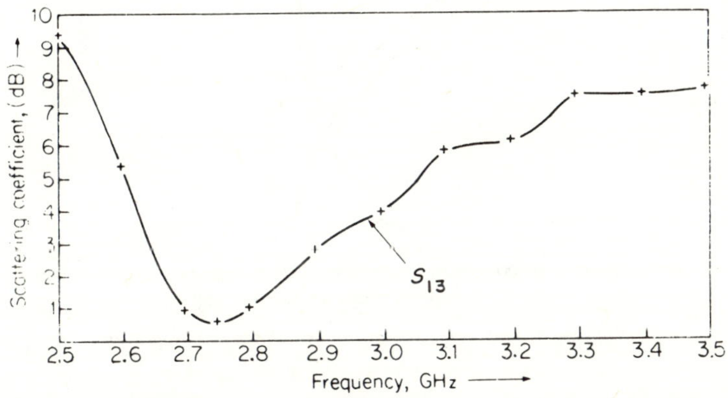

Figure 18.8.   *Scattering coefficient S_{13} versus frequency for fixed ferrite diameter for 4-port radial mode junction (Ref. 7).*

18.2.2.   *Adjustment of Symmetrical Axial Mode of Junction.*   Once the $n = \pm 1$ radial modes have been established within the junction the $TM_{0,1,1}$ axial mode is tuned by varying the length of the metal post. This circulation condition is again satisfied provided

$$S_{11} = S_{12} = S_{13} = S_{14} = \tfrac{1}{2}$$

Figure 18.9 gives the scattering coefficients of the junction as a function of the length of the metal post. The length of the metal post is a quarterwave in the ferrite medium as before.

18.2.3.   *Splitting of Degenerate Radial $n = \pm 1$ Modes.*   The final adjustment is now met by rotating the standing wave formed by the $n = \pm 1$ modes through 45° by magnetizing the junction. The variation of the

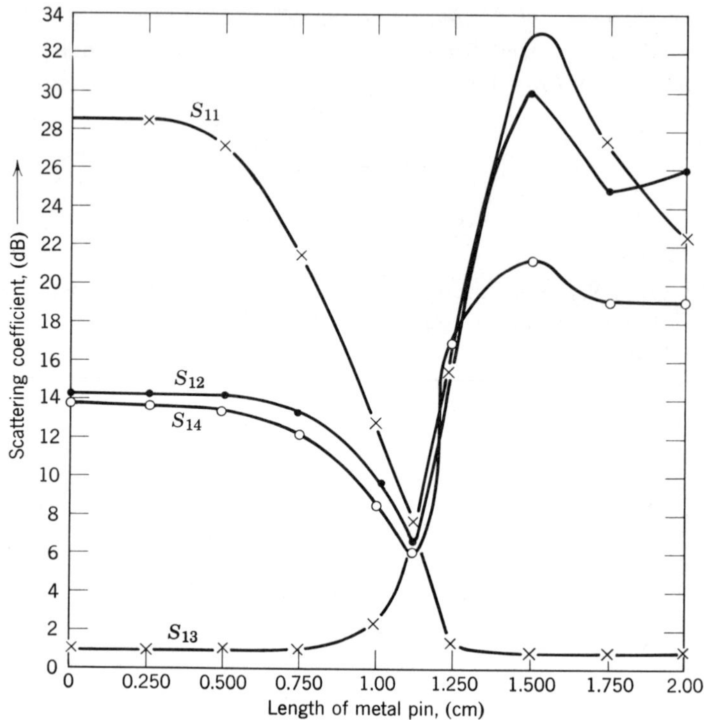

Figure 18.9.   *Scattering coefficients versus length of metal pin for 4-port radial mode junction* (*Ref.* 7).

scattering coefficients with the direct magnetic field is shown in Figure 18.10. An ideal circulator is obtained when

$$S_{11} = S_{13} = S_{14} = 0$$

$$S_{12} = 1$$

This condition is satisfied at a field of 2780 At/m.

18.3.   4-PORT SINGLE JUNCTION CIRCULATOR USING RADIAL MODES

A 4-port single junction circulator is also obtained by taking a linear combination of radial $n = \pm 1$ and 0 modes. The field patterns in this case are shown in Figures 18.11a and b. The adjustment of this device follows closely that of the first two geometries already described except that the

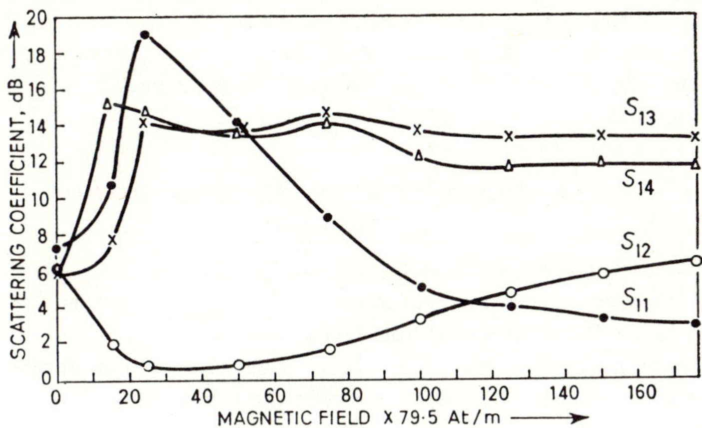

Figure 18.10. *Scattering coefficients as a function of magnetic field for 4-port radial mode junction (Ref. 7).*

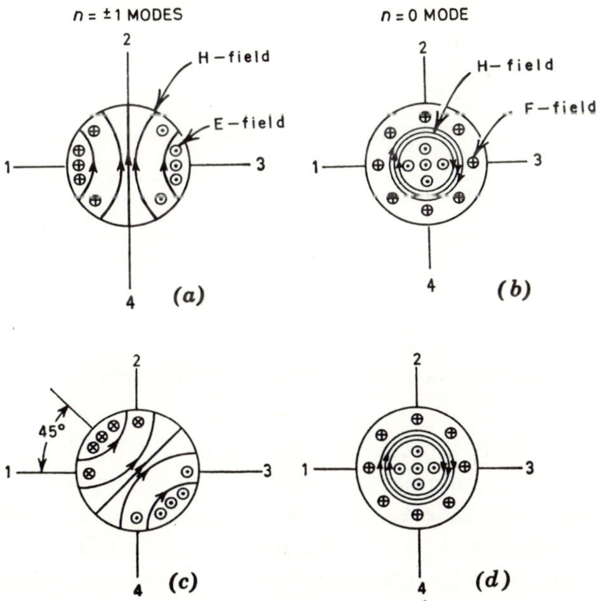

Figure 18.11. *(a) $n = \pm 1$ field patterns for unmagnetized ferrite post (Ref. 19). (b) $n = 0$ field patterns for unmagnetized ferrite post (Ref. 19). (c) $n = \pm 1$ field patterns for magnetized ferrite post (Ref. 19). (d) $n = 0$ field patterns for magnetized ferrite post (Ref. 19).*

$n = 0$ mode is tuned to the frequency of the $n = \pm 1$ modes with the help of a thin nonresonant capacitive post at the center of junction instead of a resonant one. Figures 18.11c and d give the field patterns of the magnetized junction.

18.4. *STRIPLINE 4-PORT SINGLE JUNCTION CIRCULATOR*

The purpose of this section is to describe the construction and theory of a 4-port stripline circulator. A schematic of this component is given in Figure 18.12. It consists of simple ferrite disks at the junction of four 50 Ω stripline transmission lines. The direct magnetic field is applied perpendicular to the plane of the ferrite disks in the usual way.

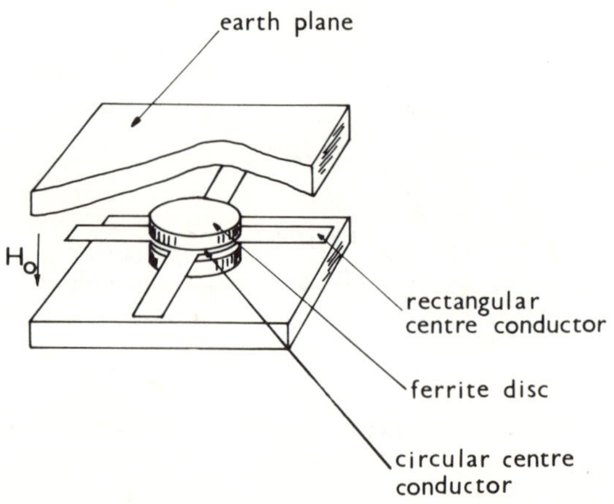

Figure 18.12.   *Schematic of single junction 4-port stripline circulator.*

The modes used in this case are the $n = 0$ and ± 3 radial ones in a simple ferrite disk.* The experimental adjustment of the circulator relies on the fact that the $n = 0$ mode already lies approximately midway between the split $n = \pm 3$ ones. It is, therefore, possible in this instance to omit the adjustment that requires the three isotropic resonant frequencies to coincide. Figures 18.13a and b show the field patterns for these modes. The radial wavenumbers for such demagnetized ferrite disks are $kR = 3.83$ for $n = 0$ and $kR = 4.20$ for $n = \pm 3$. A radial wavenumber of between $3.86 < kR < 4.10$ has been used on a waveguide junction also. Figures 18.13c and d

---

*It is observed in passing that the $n = \pm 3$ modes are not suitable for the construction of 3-port symmetric circulators.

give the field patterns of the modes with the $n = \pm 3$ ones rotated by 15° by magnetizing the junction with a direct magnetic field. If the magnitudes of the electric fields for the two patterns are equal (after rotation) at the four ports, transmission takes place between ports 1 and 2 and ports 3 and 4 are isolated.

For the 4-port circulator described here the electric field distribution around the periphery of the ferrite disk is

$$E_z = a_0 + a_3 \cos 3(\tau_3 + \phi)$$

where a_0 and a_3 are arbitrary constants and τ_3 is the phase angle through which the $n = \pm 3$ modes are rotated.

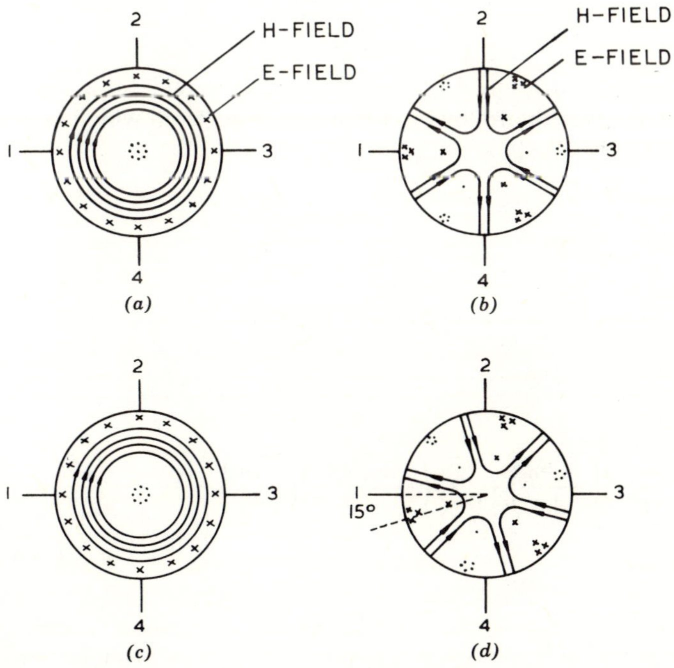

Figure 18.13.   (a) $n = 0$ field patterns for unmagnetized disk (Ref. 18). (b) $n = \pm 3$ field patterns for unmagnetized disk (Ref. 18). (c) $n = 0$ field patterns for magnetized disk (Ref. 18). (d) $n = \pm 3$ field patterns for magnetized disk (Ref. 18).

The boundary conditions at the 4 ports are

$$E_z(\phi=0)=a_0+a_3\cos 3\tau_3=+1$$

$$E_z(\phi=\pi/2)=a_0+a_3\cos 3\left(\tau_3+\frac{\pi}{2}\right)=+1$$

$$E_z(\phi=\pi)=a_0+a_3\cos 3(\tau_3+\pi)=0$$

$$E_z(\phi=3\pi/2)=a_0+a_3\cos 3\left(\tau_3+\frac{3\pi}{2}\right)=0$$

The result is

$$\tau_3=15°$$

$$a_0=\frac{1}{2}$$

$$a_3=\frac{1}{\sqrt{2}}$$

In the case of the stripline 4-port single junction circulator a solution to the *e.m.* problem gives the following three equations from which $\sin\psi, kR,$ and K/μ may be obtained

$$R^2+S^2=Q^2$$

$$R^2-S^2=-PQ$$

$$S(Q-P)=RZ_0$$

where

$$P=\left(\frac{Z_{\text{eff}}}{\pi}\right)\left(\frac{\Psi B_0}{A_0}\right)+\sum_{n=1}^{\infty}\left(\frac{2\sin^2 n\Psi}{n^2\Psi}\right)\left(\frac{Z_{\text{eff}}}{\pi}\right)\frac{A_n B_n}{\left[A_n^2-\left(\frac{K}{\mu}\right)^2\left(\frac{nB_n}{kR}\right)^2\right]}$$

$$Q=\left(\frac{Z_{\text{eff}}}{\pi}\right)\left(\frac{\Psi B_0}{A_0}\right)+\sum_{n=1}^{\infty}\left(\frac{2\sin^2 n\Psi}{n^2\Psi}\right)\left(\frac{Z_{\text{eff}}}{\pi}\right)\frac{A_n B_n\cos n\pi}{\left[A_n^2-\left(\frac{K}{\mu}\right)^2\left(\frac{nB_n}{kR}\right)^2\right]}$$

$$R = \sum_{n=1}^{\infty} \frac{2\sin^2 n\Psi}{n^2\Psi} \left(\frac{Z_{\text{eff}}}{\pi}\right) \frac{\left(\dfrac{K}{\mu}\right)\left(\dfrac{nB_n^2}{kR}\right)\sin\left(\dfrac{n\pi}{2}\right)}{\left[A_n^2 - \left(\dfrac{K}{\mu}\right)^2\left(\dfrac{nB_n}{kR}\right)^2\right]}$$

$$S = \left(\frac{Z_{\text{eff}}}{\pi}\right)\left(\frac{\Psi B_0}{A_0}\right) + \sum_{n=1}^{\infty}\left(\frac{2\sin^2 n\Psi}{n^2\Psi}\right)\left(\frac{Z_{\text{eff}}}{\pi}\right)\frac{A_n B_n \cos\left(\dfrac{n\pi}{2}\right)}{\left[A_n^2 - \left(\dfrac{K}{\mu}\right)^2\left(\dfrac{nB_n}{kR}\right)^2\right]}$$

$$A_n = J_n'(kR)$$

$$B_n = J_n(kR)$$

One solution is $\sin\Psi = 0.40$ radians, $kR = 3.706$ and $K/\mu = 0.309$ with $Z_{\text{eff}}/Z_0 = 0.300$.

REFERENCES

1. L. E. Davis, M. D. Coleman, and J. J. Cotter, "Four-port crossed waveguide junction circulators," (corresp.) *IEEE Trans. on MTT*, **MTT-12**, 43–47 (1964).

2. B. Owen, "The Identification of Modal Resonances in Ferrite Loaded Waveguide *Y*-junctions and Their Adjustment for Circulation," *Bell Syst. Tech J.*, **51** (3), 595–627 (1972).

3. J. Helszajn and F. F. Tan, "Radial Line Waveguide Junction Circulators, submitted to TRANS IEEE on Microwave Theory & Technique. (to be published.)

4. Markuvitz *Microwave Handbook*, McGraw-Hill, New York.

5. L. Lewin, *Advanced Theory of Waveguides*, Iliffe and Sons, Ltd., London, 1951.

6. J. Helszajn, "Three Resonant Mode Adjustment of the Waveguide Circulator," *Radio Electron. Eng.*, **42** (5), 357–360 (1972).

7. J. Helszajn and C. R. Buffler, "Adjustment of the 4-Port Single-Junction Circulator," *Radio Electron. Eng.*, **35** (6) (1968).

8. J. Helszajn, "The Adjustment of the *m*-Port Single Junction Circulator," *IEEE Trans. Microwave Theory Tech.*, **MTT-18**, 705–711 (1970).

9. S. Yoshida, "*X*-circulator," *Proc. Inst. Radio Eng.*, **47**, 1150 (1959).

10. D. N. Landry, "A Single Junction Four-Port Coaxial Circulator," Wescon, Part 5, Session 4.2, 1963.

11. C. E. Fay and W. A. Dean, "The Four-Port Single Junction Circulator in Stripline," PG-MTT *Digest Int. Symp. IEEE*, 1966.

12. S. R. Longley, "Experimental 4-Port *E*-Plane junction circulator," *IEEE Trans. Microwave Theory Tech.*, (corresp.), **MTT-15**, 378–380 (1967).

13. A. G. Bogdanov, "Design of Waveguide X-circulators," *Radio Eng. Electron. Phys.*, **14** (4) (1969).

14. J. B. Davies and P. Cohen, "Theoretical Design of Symmetrical Junction Stripline Circulators," *IEEE Trans. Microwave Theory Tech.*, **MTT-11**, 506–512 (1963).

15. H. Bosma, "On Stripline Circulation at U. H. F.," *Trans. IEEE Microwave Theory Tech.*, **MTT-12**, 61–72 (1964).

16. J. Watkins, "Circulator Resonant Structures on Microstrip," *Electron. Tech. Lett.*, **5**, 524–525 (1969).

17. W. E. Courtney, "Analysis and Evaluation of a Method of Measuring the Complex Permittivity and Permeability of Microwave Insulators," *IEEE Trans. Microwave Theory Tech.*, **MTT-18** (8), 476–485 (1970).

18. J. Helszajn, "Waveguide and Stripline 4-Port Single Junction Circulators," *IEEE Trans. Microwave Theory Tech.*, Vol 21 **MTT-20**, 630–633 (1973).

19. C. E. Fay and R. L. Comstock, "Operation of the Ferrite Junction Circulator," *IEEE Trans. Microwave Theory Tech.* **MTT-13**, 15–27 (1965).

20. W. H. Ku and Y. S. Wu, "On Stripline Four Port Circulator," 1973 *IEEE Microwave Theory Tech. Int. Microwave Symp. Digest*, IEEE Cat. No. 73, chap. 736–9 MTT, 86–88.

CHAPTER NINETEEN

Magnetic Parameters of Junction Circulators

The influence of the ferrite parameters on the performance of the junction circulator is treated in this chapter. This is usually stated in terms of the insertion loss and bandwidth of the device. In its matched condition, the junction exhibits many of the properties of a transmission line cavity filter. The behavior of the device may, therefore, be characterized by its loaded and unloaded Q factors. The loaded Q-factor is determined by the bandwidth of the device, as indicated in Chapter 12. It also defines the magnetic splitting factor K/μ of the ferrite material. The unloaded Q factor is determined by the dielectric and magnetic losses of the ferrite material. The magnetic loss tangent is connected to the real and imaginary parts of the effective permeability μ_e. A similar statement applies to the dielectric loss tangent. It is found that the insertion loss is proportional to the effective linewidth of the ferrite material provided that the dielectric loss is negligible. The purpose of this chapter is to relate the loaded and unloaded Q factors to the ferrite parameters with the ferrite biased either below or above the main resonance. The important material parameters being the saturation magnetization M_0, the linewidth ΔH, the applied direct field H_0, the d.c. demagnetizing factors, N_x, N_y, N_z, the Curie temperature T_c, and the operating frequency ω.

In practice, most of the material parameters, display some dependence on such variables as frequency, temperature, and direct magnetization field. Throughout this text it has also been assumed that the ferrite material is saturated. If the latter is not true, the ferrite will exhibit so-called low-field losses below some frequency. The onset of this loss being determined by the shape demagnetizing factors and the saturation magnetization. In this region the initial permeability may differ from the effective one μ_e. Finally it has been assumed so far, that the ferrite material is linear. In practice, however, ferrite materials are known to exhibit nonlinear losses if the microwave signal level exceeds some critical threshold field h_{crit}. Some comments on the nonlinear behavior of ferrites at large signal level are, therefore, also included in this chapter for the sake of completeness.

19.1. BANDWIDTH OF JUNCTION CIRCULATOR IN TERMS OF MAGNETIC VARIABLES

It has been indicated in a number of places in this text that the ratio of the off-diagonal to diagonal components of the tensor permeability is related to the loaded Q factor or the junction specifications r and $2\delta_0$ by

$$Q_L = \frac{1}{\sqrt{3}}\frac{\mu}{K} = \frac{(r-1)}{2\delta_0\sqrt{r}} \tag{19.1}$$

In terms of the original variables defined in Chapter 1 K/μ is

$$\frac{K}{\mu} = \frac{p}{\sigma^2 + p\sigma - 1} \tag{19.2}$$

and

$$p = \frac{\gamma M_0}{\mu_0}/\omega \tag{19.3}$$

$$\sigma = \gamma\left(H_0 - \frac{N_z M_0}{\mu_0}\right)/\omega \tag{19.4}$$

Here p is the normalized magnetization, σ is the normalized internal d.c. field, M_0 is the saturation magnetization, in Wb/m^2, γ is the gyromagnetic ratio (measured in rad/s per A/m), μ_0 is the free space permeability in H/m, N_z is the d.c. demagnetizing factor, and ω is the frequency in rad/s.

The magnetic splitting K/μ is shown in Figure 19.1 as a function of σ for parametric values of p. This indicates that it is possible to operate the junction either above or below the main resonance. The range which K/μ may take for below resonance circulators is discussed in Chapters 12 and 14.

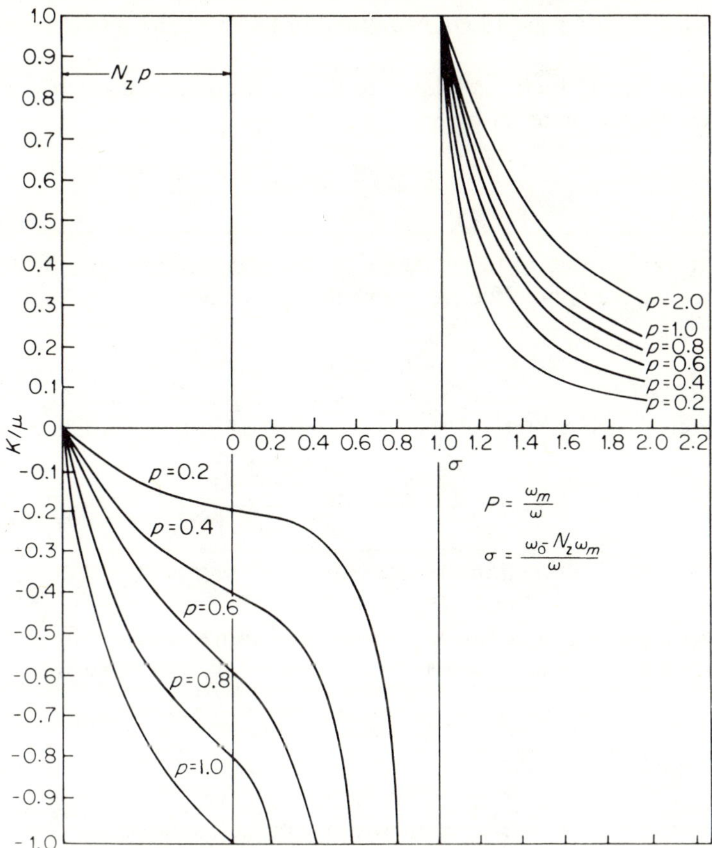

$$P = \frac{\omega_m}{\omega}$$

$$\sigma = \frac{\omega_\sigma\, N_z \omega_m}{\omega}$$

Figure 19.1.   *Variation of μ_e as a function of δ (Ref. 15).*

19.2.   INSERTION LOSS OF THE JUNCTION CIRCULATOR IN TERMS OF MAGNETIC VARIABLES

In Chapter 7 the insertion loss of circulators was related to the ratio of the loaded to unloaded Q factors of the junction by

$$L(dB) = 20\log_{10}\left(1 + \frac{Q_L}{Q_U}\right) \tag{19.5}$$

If $Q_L/Q_U \ll 1$ this last equation becomes

$$L(dB) \approx 8.686\left(\frac{Q_L}{Q_U}\right) \tag{19.6}$$

The loaded Q factor used here is defined in terms of the ferrite parameters in the last section.

The unloaded Q factor is related to the magnetic and dielectric parameters of the ferrite material by

$$\frac{1}{Q_U} = \frac{1}{Q_m} + \frac{1}{Q_e} \tag{19.7}$$

The contribution due to the magnetic losses is determined by the real and imaginary parts of the effective permeability of the ferrite material.

$$Q_m = \frac{\mu_e'}{\mu_e''} \tag{19.8}$$

The effective permeability of the ferrite material is given in Chapter 1 by

$$\mu_e = \frac{\mu^2 - K^2}{\mu} = \frac{(p+\sigma)^2 - 1}{\sigma^2 + p\sigma - 1} \tag{19.9}$$

It is illustrated as a function of σ for parametric values of p above or below the main resonance in Figure 19.2.

Magnetic losses can be introduced into the above equation by adding an imaginary frequency $j\alpha$ to σ, where α is the normalized linewidth

$$\alpha = \frac{\gamma \Delta H_{\text{eff}}}{2\omega} \tag{19.10}$$

and ΔH_{eff} is the effective linewidth in ampere per meter. Away from the main resonance, the effective linewidth can be replaced by that of the spinwave ΔH_k.

The contribution to the unloaded Q factor due to the dielectric losses is similarly given by

$$Q_e = \frac{\epsilon'}{\epsilon''} = \frac{1}{\tan\delta} \tag{19.11}$$

where $\tan\delta$ is the dielectric loss tangent. For most ferrite materials, $\tan\delta \approx 0.0002$, so that the dielectric losses are usually negligible compared to the magnetic ones and so are neglected here.

19.3.   *MAGNETIC PARAMETERS OF JUNCTION CIRCULATOR BIASED BELOW THE MAIN RESONANCE*

In the case of the circulator biased below the main resonance it is assumed, as an approximation, that the ferrite material is just saturated. The real and imaginary parts of μ_e are then found by replacing σ in Eq. 19.9 by

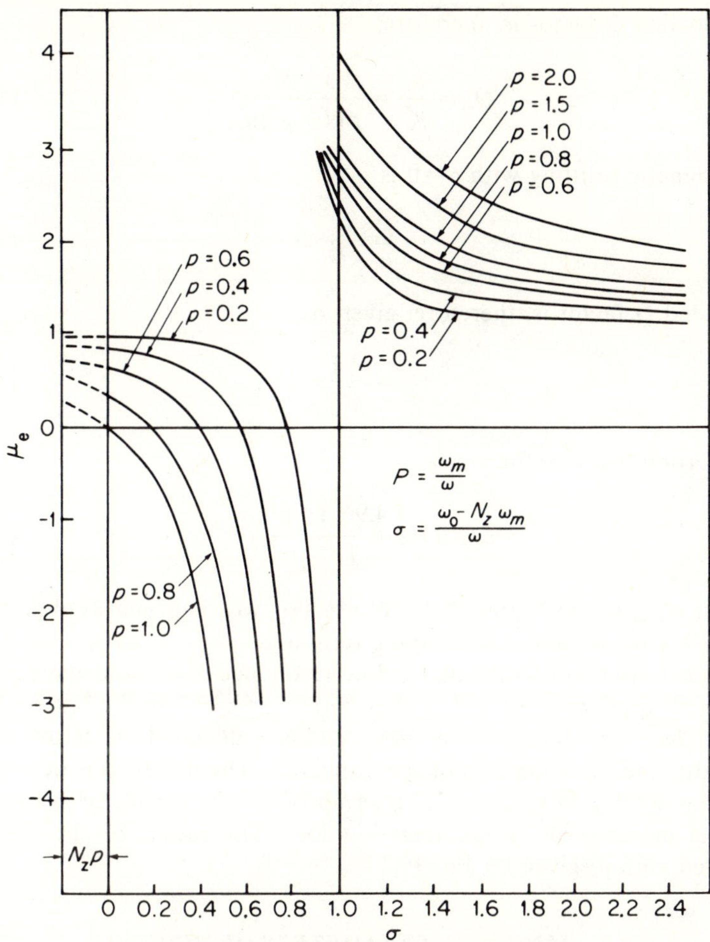

$$P = \frac{\omega_m}{\omega}$$

$$\sigma = \frac{\omega_0 - N_z \omega_m}{\omega}$$

Figure 19.2.   *Variation of K/μ as a function of p (Ref. 15).*

$\sigma + j\alpha$, and then setting σ to zero. The result for μ_e is

$$\mu_e' - j\mu_e'' = \frac{(p+j\alpha)^2 - 1}{(j\alpha)^2 + j\alpha p - 1} \tag{19.12}$$

Hence,

$$\mu_e' \approx 1 - p^2 \tag{19.13}$$

$$\mu_e'' \approx p(1+p^2)\alpha \tag{19.14}$$

The unloaded Q factor is, therefore,

$$Q_m = \frac{\mu_e'}{\mu_e''} = \frac{1-p^2}{p(1+p^2)\alpha} \tag{19.15}$$

The magnetic splitting with $\sigma = 0$ is

$$\frac{K}{\mu} = p \tag{19.16}$$

The loaded Q factor is, therefore, given by

$$Q_L = \frac{1}{\sqrt{3}\,p} \tag{19.17}$$

The insertion loss now becomes

$$L(dB) \approx \frac{4.96(1+p^2)\alpha}{(1-p^2)} \tag{19.18}$$

This last equation indicates that the insertion loss is primarily determined by the effective normalized damping term α.

For most circulators operated below resonance $p^2 \ll 1$ and the insertion loss can be approximated still further by setting $p^2 = 0$ in the last equation. The magnetic parameters are now readily calculated in terms of the bandwidth and insertion loss of the circulator. The d.c. field is determined by setting $\sigma = 0$, p is completely determined by the circulator bandwidth, and α is determined by the insertion loss. The radius of the ferrite is calculated with μ_e given by Eq. 19.13.

19.4. *MAGNETIC PARAMETERS OF JUNCTION CIRCULATOR BIASED ABOVE THE MAIN RESONANCE*

In the case of the junction biased above the main resonance, it is assumed as an approximation that $p > 1$ and $\sigma > 1$. Using this approximation in Eq. 19.9 gives

$$\mu_e \approx 1 + \frac{p}{\sigma} \tag{19.19}$$

The real and imaginary parts of this last equation are now obtained by replacing σ by $\sigma + j\alpha$. The result is

$$\mu_e' - j\mu_e'' = 1 + \frac{p}{(\sigma + j\alpha)} \tag{19.20}$$

where

$$\mu_e' \approx 1 + \frac{p}{\sigma} \tag{19.21}$$

$$\mu_e'' \approx \frac{\alpha p}{\sigma^2} \tag{19.22}$$

Hence

$$Q_m = \frac{\mu_e'}{\mu_e''} = \frac{\sigma(1 + p/\sigma)}{\alpha(p/\sigma)} \tag{19.23}$$

The magnetic splitting with $p > 1$ and $\sigma > 1$ is

$$\frac{K}{\mu} \approx \frac{p/\sigma}{\sigma(1 + p/\sigma)} \tag{19.24}$$

The loaded Q factor is, therefore,

$$Q_L \approx \frac{\sigma(1 + p/\sigma)}{\sqrt{3}\,(p/\sigma)} \tag{19.25}$$

and the insertion loss is

$$L(dB) \approx 4.96\alpha \tag{19.26}$$

Equation 19.26 again suggests that the insertion is proportional to the damping term α. Here, the material parameters are determined by the necessary magnetic splitting required to obtain the bandwidth. The ferrite radius is then determined by straightforward calculation with μ_e given by Eq. 19.21.

19.5.   LINEWIDTH DEPENDENCE UPON DIRECT CURRENT FIELD

In practice, the uniform linewidth exhibits some dependence upon the internal direct magnetic field. In particular, there is a well-known broadening of the uniform mode linewidth as the uniform precession frequency varies with respect to the spinwave manifold. The spinwave modes are higher-order modes of the equation of motion, which (unlike the uniform mode) vary spatially over the sample.

The uniform mode can be made to intersect different portions of the spinwave manifold by varying the d.c. field. The linewidth goes through a maximum when the uniform modes are degenerate with the largest number of spinwave modes. For coupling between the uniform mode and the spinwave reservoir to occur a suitable coupling mechanism must be identified. Such coupling is possible as a result of irregularities in the magnetic lattice and surface imperfections. These can act as centers for scattering of the uniform mode to spinwave modes of wavelength comparable to the dimensions of the inhomogeneity. Figure 19.3 illustrates the effective linewidth for three magnesium ferrites as a function of the applied direct magnetic field.

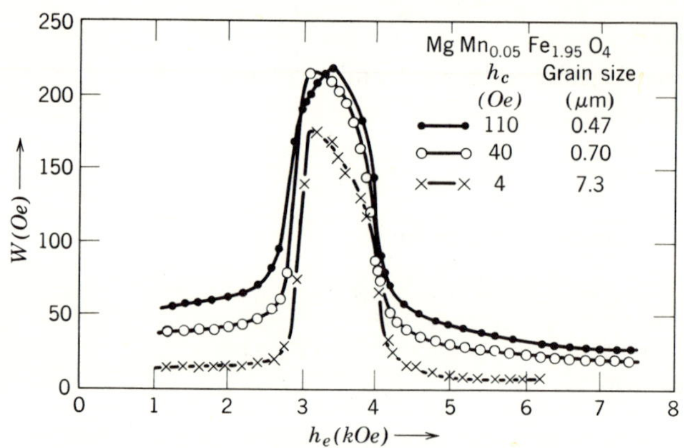

Figure 19.3.   *Variation of ΔH_{eff} as a function of external direct magnetic field (Ref. 18).*

19.6.   DEMAGNETIZED PERMEABILITY

The demagnetized permeability may be obtained from a theorem which applies to a partially magnetized ferrite and which is independent of the details of the domain configuration:

$$\mu_p^2 - K_p^2 = \mu^2 - K^2 \tag{19.27}$$

Here μ_p and K_p refer to the partially magnetized components of the permeability tensor and μ and K refer to the ones in the magnetized state. The right-hand side of Eq. 19.27 passes through a minimum in the unmagnetized state for which K_p^2 vanishes. Hence

$$\mu_{\text{demag}} = \left(\mu^2 - K^2\right)^{1/2} \tag{19.28}$$

In terms of the original variables introduced in Chapter 1 this becomes

$$\mu_{\text{demag}} = \left[\frac{\omega^2 - (\omega_a + \omega_m)^2}{\omega^2 - \omega_a^2} \right]^{1/2} \tag{19.29}$$

Here it is assumed that the effective magnetic field inside the domains is the anisotropy field H_a (and $\omega_a = \gamma H_a$). For the completely demagnetized state the random orientation of the domains can easily be taken into account in an approximate fashion. In this case the off-diagonal components of the permeability tensor vanish. If only up and down domains are present the diagonal components corresponding to the x- and y-directions are given by Eq. 19.29 and that corresponding to the z-direction equals unity. The average permeability for random orientation may then be expected to be given approximately by the average of the permeabilities applicable in the three principal directions. This yields as the average (isotropic) permeability in the demagnetized state

$$\mu_{\text{demag}} = \frac{2}{3} \left[\frac{\omega^2 - (\omega_a + \omega_m)^2}{\omega^2 - \omega_a^2} \right]^{1/2} + \frac{1}{3} \tag{19.30}$$

The permeability defined by the last equation is real for $\omega > (\omega_m + \omega_a)$ and imaginary for $\omega_a < \omega < (\omega_m + \omega_a)$. It thus accurately describes the onset of low field loss, which is to be derived in the next section.

For ω_a small compared to ω_m and ω the following simple form applies:

$$\mu_{\text{demag}} = \frac{2}{3} \left[1 - \left(\frac{\omega_m}{\omega} \right)^2 \right]^{1/2} + \frac{1}{3} \tag{19.31}$$

Figure 19.4 gives the variation of the demagnetized permeability as a function of the normalized saturation magnetization. The coordinate points plotted in this illustration are experimental ones.

19.7.   *LOW FIELD LOSSES IN UNSATURATED MEDIUM*

The general form of Kittel's resonance predicts a contribution to the natural resonance frequency arising from crystalline anisotropy and shape demagnetizing fields within a single crystal domain even in the absence of a d.c. field. This determines the lowest possible operating frequency. In an

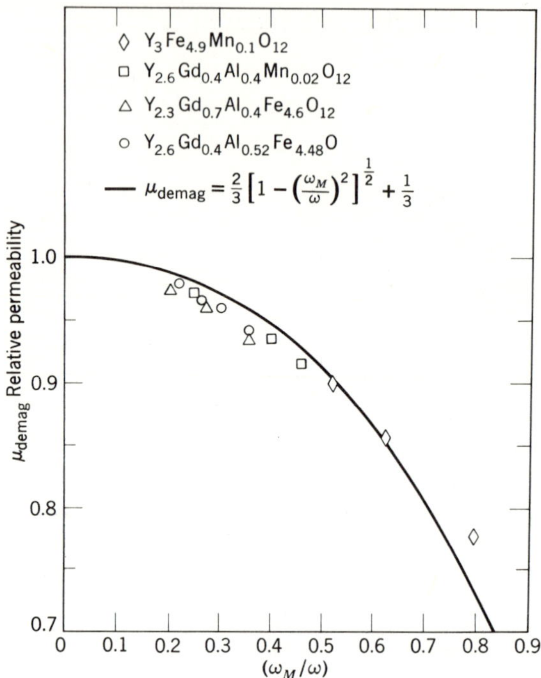

Figure 19.4.   *Variation of demagnetized permeability as a function of p for various garnet compositions* (*Ref.* 19).

unsaturated ferrite, however, resonance absorption losses occur at frequencies up to at least twice the value predicted by Kittel's resonance equation because the domain structure of an unsaturated ferrite contributes to the effective demagnetizing field. The effect of low-field loss upon the shape of the resonance curve at microwave frequencies is indicated in Figure 19.5. It is possible to calculate the contribution to the total demagnetizing field of the Weiss domains by considering a nonsaturated specimen, which is still divided into Weiss domains. However, it may be obtained more directly from equation 19.30 by observing that the permeability is real for

$$\omega > \omega_m + \omega_a \tag{19.32}$$

and imaginary for

$$\omega_a < \omega < (\omega_m + \omega_a) \tag{19.33}$$

This last condition establishes the interval over which the main rotation resonance is possible. Hence, the maximum frequency at which domain

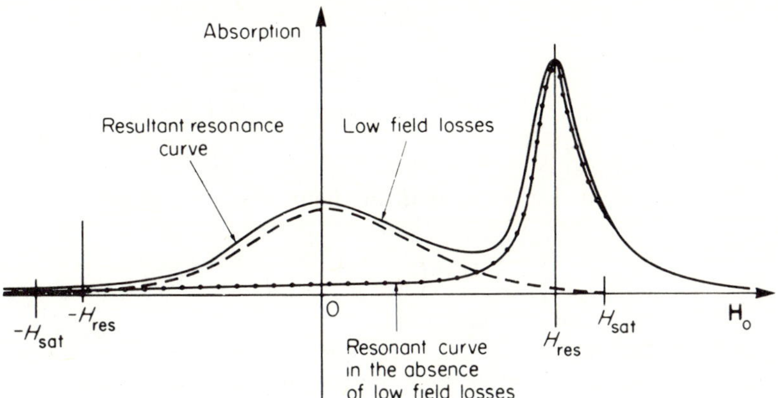

Figure 19.5.   *Effect of low field losses upon the shape of the resonance curve at microwave frequencies (Ref. 6).*

rotation resonance occurs is

$$\omega_{max} = \omega_m + \omega_a \tag{19.34}$$

and the minimum one is

$$\omega_{min} = \omega_a \tag{19.35}$$

19.8.   *LOW FREQUENCY LIMIT OF RESONANCE IN SATURATED MEDIUM*

In the previous section it has been observed that low field losses can occur in an unsaturated ferrite material below a maximum frequency given by $\omega_{max} = (\omega_a + \omega_m)$. To operate below this frequency it is necessary to ensure that the ferrite material is saturated. The minimum frequency at which this condition is satisfied at resonance is determined by the shape demagnetizing factors, the saturation magnetization and the equivalent anisotropy field. If the anisotropy field is negligible the minimum operating frequency is obtained with a thin disk shaped geometry.

To understand why this is the case consider the effect of the shape demagnetizing factors on the internal field, H_i. This field is given by

$$H_i = H_0 - N_z \frac{M_0}{\mu_0} \tag{19.36}$$

Kittel's resonant frequency in the case of a spheroid is from Eq. 1.71 given by

$$\omega_r = \gamma \left(H_0 - N_z \frac{M_0}{\mu_0} + N_t \frac{M_0}{\mu_0} \right) \qquad (19.37)$$

where for a spheroid $N_x = N_y = N_t$.

Eliminating H_0 in Eq. 19.36 with the help of Eq. 19.37 gives

$$H_i = \left(\frac{\omega_r}{\gamma} - N_t \frac{M_0}{\mu_0} \right) \qquad (19.38)$$

For low-loss resonance, the ferrimagnetic sample must be saturated, and the internal magnetic field H_i must be positive. Hence,

$$\omega_r \geqslant N_t \omega_m \qquad (19.39)$$

For a sphere, $N_t = 1/3$, and hence

$$\omega_r \geqslant \frac{\omega_m}{3} \qquad (19.40)$$

For a disk having a diameter 30 times its thickness, N_t is approximately given from Eq. 1.64 by 0.016, and

$$\omega_r \geqslant 0.016\omega_m \qquad (19.41)$$

Thus, it can be seen from Eqs. 19.40 and 19.41 that, for a disk having a diameter-to-thickness ratio of 30 to 1, the lower frequency limit of operation can be extended downwards by a factor of approximately 20, for the same ferrimagnetic material.

Figure 19.6 depicts the minimum value of ω_r of spheroidal geometries as a function of M_0 with the ratio of principal axes as a parameter.

If the material becomes unsaturated at the main resonance the linewidth increases because of the onset of the magnetic domain structure. This is shown in Figure 19.7.

19.9.   *NONLINEAR LOSS AT LARGE SIGNAL LEVEL*

The onset of nonlinear loss at large signal level is due to the exponential growth of certain pairs of spinwaves from their thermal excitation by coupling of the spinwaves to the uniform mode through demagnetizing and exchange fields.

Spinwave instabilities in ferrite samples under perpendicular pumping experimentally involves two separate effects. First, it is found that an

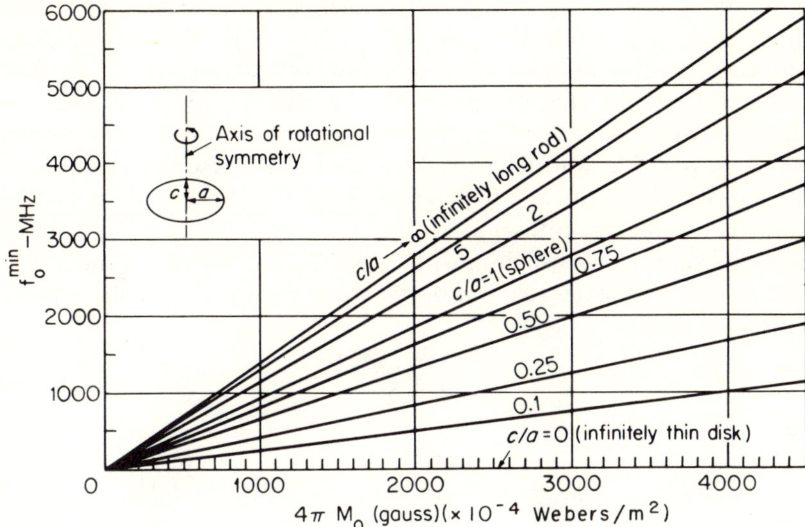

Figure 19.6.   *Theoretical minimum resonant frequencies of spheroids (Ref. 20).*

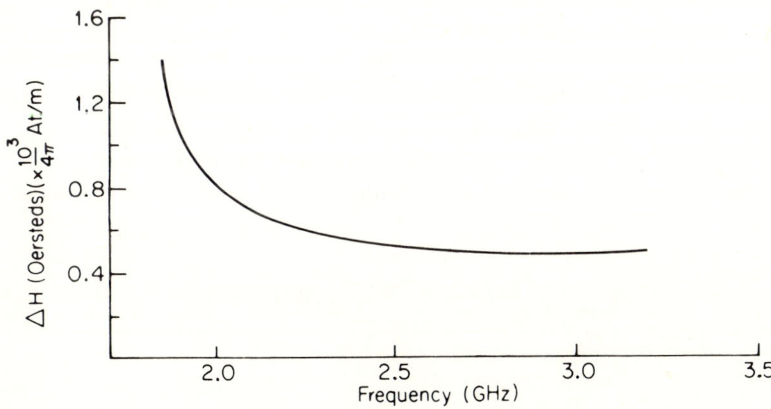

Figure 19.7.   *Linewidth of polished YIG sphere as a function of frequency (Ref. 9).*

additional subsidiary resonance peak appears below the magnetic field required for the main resonance. Secondly, the main resonance line exhibits substantial lowering and broadening as the r.f. excitation increases above some critical threshold. The latter instability is a second order one which is not particularly important in circulator design.

The onset of spinwave instability at the subsidiary resonance peak arises from coupling between the first order of the amplitude of the uniform mode and the spinwaves, provided the incident signal exceeds some critical threshold field, h_{crit}, is therefore known as a first-order nonlinear process. This process involves the excitation of spinwaves that propagate at about 45° with respect to the direction of the direct field and that are synchronous with half the r.f. excitation frequency. Under some circumstances it is possible for the subsidiary and main resonances to occur simultaneously. The critical r.f. magnetic field threshold is then particularly low (coincidence-limiting).

The first requirement, which must be satisfied to observe nonlinear loss, is to satisfy the frequency relation between the frequencies of the uniform and spinwave modes

$$\omega_k = \frac{\omega}{2} \tag{19.42}$$

where

$$\omega_k = \left[(\omega_0 - N_z\omega_m + \omega_{\text{ex}})(\omega_0 - N_z\omega_m + \omega_{\text{ex}} + \omega_m \sin^2\theta_k) \right]^{1/2} \tag{19.43}$$

Here θ_k is the angle that the spinwaves make with the direction of the direct magnetic field and ω_{ex} is an exchange frequency that is derived from the exchange energy. The latter energy is an energy of electrostatic origin and has no classical counterpart.

The second necessary requirement for the onset of nonlinear loss is for the magnetic field intensity to exceed some critical value. This field is a function of the direct magnetic field, the saturation magnetization, the frequency, and the uniform mode and spinwave modes linewidths.

For the circularly polarized wave which excites the main resonance the result is

$$\frac{h_{\text{crit}}}{\Delta H_k} = \frac{\left[(\omega_r^2 - \omega^2)^2 + 4\omega_r^2(\gamma\Delta H/2)^2 \right]^{1/2}}{\omega_m\omega(1 + e_k)\sin\theta_k \cos\theta_k} \tag{19.44}$$

where

$$e_k = \left[1 + \left(\frac{\omega_m^2}{\omega} \right)^2 \sin^4\theta_k \right]^{1/2} - \left(\frac{\omega_m}{\omega} \right)\sin^2\theta_k \tag{19.45}$$

Here, ΔH_k is the spinwave linewidth and e_k is the spinwave ellipticity. Figure 19.8 shows h_{crit} as a function of ω_0/ω in the case of a sphere.

Each curve consists of two branches that correspond to the two regions in which the unstable spinwaves have either $k \neq 0$ or $k = 0$. The latter condition corresponds to the right-hand branch. The waves most likely to

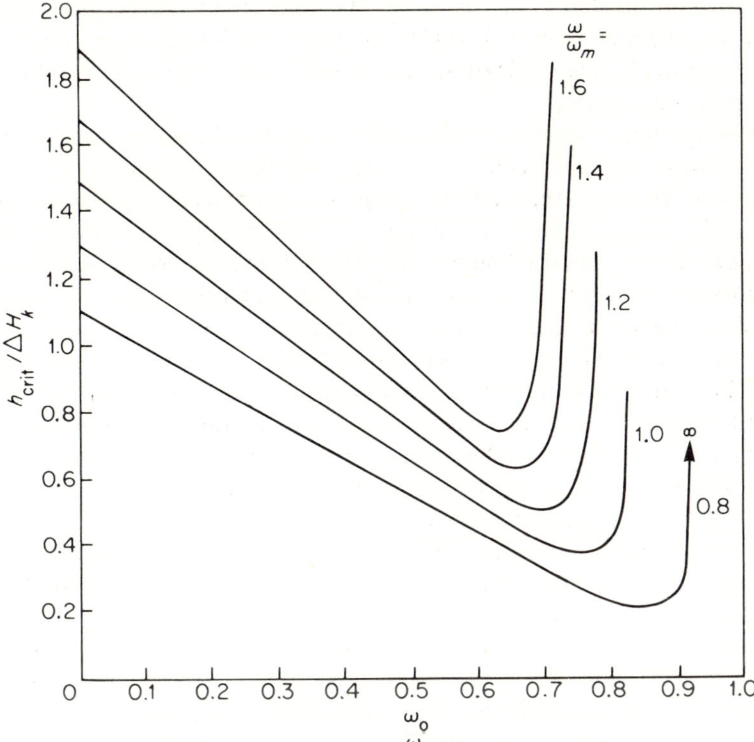

Figure 19.8.   *Threshold magnetic field for nonlinear loss as a function of ω_0/ω (Ref. 12).*

be unstable at the subsidiary resonance have $k=0$ and $\theta_k \approx 45°$.

As the direct field is increased from zero, the dispersion relation consists of two regions where $k \neq 0$ or $k=0$. The dispersion relation ceases to yield a real value for k with $\theta_k = 0$ at a direct magnetic field given by

$$\omega_0 = \frac{\omega}{2} + N_z \omega_m \tag{19.46}$$

Equation 19.46 provides a convenient rule for the existence of the subsidiary resonance.

19.10.   *SWITCHING COEFFICIENT OF MAGNETIZATION*

The change of magnetization in a ferrite core consists usually in reversal of the magnetization, for example from negative remanence to the positive remanence corresponding to the magnetic field applied. The ultimate state of magnetization that is set up (after the passage of the current pulse) is here always symmetrical with respect to zero. It is found that in most cases the change in the magnetization produced by this field cannot follow the increase in the current.

For an applied magnetic field slightly in excess of the coercive force, H_c, domain wall motion will, in general, be the predominant reversal mechanism. In this case, the flux change is accomplished by the motion of Bloch walls, which separate the domains of differently oriented magnetization. For suitably oriented single crystals of ferrite, a very simple domain configuration may be achieved, which makes it possible to obtain information on the behavior of a moving domain wall.

Theoretical studies and experiments performed with single crystals of ferrite have demonstrated that the wall velocity depends linearly on the applied magnetic field. This leads to a linear relation between the direct field and the reciprocal of the switching time. Such a relationship is also found experimentally for polycrystalline ferrites although the actual domain configuration is not known.

The switching time is usually measured by using a core with two windings as shown in Figure 19.9. The output voltage pulse appearing at the terminals of the secondary winding exhibits a characteristic shape with two separate maxima when a current pulse is passed through the magnetizing winding. Figure 19.10 shows three secondary voltage pulses as measured on ferroxcube 6A for three different field strengths. For ferrites with rectangular hysteresis loops the first maximum in the output voltage represents a small percentage of the total area under the curve and hence of the change in magnetization. The duration T of a voltage pulse is defined as the time (counted from the beginning of the current pulse) that elapses before the voltage has dropped to 10% of the maximum value; for the maximum value the second peak is considered. The dependence of the switching time on the magnetizing force producing flux reversal is most clearly presented by plotting $1/T$ as a function of the magnetic field strength H, as shown in Figure 19.11. Over the majority of the range shown, $1/T$ has a linear dependence on H. Which may be adequately represented by

$$T(H - H_0) = S \tag{19.47}$$

In this expression S is known as the switching coefficient, and H_0, which is of the same order of magnitude as the coercive force H_c of the material, may be termed the threshold field for irreversible magnetization. It should

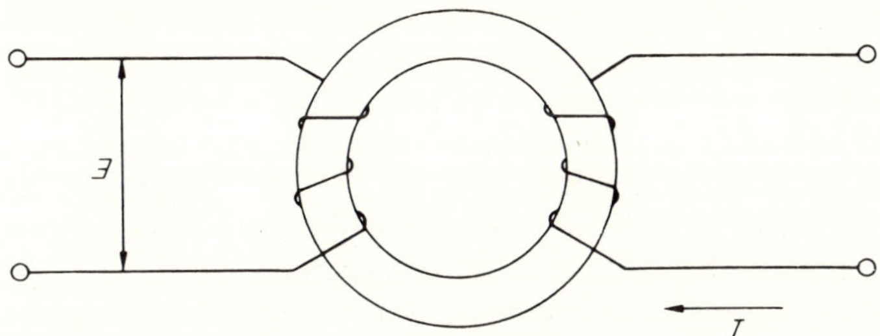

Figure 19.9.   *Ferrite core with two windings for measuring switching time.*

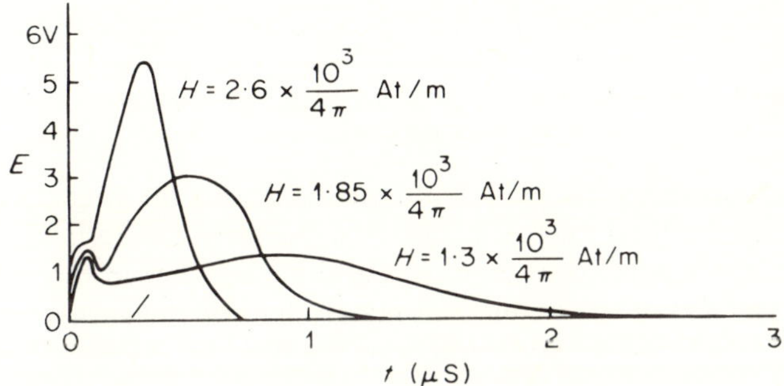

Figure 19.10.   *Voltage pulses measured on ferroxcube 6A (Ref. 21).*

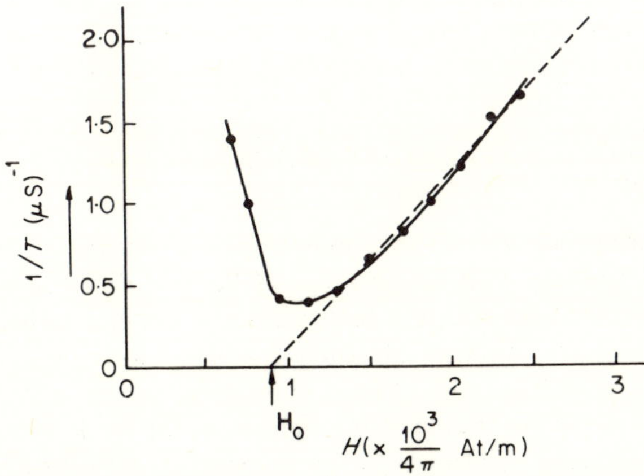

Figure 19.11.   *Reciprocal reversal time $1/T$ as a function of H for ferroxcube 6A (Ref. 21).*

337

be noted that, although the curve is continued to values of H less than H_0, the switching of the core under these conditions produces a smaller hysteresis loop; that is, the material is not driven to magnetic saturation, and such operation is not desirable. The optimum squareness ratio R_s occurs for values of H very little different from H_0, but it is usual to adopt magnetizing fields of at least $2H_0$, the slight deterioration in squareness being accepted in the interests of faster switching.

It is found that the great majority of ferrites with the squareness and the coercive force usual for switching elements all have a "reversal constant" $S = (H - H_0)T$ of the same order of magnitude. In the case of ferroxcube 6A considered above, the resultant values are $H_0 = 80$ A/m and $S = 80$ (μsec)(A/m).

19.11. *THE MAGNETIZATION VALUES OF VARIOUS FERRITES*

The ferrite parameters throughout this text have so far been discussed in terms of the normalized magnetization p of the ferrite material. In practice, the saturation magnetization of ferrites and garnets lies in the range 0.015 to 0.50 Wb/m^2 for most commercial materials. For the garnet materials it lies between 0.015 and 0.18 Wb/m^2. For magnesium manganese ferrites it lies between 0.08 and 0.30 Wb/m^2. For nickel ferrites the magnetization lies between 0.10 and 0.50 Wb/m^2, and for lithium ferrites it lies between 0.015 and 0.50 Wb.m^2. Figure 19.12 illustrates the variation of magnetization of a yttrium gadolinium aluminium iron garnet with aluminium substitution for iron. This illustration also shows that the magnetization varies with temperature becoming zero at the curie temperature T_c.

19.12. *TEMPERATURE DEPENDENCE OF MAGNETIZATION*

The temperature dependence of the magnetization of ferrites is dependent on the change in moment with temperature of the two or more oppositely directed sublattices that constitute a ferromagnetic material. The majority of ferrites contain ferric ions as the only magnetic ions, and the curie temperature, magnetization, and change of magnetization with temperature depends only upon the degree of interaction of the ferric ions within a sublattice. This interaction varies with the proximity and crystallographic orientation of the magnetic ions and can be greatly diminished by the substitution of nonmagnetic ions on the same crystallographic site as the magnetic ions. The net magnetization of the material can, therefore, be raised or lowered by this means, depending on whether the substituted

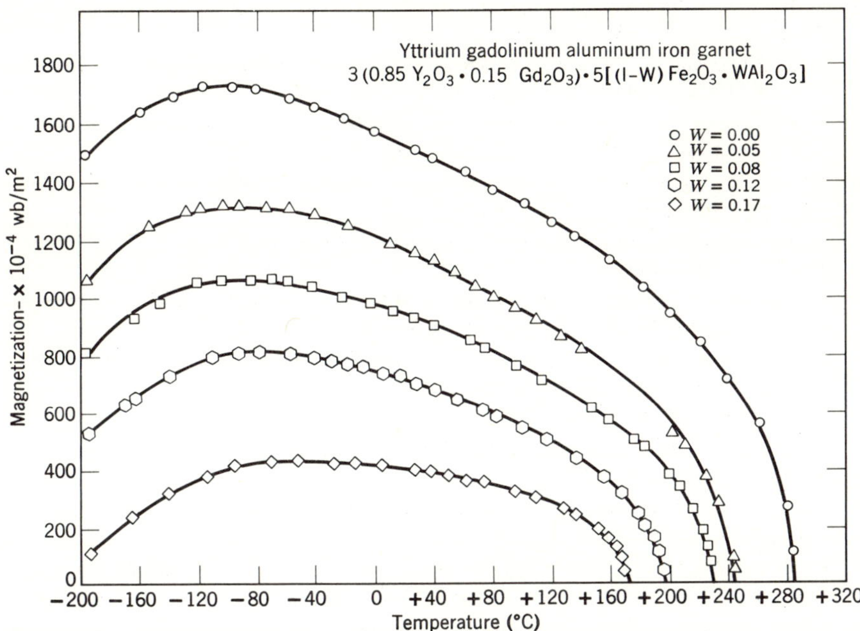

Figure 19.12.   *Magnetization versus temperature for yttrium gadolinium aluminium iron garnet.*

sublattice is initially dominant or not. The curie temperature is almost invariably lowered by the introduction of nonmagnetic ions, although the room temperature magnetization can be raised or lowered.

The temperature dependence of the magnetization may be further modified by the introduction of a different magnetic ion, for example, gadolinium in garnet, nickel in spinel ferrite. Nickel ions in spinel add to the net magnetization because of their orientation, and also increase the curie temperature. Gadolinium, while not affecting the curie temperature, opposes the net room temperature ferric moment (garnets have three sublattices). The temperature dependence of the moment of gadolinium is such that it dominates the net ferric moment at low temperature. The result is a low temperature compensation point, where the magnetization passes through zero. The effect near room temperature is to create a relatively flat region of temperature versus magnetization.

REFERENCES

1. G. Rado, "Theory of Microwave Permeability Tensor and Faraday Effect in Non-saturated Ferromagnetic Materials," *Phys. Rev.*, **89**, 529 (1953).

2. C. Kittel, "On the Theory of Ferromagnetic Resonance Absorption," *Phys. Rev.*, **73**, 155–61 (1948).

3. C. R. Buffler, "Resonance Properties of Single Crystal Hexagonal Ferrites," *J. Appl. Phys. Suppl.*, **33** (3), 1360–1362 (1962).

4. H. S. Belson and C. J. Kreissman, "Microwave Resonance in Hexagonal Ferromagnetic Single Crystals," *J. Appl. Phys. Suppl.*, **30** (4), 175S (1959).

5. E. Schlömann and R. V. Jones, "Ferromagnetic Resonance in Polycrystalline Ferrites with Hexagonal Structure," *J. Appl. Phys. Suppl.*, **30**, 177S (1959).

6. B. Lax, "Frequency and Loss characteristics of microwave ferrite devices," *Proc. IRE*, **44**, 1368–1386 (1956).

7. J. O. Artman, "Microwave Resonance Relation in Anisotropic Single-Crystal Ferrites," *Phys. Rev.*, **105**, 62 (1957).

8. L. Landau and E. Lifshitz, "On the Theory of the Dispersion of Magnetic Permeability in Ferromagnetic Bodies," *Phys. Z. Sowjetunion*, **8**, 153 (1935).

9. F. C. Rossol, "Subsidiary Resonance in the Coincidence Region in Yttrium Iron Garnet," *J. Appl. Phys.*, **31**, 2273 (1960).

10. J. Smit and N. H. P. Wijnn, *Ferrites*, 83–84, Phillips Technical Library, Holland, 1959.

11. Y. Konishi, "Lumped Element *Y* Circulator," *IEEE Trans. Microwave Theory Tech.*, **MTT-13**, No 6 852–864 (1965).

12. H. Suhl, "The Non-linear Behaviour of Ferrites at High Signal Levels," *Proc. IRE*, **44**, 1270 (1956).

13. H. Bosma, "A General Model for Junction Circulators: Choice of Magnetization and Bias Field," *IEEE Trans. Magn.*, **MAG. 4** (3) 587–596 (1968).

14. J. Edrich and R. C. West, "Very Low Loss *L*-Band Circulator with Gallium Substituted YIG," *IEEE Intermag. Conf.*, 1969

15. C. E. Fay and R. L. Comstock, "Operation of the Ferrite Junction Circulator," *Trans IEEE Microwave Theory Tech.*, **MTT-13**, 15–27 (1965).

16. R. C. LeCraw and E. G. Spencer, "IRE Convention Record. Part 5," 66, *J. Appl. Phys.*, **28**, 399 (1957).

17. J. Helszajn, "Ferrite Parameters of the Junction Circulator," *Electronic Components*, July 1970.

18. Q. H. F. Vrehen, G. H. Beljers, and J. G. M. de Lau, "Microwave Properties of Fine-Grain Ni and Mg Ferrites," *IEEE Trans. on Magn.*, **MAG-5**, 617–621 (1969).

19. W. E. Courtney, "Analysis and Evaluation of a Method of Measuring the Complex Permittivity and Permeability of Microwave Insulators," *IEEE Trans. Microwave Theory Tech.*, **MTT-18** (8), 476–485 (1970).

20. F. R. Arams, M. Grace, and S. Okwit, "Low Level Garnet Limiters," *Proc. IRE*, **49**, 1308 (1961).

21. Van Der Heide et al., *Philips Tech. Rev.*, **18**, 339 (1956).

22. E. G. Spencer, R. C. LeCraw, and C. G. Porter, "Ferromagnetic Resonance in Yttrium Iron Garnet at Low Frequencies," *J. Appl. Phys.*, **29**, 429 (1958).

23. R. W. Damon, "Relaxation Effects in Ferromagnetic Resonance," *Rev. Mod. Phys.*, **25**, 239–245 (1953).

24. E. Schlomann, "Ferromagnetic Resonance at High Power Levels," Raytheon Technical Report R-48, 1959.

25. J. Kemanis and S. Wong, "Analysis of High Power Effects in Ferrimagnetics from the Point of View of Energy transfer. I. First-Order Instabilities," *J. Appl. Phys.*, **35**, 1465–1470 (1964).

26. J. Helszajn and J. McStay, "Simplified Theory of Nonlinear Phenomena in Ferrimagnetic Materials," *Proc. IEE*, **114**, No. 11, 1585–1591 (1967).
27. J. Helszajn and J. McStay, "Parallel Pumping in Hexagonal Ferrites with the d.c. Field Off the Easy Plane," *IEEE Trans. Microwave Theory Tech.* (1970).